Cancer

Nutrition and Survival

Cancer

Nutrition and Survival

A microevolutionary approach to the causes, prevention and treatment of cancer

Steve Hickey PhD

and

Hilary Roberts PhD

Dedicated to Lyn Mesham,
for her help and support over the years.

Also to Jill and Michael Roberts,
for every reason, with love.

About the authors

Hilary Roberts' PhD research on early life undernutrition was carried out in the Department of Child Health at the University of Manchester. Steve Hickey gained his PhD in Medical Biophysics from the University of Manchester. He is now a member of the Department of Biological Sciences, Manchester Metropolitan University. Previous books by the authors include *Ascorbate: The Science of Vitamin C* and *Ridiculous Dietary Allowance.*

Contents

Acknowledgements

We are grateful to all those who have helped us in writing this book. In particular, the late Dr Hugh Riordan's research was an inspiration to us.

We wish to thank Michael Roberts, FRCS, who provided medical information, read the text and provided detailed feedback. Dr Michael Gonzalez provided details of his research and Dr Jorge Miranda-Massari supplied information about research into ascorbate and cancer. We received encouragement and feedback from Dr Selva Kumar on the use of vitamin C in cancer patients. We are grateful to the staff of the Department of Biological Sciences at Manchester Metropolitan University and, in particular, Professor Terry Looker and Dr Olga Gregson.

Dr Reginald Holman provided valuable information about his early experiments on hydrogen peroxide. Dr Abram Hoffer gave details of his experiences with vitamin C and cancer. Professor Joel Kauffman helped with discussion about the redox cycling of alpha-lipoic acid. Dr John Ely inspired us to emphasise our description of the interactions between ascorbate and glucose. Dr Devarajan Karunagaran supplied background information on the use of curcumin in cancer. Dr Andrew Saul has given frequent assistance and moral support during the writing of this book; he also suggested the title. Dr Ralph Moss and Anne Beatie helped clarify some points in Moss's books.

The publication of this book gives us the opportunity to acknowledge the continuing work of Dr Mark Levine and his vitamin C research group at the US National Institutes of Health. Dr Levine has helped bring the actions of high dose vitamin C against cancer to the attention of mainstream medicine.

Bill Sardi sent frequent emails, with details of developments in the area, and suggested we consider the anticancer action of IP6. Owen Fonorow and the Vitamin C Foundation have also supported our work, by providing useful information and by encouraging us to revisit the work of John Beard on development and cancer. Professor Harry Rubin kindly provided information about his work on magnesium and the cell cycle.

Pauline Rose read the manuscript and provided feedback. Andrew Hickey provided detailed comments on the text.

The drawing of a typical animal cell (page 30), is from the US National Institutes of Health and is in the public domain. Pictures from the US National Cancer Institute, also in the public domain, include the image of normal and cancer cells (author Pat Kenny, page 57); the karyotype (page 33); the breast images (MRI and X-ray by Dr Mitchell D. Schnall, University Of Pennsylvania, page 117); the SEM image of a malignant cell (by Dr. Raouf Guirgus of Dr. Liotta's Laboratory, page 69); and the cover image of a breast cancer cell, by an unknown photographer.

Preface

There are few accessible accounts of the biology of cancer,[1] although many on its treatment. Self-help books, often proposing unconventional treatments that lack scientific backing, are common. Some writers are scornful of these, while assuming that conventional treatments have substantial scientific support. It is our view that both alternative and conventional approaches require further investigation.

The aim of the present book is to evaluate existing research on the use of simple nutrients in the treatment of cancer. Our purpose is educational, to report the science, rather than to promote medical therapies. Cancer is deadly and discussion of its treatment is fraught with difficulties. However, ethical problems should not be used to suppress the dissemination of basic scientific ideas.

It will become clear to the reader that it is possible to self-treat cancer, using nutritionally based therapies. However, we are not promoting this idea: rather we are pointing out the possibilities for future research. Cancer is a serious disease, which should be treated by competent physicians. Anyone considering nutritional options should discuss the possibilities with their doctor. Several physicians who specialise in treating cancer using nutritional therapies are mentioned in the book.

This book is not intended to provide medical advice; the authors disclaim any responsibility for use of the information contained within for such a purpose. We are not promoting any therapy, nor do we encourage individuals to use the information in this book to go against the advice of their physician. People electing to self-treat cancer do so at their own risk. Of course, patients may choose to bring the book to the attention of their physicians, to help in demanding appropriate treatment.

The book challenges the prevailing idea that radiation and toxic chemotherapy are the most generally appropriate treatments for metastatic cancer. For decades, these methods have been employed with limited success. Dr Ralph Moss has provided an extensive account of this failure in his books on cancer. In view of the potential for a poor outcome, medical practitioners have an obligation to examine all possible options. However, the lack of a common literature for dissemination of information on the biology of cancer makes this difficult. Those defending the current position should not rely on historical grounds: the statement that we have 50 years experience of chemotherapy carries little scientific weight. By way of comparison, we could ask how much experience of blood letting was required before it was realised that the procedure was killing the patients.

We have written this book to be accessible to the intelligent general reader, who may have little knowledge of biology, chemistry or medicine. This involves some simplification, although we have tried to avoid being inaccurate. Readers who lack a scientific education can read the text and ignore footnotes and references. If the text becomes difficult, because of biochemical words or jargon, please read on, these details are not pertinent to the overall story. As far as possible, we have tried to exclude jargon or consign it to footnotes. A glossary is provided towards the end of the book, to clarify the language.

Physicians should find the book easy to read, but we hope they will find the ideas sophisticated and challenging. Biological scientists should feel quite at home, as they are used to an evolutionary approach. We have included references in the text, for the benefit of readers who wish to follow up the ideas. Often we have cited one example, where other references would have been equally valid. We have used such references to indicate additional sources of information, rather than as specific justification for the points made. However, it should be possible to validate our statements by following the references or by an elementary internet or library search.

Throughout the book, we describe the international organisation of medicine as the *medical establishment* or *medical orthodoxy*. This is not intended as a derogatory term, nor do we intend it to be applied to individual physicians. Modern medical research requires large amount of funding; consequently, there is a risk that undue control could be exerted by the pharmaceutical giants and their financial interests. Even the most outspoken and diligent physician can have limited influence on the diverse and international organisation of such research.

References to vitamin C, ascorbic acid and ascorbate are used interchangeably. Generally, we do not distinguish L-ascorbate from racemic ascorbate in the text, although L-ascorbate enters cells more effectively and is the preferred supplement for therapy. References to intravenous vitamin C refer only to sodium ascorbate and not to other forms, such as ascorbic acid, which have side effects with this mode of administration. Most chemical equations are relegated to footnotes. Chemical species are used in the text and diagrams, to illustrate the mechanisms involved, and should not be assumed to be intended to represent a balanced chemical reaction. In our previous books, *Ascorbate: The Science of Vitamin C*,[15] and *Ridiculous Dietary Allowance*,[2] we invited readers who found errors to email us at radicalascorbate@yahoo.com. More than two thousand copies of Ridiculous Dietary Allowance were provided by free internet download, to ensure that factual errors and anything else we had missed were corrected. This is a most constraining form of peer review and may provide the readers with reassurance that the ideas presented are factually correct and logically consistent. Readers are encouraged to email details of errors in this text to the same address, so we may correct later editions.

Introduction

"Science is built up of facts, as a house is with stones. But a collection of facts is no more a science than a heap of stones is a house." Henri Poincaré

The aim of this book is to take a fresh look at cancer, explaining what it is, why it develops, and how it might be overcome. As far as most people are concerned, cancer is a death sentence. We aim to show that this need not be the case. Although cancer is an almost inevitable feature of multicellular animals, there are ways to control it.

To recognize the strategies needed to win the fight against cancer, we must first understand its biology. This will help us explain why people get the disease, and why established cancers can be so intractable. A biological perspective also suggests promising approaches to treatment. We may not yet be able to eradicate cancer, but we can predict methods to restrict its development and spread. The aim of such treatments would be to allow patients to live out their natural lifespan, unaffected by the presence of cancer cells within their bodies. There is solid scientific evidence that this is possible.

Cancer is the name for a group of diseases, characterised by uncontrolled growth of abnormal cells in the body. Although cancer has been with us throughout history, it is often claimed that major improvements in cancer treatment have been made over the last 50 years. These include more accurate and earlier diagnosis, selective surgery, chemotherapeutic drugs and radiation therapy. People generally assume that these advances have decreased the number of cancer deaths in developed countries. Despite this hope, cancer remains a feared disease. Many conventional treatments have severe side effects, and progress in fighting the worst cancers is slow.

Any approach to cancer should depend on scientific answers to fundamental questions. We are not interested in opinion, conjecture or conspiracy theories, but in demonstrable experimental data and its immediate implications. Such evidence suggests that far more people could survive cancer than do at present.

An ideal treatment

Suppose a medical breakthrough could prolong the lifespan of terminal cancer patients. If the therapy could extend a patient's life tenfold, a patient previously given three years to live could expect 30 years of relative health. These patients might expect to die from other causes. They would have cancer, but it would not reduce their quality of life or expected lifespan.

Now imagine the treatment is non-toxic. The patient would not be poisoned by chemotherapy or radiation. Indeed, their other tissues would be invigorated by the therapy. The treatment would involve little more than a change of diet, taking tablets and attending outpatient checks at the hospital. Some people might achieve a complete cure, although others would need to stay on the medication indefinitely.

According to prevailing theories, this imagined therapy is too good to be true. Unless medicine undergoes a paradigm shift, mainstream efforts appear unlikely to achieve such a breakthrough in the next century. Surprisingly, however, the background for non-toxic therapies has already been established. These treatments may already exist, although they have not been properly investigated in clinical trials. Based on known scientific results, such a therapy might be achieved within a decade. Although cancer may not become a disease of the past, we might consider it manageable and much less frightening.

Cancer as microevolution

Current medical models do not explain why cancer develops. Furthermore, they give only a limited appreciation of the underlying processes. These theories of cancer and its development may have acted like a mental straightjacket, restricting productive lines of research.

This book presents a new model, describing the development of cancer as the *microevolution* of a tumour within the body. This provides a guide for understanding the progression of the disease and the effects of treatment. The new approach considers cancer in terms of the microevolution of cells and their descendants. While it has long been recognised that the progression of cancer has evolutionary components,[3,4,5,6,7] we take the process further. In our view, microevolution is the driving force behind the development of cancer, and can be used to explain the varied features of the disease. This defining mechanism of cancer is an essential part of life itself. By understanding the importance of these evolutionary processes, we can gain insights that can be used to predict effective treatments.

The formation and spread of malignant cancer is an inevitable result of faulty cell division. The theory of evolution, applied to a set of cells that are characterised by erroneous growth, explains why cancers become malignant. Any set of cells with these characteristics would eventually lead to the development of a malignant tumour. This simple model predicts many of the properties of cancer, as well as the nature and effectiveness of treatments. Our approach implies it should be relatively easy to develop non-toxic remedies, offering great benefit to cancer sufferers. The model also explains why conventional cancer treatments have failed to eliminate the disease.

Ineffective treatments

Conventional treatments have had limited success against some rare cancers, such as testicular carcinoma. Unfortunately, most cancer therapies have an unacceptably low success rate and too few people live longer than five years after diagnosis. Over recent years, much apparent improvement in treatment has been a result of other factors, such as earlier diagnosis. Physicians avoid the word cure when discussing cancer; one of the main reasons is that, with the exception of surgery, conventional treatments are rarely curative and often do not extend lifespan.

Nutritional therapies

Over millions of years, multicellular organisms, such as animals and plants, have evolved mechanisms to control abnormal cell growth and cancer. This implies the existence of nutrients and naturally occurring substances, which destroy cancer cells without being toxic to healthy tissues.

The treatments described in this book are based on such non-toxic nutrients. Food includes many components that can kill cancer cells selectively, when taken in sufficient quantities. Despite the commonly held belief that anticancer agents are rare, we will show that many dietary antioxidants destroy cancer cells, while leaving healthy cells unharmed. The effectiveness of these nutrients has been established for decades. However, perhaps because of increasing specialisation and the lack of a common literature pool, physicians and specialists are unlikely to come across this information. This approach has therefore been largely overlooked by the medical establishment.

Potential treatments neglected

People might assume that conventional medical researchers would be striving to follow any path that might lead to a solution. This is not the case: medical scientists could learn a lot from ideas they have discounted. Many alternative therapies have been discarded as scientifically unsound. However, it will become apparent that often such treatments have been arbitrarily rejected. A notable exception is the investigation of therapies that might lead to direct financial profit, for which research funding is more readily available.

There is something wrong with the medical establishment's response to cancer. Cancer has been inadequately researched for decades and it is time for this to change.

Is a cure possible?

Cancer is sometimes described as a number of different diseases, with little in common apart from unrestrained cell growth. Because of the diversity of its different forms, many experts consider the idea of a single cure to be almost impossible. Clearly, if researchers do not believe there is any potential for a cure, it is unlikely to be found.

By contrast, the microevolutionary model suggests that cancers have a great deal in common; they develop in a consistent way, which is well described by standard biology. The biological features of cancer make it difficult to eradicate from the body, but also suggest ways to fight it. The model implies that even patients considered terminal may have a real chance of a longer life or, occasionally, a complete cure, through properly administered, non-toxic nutritional therapies.

In this book, we describe how cancer reflects the differing evolutionary pressures on single and multicellular organisms: it is caused by the microevolution of injured cells. Given enough time, cancer will develop whenever there is a proliferation of damaged cells. The role of antioxidants and free radicals is also central to the development, control and elimination of cancer. Plants contain large numbers of non-toxic, anticancer agents. The presence of these can be explained using basic biological principles.

Because of its fundamental biological nature, a "cure" for cancer is probably impossible. However, non-toxic treatments, based on massive oral or intravenous doses of vitamin C and other antioxidants, can destroy cancer cells and prevent them from growing. This is especially true if glucose is in short supply. There is an overwhelming preponderance of scientific evidence that even "terminal" cancer is treatable, using this approach.

Based on this understanding, we can conclude that cancer is, potentially, a controllable disease. In the majority of cases, it should not lead to premature death.

What makes you think they really want a cure?

"The combined profits for the ten drug companies in the Fortune 500 ($35.9 billion) were more than the profits for all the other 490 businesses put together ($33.7 billion)." Marcia Angell

Many doctors and medical scientists chose their career because they wanted to find a cure for cancer. Other youngsters might want to be a footballer, pop star or movie icon, but these had more altruistic aims. This ambition is hardly surprising: cancer is perhaps the most dreaded disease and finding a cure is a central aim of medical science.

Unexpectedly, while researching this book, we gained the impression that a cure for cancer would perhaps be less welcome than one might think. Where were the scientists who were searching for a cure, rather than a slightly improved treatment or minor variation to clinical practice? Despite all the hype, we could find little evidence that a cure was considered an achievable aim. Charities make pleas for money to cure cancer, but areas of research that appear to have great potential have been sidelined. To explain this finding, we need to describe the conflicts of interest that lie at the heart of modern medical science.

Consider the case of an eminent scientist, who spent years working to find a cure for cancer.[a] Decades ago, he made substantial progress in finding an effective treatment, but received little encouragement. Numerous originators of "cancer cures", often without foundation, make similar claims that their treatments are ignored. This was different. The report was from a leading scientist, a physician we have known for many years. His work was pointed out to us by a senior surgeon, who told us that the cytotoxic mechanisms presented our book on vitamin C were similar to this doctor's early claims.[15] Although initially sceptical, we were struck by the similarity of the two mechanisms. The scientist's early research had been published in leading journals, before apparently being forgotten.

---

[a] We have elected to respect the scientist's anonymity, for reasons explained later in this chapter.

We were incredulous. Why, we asked, if he had found a cure for cancer, did he not have a Nobel Prize? And why are people still dying? His shocking reply was, "*What makes you think they really want a cure?*" He explained that he had a wife, children and a career, and had not been willing to give these up for the sake of people who would not listen. On a happier note, as we describe later, his ideas fit well with the microevolutionary model of cancer.

To people unaware of the pressures on medical scientists, the suggestion that anyone might not want a cure for cancer appears ludicrous. Despite this, the suppression of new ideas because of financial interest is so often overstated that it is almost a cliché. The classic film, *The Man in the White Suit*, is a good example. Alec Guinness plays Sidney Stratton, a scientist who invents a wonderful cloth that does not wear out or get dirty. As soon as the financial implications of the new cloth are realised, there is a concerted effort by all concerned to stifle its exploitation.

Occasionally, reality is stranger than fiction. We began to suspect that the criterion used for investigating cancer treatments is the potential for financial profit.[8] Modern medicine is structured around the development of new treatments, while the less profitable aim of prevention is a secondary consideration. If a new treatment will generate profits, it is investigated. However, if the treatment provides no mechanism for financial gain, it is discarded. Moreover, if a therapy provides the possibility of a cure without profit, then not only will it not be investigated, it might even be actively suppressed.

The financial corruption of medicine has been described by others.[8,9,10] We make only brief and passing comment, to put the research on cancer into context. Corruption in medical science is now a mainstream consideration for academic researchers. Dr Marcia Angell, formerly Editor-in-Chief of the prestigious New England Journal of Medicine, recently described how drug companies were biasing results in medical studies.[11] Angell describes how the pharmaceutical industry,

"*...corrupted by easy profits and greed, has deceived and exploited the American people.*"

Such strong statements by a leading physician make the point clearly. Pharmaceutical companies regard medical research as a branch of marketing. As a result, the medical literature has been contaminated with misleading reports and biased studies.

The popular impression is that cancer researchers are forging ahead at the forefront of science, leaving no stone unturned in their fight to conquer the disease. Sadly, this description is naïve. It appears that the medical establishment is structured to prevent a breakthrough in cancer research. We can compare our modern systems to the time when the development of penicillin was fast-tracked, to keep soldiers fighting during the war. That probably could not happen nowadays - medicine is administered in such a way as to make the necessary recognition, experimentation and use of such a new treatment almost impossible.

Suppose a researcher discovers a simple chemical substance that kills cancer cells but leaves healthy cells undamaged. Supporting experiments confirm the finding in test tube studies and tissue models. There are reports of successful treatments and even cures.[15] The public might expect that the "cure" would be tried immediately, to see which cancers it would be effective on, so it could be given to patients in desperate need. However, modern medicine demands extensive tests, clinical trials, regulatory approval, and so on. Such trials are exceedingly expensive and, unless funding is available, might never be performed. Most current patients would be dead before they got the chance to try the treatment. As we show later, this claim is not simply rhetorical: such anticancer substances already exist.

Professional suicide

This book was written to try to make the study of cancer more scientifically and ethically respectable. Research into cancer and its treatment should be open and refutable. A scientist claiming that, say, carrot juice can cure cancer, has a right to generate the idea as a hypothesis, providing it is consistent with the available evidence.[b] It is not good science to suppress such an idea on the basis that it conflicts with current guidelines, gives false hope, or may prevent patients from getting conventional treatment. To claim that the idea has no scientific proof is meaningless.[15] Such a constraint on the generation or investigation of scientific ideas is counterproductive. If someone objects to a particular hypothesis, the scientifically valid response is to test it, and show that it is wrong.

Clinicians are generally not trained as scientists. Indeed, during a long and arduous training, they have to learn and accept mountains of facts, based on recognised medical opinion. This training teaches them to

[b] Carrot juice is unlikely to have more than a slight effect on the disease.

be conservative and to conform to best practice, unless they have solid reasons for deviation. The result is a medical profession that is essentially conformist, which is what society asks of people who are involved in life and death decisions. Patients expect an established and appropriate treatment for a health problem, not a therapy based on the whim of the doctor.

The position in science is different. Scientists are iconoclasts. Their personal attributes are almost the opposite of those required of a physician. A scientist is required to challenge every idea and try to find improved explanations. Scientists are supposed to generate new ideas and investigate them. The idea of accepting current ideas and common sense should be an anathema to a scientist. Indeed, ideas in physics, the most solid of disciplines, may be rejected *because they are not crazy enough*! However, medical research seems to operate by different rules from the rest of science. If a medical scientist's ideas challenge the current orthodoxy, the researcher's future career may suffer.

An extreme example of scientific suppression in medicine can be found with AIDS. Peter Duesberg was a high flier in virus research. From 1968-1970, he demonstrated that influenza virus genes are segmented into parts, clarifying the virus's ability to vary rapidly, by re-assorting its genes. This explains why flu can change from year to year, and why new vaccines are always needed. Shortly after, Duesberg isolated the first cancer gene, through his work on retroviruses. These are the class of virus that include the human immunodeficiency virus (HIV), claimed to cause AIDS. Peter Duesberg was among the first to study the structural proteins of these viruses. This outstanding work led to his election to the US National Academy of Sciences, in 1986. Between 1985 and 1992, he was awarded a seven-year Outstanding Investigator Grant, from the National Institutes of Health.

Duesberg's mistake was to challenge the idea that the HIV virus is the cause of AIDS.[c] He argued this in a series of papers, published in prestigious, high impact, scientific journals. He proposed an alternative hypothesis: that the various AIDS diseases are brought on by long-term consumption of recreational drugs and anti-HIV medication, such as AZT, which is prescribed to prevent and treat AIDS.

Duesberg had been expected to gain a Nobel Prize, but instead his research funding was terminated and his reputation attacked. He had nothing to gain by proposing an alternative hypothesis. Indeed, since he

[c] Note that we are making no statement on the role of HIV in AIDS.

had done outstanding work on retroviruses, his status would have been enhanced if HIV were shown to cause AIDS. However, it is normal practice for scientists to propose hypotheses for testing. It does not matter whether Duesberg's ideas were ultimately correct; he had a right, even a duty, to state his objections to the HIV hypothesis. This is a basic rule of science, and any action that restricts the scientific method is ultimately harmful.[15]

Medical scientists who publish radical ideas are often suppressed, and Duesberg had crossed the line. The medical establishment, having decided that HIV was the cause of AIDS, believed it was irresponsible of Duesberg to express his doubts. After all, if people listened to him, they might ignore recommendations for preventing the spread of the virus. Although Duesberg was a leading virologist, speaking in his area of expertise, his ideas were condemned.

After publication of Duesberg's initial scientific paper,[12] which cast doubts on the theory that HIV causes AIDS, Chuck Klein, of the US Department of Health and Human Services, wrote a media alert to the heads of associated government departments, including the White House. He said,

"This obviously has the potential to raise a lot of controversy (if this isn't the virus, how do we know the blood supply is safe? How do we know anything about transmission? How could you all be so stupid and why should we ever believe you again?) and we need to be prepared to respond. I have already asked NIH public affairs to start digging into this."[13]

This is hardly an appropriate response to a scientific paper, published in the respected journal, Cancer Research. A memorandum by Florence Karlsberg, also from the Department of Health and Human Services, describes damage limitation actions and an attempt to "educate" Duesberg.[13] Duesberg alleges that government officials tried to bribe him to recant his position. Moreover, he claims 22 of his grant proposals were rejected sequentially.[14] It can take weeks to put together a proposal for research, so this may correspond to several years' work being wasted. Duesberg's research funding effectively ceased, damaging his future career. However, he remains a professor of molecular and cell biology at the University of California.

If the establishment can destroy the career of a world-leading scientist like Peter Duesberg, only the bravest physicians will champion unwelcome theories. In medicine, scientific theories are often viewed with scorn. The gold standard is the "*large scale, randomised, double-blind clinical trial*", an experimental method that is a particularly weak kind of science.[15] Large-scale clinical trials are appropriate for finding rare diseases or side effects, which occur with a low frequency in a large population. In science, small, easily reproducible experiments are preferred.

Biology

"All science is either physics or stamp collecting." Ernest Rutherford

The science of biology is concerned with the variation and diversity of life.[16] By contrast, one of the features of modern medicine is its concentration on the particularities of the human body. In medicine, humans are considered distinct and results found in another animal, such as the guinea pig, are viewed with suspicion. Such experimental animals may react quite differently to a drug or treatment. For the most part, this concentration on human specifics is a sensible approach, recognising biological variation. However, a second, apparently contradictory, feature of living things is their similarity.[17] It is sometimes claimed that people share more than 98% of their genes with chimpanzees and about 50% with bananas.[a] The reasoning behind this humorous statement is that the basic biochemistry of cells is common across living things. Many core human genes and enzymes are shared with plants, animals and even simple, single-celled organisms.

We will show that viewing cancer from an evolutionary perspective provides a consistent explanation for the disease. This will come as little surprise to biologists, who use a broad evolutionary framework as a context for their research. Physicians, on the other hand, are more concerned with distinctive features of human beings. Occasional approaches to cancer have described clonal development, when a single cell divides, mutates and diverges, ultimately producing a malignancy.[18,19,20] However, physicians generally view cancer as a disease, an abnormality or developmental process, rather than an evolutionary sequence of events.

This book considers cancer at the level of biological principles and microevolution. By evolution, we do not mean the development of life on a grand scale, with its philosophical and religious problems. The form we consider, microevolution, is more limited and covers the development of a population with time, together with the processes by which living things change. Knowledge of microevolution is essential to our

---

[a] The statement "we share half our genes with the banana" is sometimes attributed to an offhand comment by the British scientist, Robert May.

understanding of cancer; we also need a basic idea of cells, their structure and the ways they combine to form organisms.

Microevolution can be observed directly, so even those who reject the concept of evolution find it hard to dispute. The term does not stretch the concept of evolution or attempt to explain the origins of life. Microevolution is involved in processes such as selective breeding, to create different kinds of plants or animals; its existence can be demonstrated by comparing a Great Dane to a Chihuahua, for example. The breeding of animals for pets and flowers for our gardens can be thought of as human-directed microevolution.

The modern medical approach to disease processes tends to ignore their core biology. Medicine has become increasingly specialised, with problems being split into different perspectives. Geneticists study the genes; biochemists, the enzymes; clinicians, the methods of treatment; and epidemiologists, the distribution of the disease in the population. With cancer, this process of specialisation has caused the underlying nature of the disease to become lost in a morass of detail.

Recently, molecular biology and genetics have dominated cancer research, but this viewpoint provides a limited description of the problem. Although research at the level of molecular biology is essential to gain a full understanding of the mechanisms of cancer, such low-level research is incomplete. Currently, it is not possible to derive the workings of even a simple bacterial cell from the actions of its genes and enzymes. We therefore need to take a broader view, in order to obtain insight into this disease.

In this book, we approach the problem primarily at the level of the cell and its function. What distinguishes a cancer from a normal, healthy tissue is the behaviour of its cells. This ultimately depends on small-scale processes, such as those involving genes and proteins. We shall see that the metabolism of cancer cells, the process by which they use oxygen and generate energy, is different to that of healthy cells. Cancer cells use less oxygen to generate energy, and use antioxidants in a distinctive way.

Oxidation and reduction

Antioxidant supplements and anti-aging cosmetics are increasingly familiar. Adverts promote fruit, vegetables and other foods as healthy, because they contain antioxidants. Cosmetic creams contain "age-defying" antioxidant substances. As the name suggests, antioxidants prevent oxidation; they are also called reducing agents. Historically, the word oxidation meant combining with oxygen. For example, iron combines with oxygen to form rust. Nowadays, the term has been generalised and oxidation can be defined as the loss of one or more electrons. Similarly, reduction involves gaining electrons. Oxidation and reduction are core biological processes; ultimately, all life depends on chemical oxidation and reduction reactions.

A basic division in biology is the classification of organisms into those that use oxygen (aerobic) and those that live without oxygen (anaerobic). Humans, like other animals, breathe in order to supply oxygen, via the blood, to our tissues, which need oxygen in order to burn food and generate energy. Oxygen is so important that prevention of breathing for even a short time results in death. The brain is particularly at risk: if its blood supply fails, brain cells can be damaged or die from lack of oxygen.

Paradoxically, although oxygen is essential for survival, high concentrations of the gas are poisonous. The air that we breathe for its life-giving properties is slowly poisoning our tissues, by oxidising them. This oxidation involves free radicals and, chemically, it is like a form of slow burning. Antioxidants prevent this oxidation damage.

The term oxidation is still used for chemical reactions involving combination with oxygen. However, the scientific use of the term is more precise:

Oxidation is the loss of an electron by a molecule or atom.

Reduction is the uptake of an electron by a molecule or atom.

A simple mnemonic to remember this is *LEO the lion goes GER*. LEO is short for "Lose Electrons - Oxidize" and GER is "Gain Electrons - Reduce". Others prefer the mnemonic *OIL RIG*: "Oxidation Is Loss", "Reduction Is Gain".

The processes of oxidation and reduction are complementary. In a chemical reaction, electrons lost by one substance are gained by another. However the terms are described, these ideas take a little time to assimilate.

To a first approximation, we can consider that oxidation is essential to life, but also damages tissues. Conversely, reduction by antioxidants is generally beneficial. A healthy tissue can be well supplied with oxygen, while remaining in a reducing state; high levels of antioxidants provide electrons to prevent oxidation and free radical damage. A sick tissue, which is short of oxygen, can still be in an oxidising state, if there is a source of free radicals to consume the available electrons.

The relative level of reduction and oxidation in a tissue is called the *redox state* or *redox environment.* Redox processes are the most fundamental reactions in chemistry. We will be using the terms oxidation, reduction and redox throughout this book. To a first approximation, a high level of oxidation is bad for a cell, whereas reduction is healthy.

Oxidants and antioxidants

Oxidants and antioxidants have central roles in biology; they are involved in immune defence, metabolism (the process by which cells generate energy) and in the manufacture of chemical messengers, called hormones. They are also used to send messages between cells, to synchronise cellular behaviour. This process is called redox signalling; it plays a part in cell growth, division and death, and helps control oxygen levels.[21,22] These are central processes in the development of cancer. The redox state of a tissue is a broad measure of its health: higher levels of oxidation are associated with illness, lower levels with health.[15]

Although oxidation is essential for generating energy in our cells, it is also a source of damage. Most diseases harm our tissues by means of oxidation and free radical damage.[15] Our bodies retaliate in kind: white blood cells attack organisms, such as bacteria and viruses, using oxidants to damage and disable the invaders. Although we depend on oxidation for metabolism, immune defence and cell signalling purposes, we must also avoid the damage it can cause. For this reason, cells have a wide array of antioxidant defences. We define an antioxidant as any factor that can prevent oxidation. For our purposes, we can think of an antioxidant, such as vitamin C, as a donor of electrons.

The term free radical[b] is often associated with the use of antioxidants. A free radical is a molecule with an unpaired electron. In molecules, electrons usually occur in balanced pairs, with each of the pair spinning in the opposite direction. If oxidation strips an electron from a molecule, it leaves behind an unpaired electron. This unbalanced electron will search for a partner, to restore its equilibrium. In this way, the

---

[b] For simplicity, we use the term free radical loosely and include reactive oxygen and nitrogen species, which can act as biological oxidants, but may not have an active unpaired electron.

oxidised molecule acts as an oxidant, which needs to steal an electron from another molecule. It has become a free radical.

An oxidising free radical sets up a chain reaction, with each molecule stealing an electron from the next until, eventually, a molecule reacts with itself or changes form. When this happens, the molecule becomes damaged and is no longer able to perform its normal function within the cell. We use the terms oxidation and free radical damage to describe these injuries to cells and tissues. Oxidising free radicals are involved in almost all kinds of disease and damage to the human body. Antioxidants quench free radical chain reactions, by donating electrons.

Perhaps the best-known antioxidant is vitamin C; its main role in the body is to donate electrons. The vitamin C molecule exists in two forms: the reduced form, known as ascorbate, and the oxidised form, known as dehydroascorbate. The ascorbate form of vitamin C can donate a single electron to form the ascorbyl radical, or two electrons, to form dehydroascorbate. When vitamin C has lost electrons in this way, it ceases to be an antioxidant.

The ascorbate redox ratio, which is the proportion of reduced to oxidised vitamin C, provides an indicator of illness in the body. Diseased tissues have relatively less ascorbate and more dehydroascorbate. However, cells in the body can provide electrons to reduce dehydroascorbate back to ascorbate, restoring vitamin C to its electron-donating antioxidant state. This process is reversible. When dehydroascorbate is reduced by the cellular metabolism, it is acting as an oxidant, taking electrons from the local environment. When ascorbate donates electrons, it is acting as an antioxidant.

Ascorbate (antioxidant) is oxidised to form dehydroascorbate.

Dehydroascorbate (oxidant) is reduced to form ascorbate.

A central feature of healthy cells is that they have an ample supply of antioxidants and are in a reduced state. Sickness and damage result in production of free radicals, making the tissues more oxidising. This forms the basis for the antioxidant treatment of disease, which suggests that tissues should be maintained in an environment with plenty of electrons, to deal with harmful, oxidising free radicals. By giving massive amounts of vitamin C to a sick individual, the ratio of ascorbate to dehydroascorbate can be increased, and the tissues returned to a healthy, reducing state.[15]

Other cellular antioxidants, such as vitamin E, glutathione and coenzyme Q10,[23] exist in both oxidised and reduced forms. The proportion of these molecules in the reduced state is also a measure of tissue health. However, vitamin C is unique, in that it has low toxicity and can be given in the massive amounts needed to overcome the free radical onslaught of severe disease.[15] The doses involved can be as high as 200 grams per day, much larger than the gram level doses normally considered as large intakes.

Free radicals and oxidation are involved in all aspects of cancer. The formation of cancer often involves free radicals, such as those generated by X-rays or carcinogenic chemicals. Cancer cells need oxidants to divide and grow. Furthermore, the availability of an oxygen supply dominates the way cancer grows and develops. Cancer cells develop under anoxic conditions (without oxygen), and consequently, they are defective in the way they handle oxygen.

In addition, cancer treatments, such as radiation and chemotherapy, often work by generating free radicals in the cancerous tissue. These damage the cancer more than they do healthy tissues, because cancer cells are in a more oxidised redox state and have impaired antioxidant controls.[645] Damaged cancer cells commit suicide, in a process that also depends on free radicals. If the level of oxidants increases enough, the cancer cell will die. Furthermore, the death of cancer cells can generate further free radicals. This involvement of oxidation and reduction in cancer biology is to be expected: it is a central feature of life.

Cell types and structure

Cancer is a disease of cells. Cells are the fundamental structural unit of life, the smallest biological units of specialized function that can self-replicate. They are made up of smaller parts, such as proteins, membranes and genes. Although it is difficult to define 'life' precisely, an isolated gene or protein is not normally considered as living. People can normally distinguish living things from inanimate objects with relative ease. Most people would accept that an active cell, seen under a microscope, is alive.

Viruses are at the limits of what can be considered alive. They are smaller than cells and are not usually thought to be fully living, as they are incapable of independent growth, movement and replication. Viruses need cells to reproduce: they multiply by invading cells and taking over their biochemical machinery.

The cells of different life forms have varied characteristics; bacterial cells are usually smaller than plant or animal cells, for example. Bacteria are simple, living organisms, consisting of a single cell that lacks easily identifiable internal structures. An outer plasma membrane contains and encloses the cell, separating it from the surrounding environment. Simple cells of this type are called prokaryotes.

Typical animal cell structure

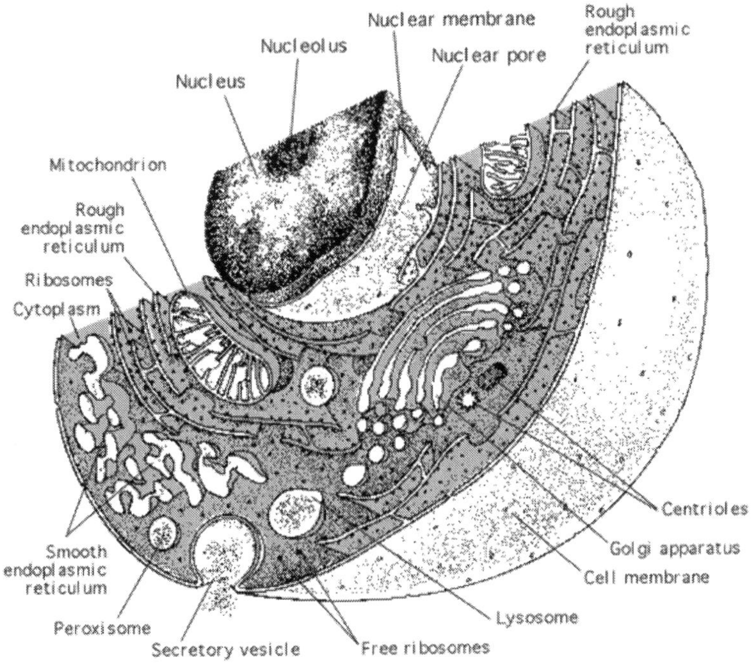

Generally, animal and plant cells have a clearly defined organization, with many internal structures. Biologists call these larger cells eukaryotes. Such cells contain a membrane-bound nucleus and other well-developed internal compartments. Most organisms, with the exception of viruses, bacteria and blue-green algae (also called cyanobacteria or pond-scum), consist of one or more of these relatively large eukaryotic cells.

Under the microscope, a typical animal cell looks as though a large cell has swallowed some smaller, cell-like objects, called the nucleus and mitochondria. Indeed, this may be exactly what happened with mitochondria, during evolution. A larger cell could have engulfed a smaller bacterium, which eventually produced the mitochondria we see today, with both cells gaining an advantage.[24] When living creatures work together to the advantage of each, the process is called mutualism and is a form of symbiosis, or living together in close association.

The most striking difference between the appearance of typical plant and animal cells is that plant cells have a thick outer cell wall. This gives plants their internal structure and strength. The wall is composed of structural materials, such as cellulose, and provides rigidity. Plant cells are often immobile, held in place by their thick cell walls. These limit cell movement and growth, giving plants a resistance to the kind of runaway malignant growth found so frequently in animals.

Animal cells are bounded only by a membrane and are sometimes capable of independent movement through the body tissues. An example of this is the migration of white blood cells towards injured tissues, to fight infection. The absence of a thick cell wall facilitates the growth and mobility of animal cells, but also contributes towards the spread of abnormal cancer cells.

A typical human cell has a thin outer membrane, which encloses a number of active membranes and internal structures. Two of these subcellular structures, the nucleus and the mitochondria, are of particular relevance to our subject. Inside animal cells, the nucleus contains the genetic material used to pass information to the cell's descendants. Like the cell as a whole, the nucleus is contained within a plasma membrane. The nuclear membrane separates it from the surrounding fluid, or cytoplasm, within the main cell body. Most of the DNA (deoxyribonucleic acid) that forms the genes is contained within the nucleus.

DNA consists of two strands, in the famous double helix structure, discovered in 1953 by Rosalind Franklin, Francis Crick, Jim Watson and Maurice Wilkins. For our purposes, a gene is a length of DNA, which codes the information needed to produce a protein. The building blocks of DNA are nucleic acids, which come in two forms: purines and pyrimadines. These small molecules are strung like beads on a chain, and the sequence of nucleic acids makes up the genetic code. In DNA, three adjacent nucleic acids code for an amino acid, a single building block of a protein.

Amino acids are small molecules that can join, or polymerise, to form long strands, or proteins. Protein chains can form an alpha helix, as described by the chemist, Linus Pauling. More typically, proteins fold into complex, three-dimensional structures. The purpose of a gene is to determine the sequence of amino acids and the length of the protein chain. In this way, genes carry the information needed to make the thousands of proteins that form our bodies and help make each of us biologically unique.

DNA is the central information store in the nucleus, specifying the proteins that provide the cell with its structure and function. The mitochondria are equally important, but for a different reason. Mitochondria are small structures within cells that oxidize food (proteins, fats and carbohydrates) to generate a usable form of energy. They produce most of the power for animal cells, providing the energy for life.

Mitochondria contain DNA, which is more easily damaged than that in the nucleus. Unfortunately, mitochondria produce free radicals and reactive oxygen species, as a by-product of energy production. Over time, the resulting oxidation may damage the mitochondria, making them function less efficiently. This damage leads them to generate more oxidants, in a self-perpetuating vicious cycle. Antioxidants, such as vitamin C and alpha-lipoic acid, help prevent damage to the cell by neutralising these free radicals. Mitochondrial damage may cause much of the decline in aging,[25] and may influence the effectiveness of cancer treatments.

Genes and chromosomes

Genes are the physical units of heredity. Gregor Mendel first suggested the existence of genes, based on his work with pea plants. His success owed much to the peas' visible characteristics, such as wrinkles or smoothness.

In a series of experiments, from 1856 to 1863, he grew 28,000 pea plants. He developed two important ideas, later named Mendel's Laws of Inheritance. The primary law was that the transmission of genes is independent. He also demonstrated that a single gene could have many forms, and that one form of a gene was inherited from each parent. Only one of each pair of inherited gene types, called alleles, was used by the offspring.

Subsequently, Mendel's experimental results caused a great deal of dispute. A leading statistician, Sir Ronald Fisher, analysed the data and found the results to be improbably close to the theoretical predictions of Mendel's model. However, this was not a case of outright scientific fraud, as his theory has been confirmed many times by later experiments. It is probable that Mendel, or one of his colleagues, massaged the data to fit the theoretical predictions more accurately.

Within a cell nucleus, each gene sits on a chromosome. The chromosomes are composed of DNA and protein; they become visible during cell division, as striped, blob-like structures. Human cell nuclei contain 23 pairs of chromosomes. Red blood cells are notable exceptions, containing neither chromosomes nor a nucleus. Red cells are specialised for transporting oxygen within the body: not having a nucleus makes them smaller and more efficient carriers of this essential cargo.

Normal karyotype: full set of human chromosomes

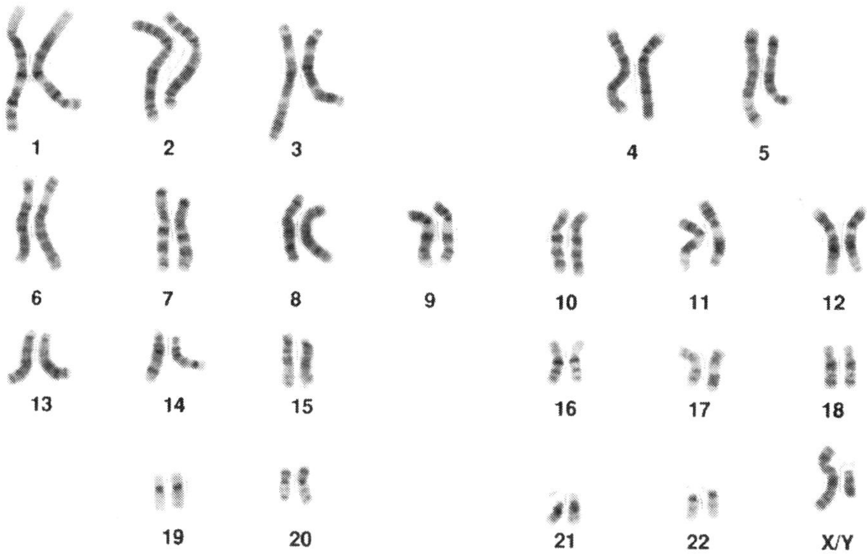

Other plant and animal cells contain different numbers of chromosomes, characteristic of their species. Chromosomes are used to package and contain DNA, which facilitates movement into two daughter cells, during cell division.[26,27] The long spiral of the DNA double helix is coiled within the chromosomes. In this "super-coiling", the DNA helix curls up like a telephone cord, shortening the molecule so it takes up less space. This spiralling into super-coils makes for efficient packing and separation of the long DNA molecules, which could otherwise look like spaghetti.

A gene consists of DNA, which provides information for the synthesis of its sister molecule, ribonucleic acid (RNA). Like DNA, RNA carries genetic information. RNA is a long molecule, similar to DNA, but is usually single-stranded. RNA molecules come in several forms. Messenger RNA carries the instructions for making proteins from the nucleus to the cell cytoplasm. The use of messenger RNA separates the DNA data storage from the processing of the information into proteins. The DNA stays behind a membrane in the nucleus, keeping it separate from the rest of the cell. This provides a controlled environment, to minimise damage to the DNA.

Evolution

Grand theories of evolution and the origins of life are scientifically, philosophically and theologically controversial. However, our concern is the more limited evolution of a cancerous cell line, within the body. Such small-scale microevolution can be followed mechanistically in simple chemistry, or even computer software.[28,29] It is possible to simulate the microevolutionary changes in a reproducing population, in a computer model.

We are interested in how changes in normal cells lead to cancer and how such cancer cells vary with time. To understand this, we will consider a population of cells within the body, viewing the process of microevolution at the level of the cells. Most changes will be traceable to variations in the genetic information, contained in the DNA, or to its expression as proteins. There is little doubt that variation in cells, like that between different animals or plants, is dominated by genetics.

Although it is essential to understand genetic changes, the properties of whole cells are what make cancer such a dangerous disease. Cancer is most easily described in terms of abnormal cell growth. The presence or absence of a gene may make a cell more or less likely to generate a cancer, but several other factors may also play a role. As we shall see, there is no "gene for cancer".[30] The start of a cancer is a more interesting process than simple gene mutation.

If we ask a typical person to define evolution, they may well reply "survival of the fittest". However, this description can lead to a circular argument, where the fittest creatures are defined as those that are more likely to survive. Evolution is more clearly described as an elimination process. For our purposes, evolution requires a population of cells that vary in their characteristics. The cells must be able to divide and multiply.

There is also a selection process, which kills cells or prevents them from multiplying. Given time, those that divide and leave more offspring will eventually dominate the population.

A fitter cell is one that avoids elimination and leaves more offspring. One cell is not more fit than another in any absolute sense, unless the second cell is sick and dying. Different organisms will respond to the local environment in distinct ways. For example, a cactus may be slow in growing to the stage where it can produce offspring, but will survive better under desert conditions than a polar bear or a fish. The fitness of an organism depends upon its habitat, local environment and current conditions.

Cell population growth

Most harmful cancers grow slowly and are limited by a number of control factors. By the time it is noticed, a cancer may have been growing for several years. Unrestrained tumour growth would soon kill the patient. However, limits to growth are a fundamental characteristic of biological populations. Using a simple bacterial colony as our model, we will illustrate how populations of cells grow. The biological principles of population growth in bacteria provide a representation for our later discussion of cancer.

Consider the growth of a single bacterial cell. Given sufficient nutrients and oxygen, the cell will divide and multiply. It will grow and split in two, forming two daughter cells. Each of these daughter cells will then continue growth and division. Different species of bacteria will divide at different rates, depending on the conditions. To keep things simple, let us assume that our bacteria divide every hour. Under favourable conditions, bacteria would increase as shown in the following table.

At this rate of growth, in just two weeks we would have more bacteria than the number of atoms in the whole universe.[c] Since this is impossible, it is obvious that something must be constraining bacterial growth. Such growth is limited by environmental factors, such as the supply of oxygen or food. The exponential growth of populations was described and related to humans by Thomas Malthus, in his work *An Essay on the Principle of Population*, published in 1798.[31] This greatly influenced Charles Darwin, who realised that mechanisms must be in place to limit the growth of living creatures. Malthus' description of the

---

[c] The number of atoms in the universe is only about 10^{80}

unconstrained growth of organisms led Darwin to his theory of natural selection.[32]

Hours	Number of Bacteria
0	1
1	2
2	4
3	8
4	16
5	32
24 (1 day)	16,777,216
168 (1 week)	3.7×10^{50}
336 (2 weeks)	1.4×10^{101}

The selection pressure that reduces the growth of populations does not necessarily come from the environment or other creatures. An internal factor could also slow growth: if the bacterial cells were to age and die, for example. Bacterial cells have a strategy of living a long time, to maximise their potential growth rates and chances of survival. In many cases, they are said to be immortal: they simply divide to form two daughter cells. This long life, relative to the rate of division, allows a single bacterial cell to leave a maximal number of descendents. For bacteria and other cells that grow in this way, long life confers an increased evolutionary advantage.

As the first bacterium in our population divides, its local environment changes from the pristine conditions that surround the original cell. Later, as descendents encircle the original, they compete with it for nutrient supply and may release waste products. The local environment for a bacterium, or other cell, in the centre of a large colony may be quite different from that at the edge. As the colony grows, it interacts with the environment, changing the local habitat, and is itself modified in turn.

As the environment changes, cells more suited to the new conditions can thrive. For this reason, a population of bacteria can be fitter if it contains a variety of differing cells. A population of indistinguishable cells is unable to respond effectively to the changing conditions, as all the cells are the same and have identical requirements for growth. Changes in DNA, caused by mutations, ensure the bacteria vary and that the changes are inherited. Through geological time scales, bacteria have had continual selection pressure to form colonies with a small amount of variation in their DNA. Bacteria that possess a mechanism for conserving, accommodating or generating variation will leave more offspring, in their changing habitats.

One way of overcoming the restrictions imposed by a particular location or habitat is to move to another. For example, a bacterium and its offspring on a single leaf of a tree may be forced to compete for nutrients. However, if it is blown by the wind to another location, it may be able to form a separate colony and leave more offspring. Bacteria that have a mechanism for dispersal can gain a selective advantage by setting up colonies in other environments. However, the process of dispersal is fraught with danger, as the new habitat may be less favourable. Many environments may be inhospitable and bacterial growth may not be possible. Over time, selection pressure will ensure a balance between dispersal and staying put. The bacteria with the optimal characteristics for dispersal will prosper and, by our definition, be biologically fitter.

In 1859, Charles Darwin proposed another evolutionary idea: the principle of divergence.[33] This suggests that a varied population can spread more easily to new locations and habitats. He provides a simple quotation, for a carnivorous animal, that illustrates the meaning,

"The more diversified in habits and structure the descendents of our carnivorous animal became, the more places they would be able to occupy".

An organism that can occupy a new habitat has the chance to find more resources for growth, allowing it to increase in number. A varied population is more likely to make this transition and survive under the new environmental conditions.[16] Some individuals will be better suited to the new habitat than will others.

The principle of divergence provides a biological explanation for why malignant cancers invade local tissues. A cancer that is able to spread to distant sites in the body, forming metastases, has increased growth and survival possibilities. This process is analogous to the invasion of other habitats by animals. This leads to the prediction that cancer cells will proliferate more quickly in secondary sites than at the primary location. In addition, cancer cell lines in the favourable habitat of experimental cell culture will grow more quickly than cells under selection pressure in the body.

Rates of cell growth in cancer cells[1]		
		Doubling Time (days)
Colon cancer	Primary tumour	700
	Metastasis	100
	Cell line	3
Breast cancer	Primary tumour	200
	Metastasis	20
	Cell line	3

New species often fail to colonise a novel environment but, occasionally, like rabbits in Australia, the population can flourish. Moreover, when two colonies occupy different habitats, they will tend to diverge with time. Increased numbers of colonies produce a total population that is less liable to extinction. If one colony's environment becomes inhospitable and the organism becomes locally extinct, it is likely that the habitat at another location may still allow population growth. This factor of increased variation and survival advantage in separated colonies is important in the treatment of spreading cancer.

For bacteria and other single cell organisms to thrive in the longer term, they need the following characteristics for fitness:

- Long life or immortality
- Rapid growth and cell division
- Inclusion of genetic variety
- Mechanisms for occasional dispersal

We will see later that these characteristics are similar to the requirements for cells to become malignant.

Microevolution in progress

Not long ago, with the introduction of antiseptics and antibiotics, it became possible to find chemicals, such as penicillin and tetracycline, which would kill bacterial cells and leave normal cells undamaged. This was the holy grail of treatment for such diseases. Although some people did suffer side effects, such as allergic sensitivity to penicillin, generally the toxicity of these substances was low. Over the following decades, antibiotics were given widely for minor infections, and even added to animal feedstuffs. Caught up by hubris, some medical experts suggested that infectious diseases, at least those caused by bacteria, were a thing of the past. It soon became clear that such optimism was misplaced. The phenomenon of resistance became apparent, shortly after antibiotics were introduced.

When a susceptible population of bacteria, such as the boil-causing *Staphylococcus aureus*, encounters an antibiotic, such as penicillin, the bacterial cells fail to grow or die. However, some bacteria will be slightly healthier in the presence of the antibiotic, because they have evolved over millions of years to exist in a variety of habitats and conditions. Penicillin evolved in a common fungus, so the bacterial ancestors could have experienced this substance many times before. An antibiotic treatment that is less than completely effective may simply select for organisms that are more resistant.

If bacteria are given an inadequate treatment with an antibiotic, some of the cells survive. Cells that die are those most susceptible to the drug. Those that prosper are the most resistant to the antibiotic. When these resistant bacteria divide and grow, they form a colony of bacteria selected for their ability to survive the antibiotic. If the antibiotic is applied again, it will be less effective. Any surviving cells may be even more resistant to its action. People reading this may consider the process leading to generation of resistance obvious, and easily remedied. However, misuse of antibiotics, by medicine, agriculture and pharmaceutical companies, continues to promote resistant organisms.

So far, we have described bacteria as reproducing asexually, by division. Bacteria also have a form of sex, during which they transfer genetic information between organisms. This means that when a bacterium gains antibiotic resistance, it can transfer that resistance to other bacteria. Indeed, the genes for resistance to multiple antibiotics can be transferred on a single loop of bacterial DNA.[34,35] Bacteria are very promiscuous and will transfer genes between widely differing species.[36]

Once medicine employed antibiotics widely, the effectiveness of these drugs was destined to decline. Scientific development and antibiotic discovery could not keep pace with bacterial microevolution and the development of resistance. Microevolution and selection processes are fast and powerful: computer models such as genetic algorithms can generate variation exponentially, by recombining genes. The development of bacterial resistance swamped the progressive and incremental nature of antibiotic drug discovery; it was also hastened by misuse of antibiotics. Antibiotic resistance is an example of microevolution in action.

Multicellular animals

So far, we have discussed single-celled organisms. However, most of the animals and plants we meet in our daily lives are multicellular. Indeed, the clumping together of cells into a single structure is one of the most important features of life on earth. A recent review placed multicellularity top of the ten most important of Mother Nature's greatest inventions.[37] A feature of multicellular animals is that they increase their range of behaviours. Single cells do not generate wings to fly, eyes that see or a complex brain: these features require large numbers of cells, working together.

The characteristics of evolutionary fitness that apply to single cells are not appropriate for cells in large multicellular organisms. In order for multicellular organisms to exist, their cells have to cooperate and communicate. Unless the cells work together for the good of the whole, the organism will die.

Consider a flock of birds. In order for the flock to exist, the birds have to agree to move in the same direction, at the same speed, while keeping a short distance apart. Otherwise, the flock will break up and the birds will go their own ways. Birds in a flock retain their capabilities for independent existence but, while they are together, their behaviour keeps the flock functioning as a single unit.

The rules for flock formation have parallels to the co-ordinating controls of multicellular organisms. The processes that provide fitness for unicellular creatures are retained within our cells. (The importance of this fact will become obvious as we progress.) However, mechanisms for signalling, cooperation and coordination are superimposed, on top of the basic biological mechanisms of cell multiplication, invasion and growth.

Single-celled organisms often divide and grow as quickly as their resources allow. Their aim is to make as many copies of themselves as possible, to increase the chance that some of their offspring will survive. Small multicellular organisms often have a similar approach to survival. Insects, for example, frequently have many offspring, in the hope that some will grow to adulthood and reproduce. Animals that suffer high infant mortality and severe selection pressures practice this evolutionary strategy. By contrast, higher animals have fewer young, each of which has a greater individual chance of survival. For this reason, the parents can pay great attention to their offspring's growth and maturation.

Multicellular animals evolved because ancestral cells that cooperated in colonies gained survival advantages. Paradoxically, these advantages are based on cell death. William Hamilton has suggested a similar situation in animal groups, known as the selfish herd.[38] In the presence of predators, herds of animals provide each individual with a greater chance of survival. There are other advantages: Israeli biologist, Amotz Zahavi, proposed that herd animals share information, such as where to find food.[39,40] Organisms gain benefits by modifying their behaviour and cooperating with others.

The presence of oxygen in the environment suggests another reason for the formation of multicellular organisms. Oxygen, as its name suggests, causes oxidation damage, which can kill cells. However, if some cells clump together to form a colony, the central cells are shielded from high levels of oxygen. The outermost cells may suffer, or even die, but the majority of the colony survives.

In times of food shortage, a multicellular colony may tend to be resistant to starvation: some of the cells may die, providing nutrients for the remainder. Even in a group of single-celled organisms, individual cell death may benefit others in the group. If members of a colony are genetically different, competition will tend to predominate. The death of a cell has no evolutionary advantage, if it benefits survivors with a different genetic makeup. However, if all the cells are genetically identical, then some can be sacrificed, for the good of the rest.

Animal cells go through complex, coordinated processes, requiring precise management of cell division and growth, to form whole bodies. Single organs, such as the eye or brain, have a level of complexity that stretches the imagination. To produce such structures, the organism needs exceptionally fine control of cell division and movement. Some of these processes rely on enhanced cell division, movement and invasion of other tissues. These properties are similar to those found in primitive organisms. They indicate that, beneath our highly developed cellular controls, the original features of evolutionary fitness, found in bacteria and other single-celled organisms, still exist.

New species

If two populations of an organism are separated, both groups will vary and, given enough time, will generate new species. When the organisms are separated geographically, the process is called allopatric speciation. Separation of the populations isolates the genes, allowing variation and natural selection to create sufficient differences between the populations to produce a new species. This process can take many generations and millions of years. Later, we use such speciation as an analogy for the formation of malignant cancer. However, in the case of cancer development, the time is short and there are several mechanisms, such as cell suicide, to inhibit the process.

When an organism is not separated geographically, the process of forming a new species is called sympatric speciation. This form of species formation is common in parasites and is heavily involved in plant speciation.[41] For genetic separation to occur, the two subpopulations must not interbreed.[d] Faulty cell division, which results in abnormal numbers of chromosomes, is one way to prevent interbreeding.[42,43] During normal sexual combination, each sperm and egg cell provides half the chromosomes. If the two cells do not contain equal numbers of chromosomes, they cannot pair up, one to one. The resulting fertilised cell has some chromosome pairs and some single chromosomes. Organisms created with such cells, even if viable, would not breed true.

---

[d] In development of cancer cells, we are not dealing with sexual reproduction but with simple division by mitosis, so our cells are already genetically isolated.

Too many chromosomes

The Kew Primrose (*Primula kewensis*), a common garden plant, has a special place in the history of biology. Crossing distantly related plants usually produces sterile hybrids. In 1912, Digby crossed two primrose plants, *Primula verticillata* and *P. floribunda* and, as expected, most progeny were sterile. However, there was a single fertile plant, the Kew Primrose.[44] This plant had double the number of chromosomes of its sterile siblings, a condition known as polyploidy. The Kew Primrose is the first recorded human-made polyploid variety.

In 1916, Hans Winkler found that mechanically damaged tissue often contained cells with too many chromosomes. He introduced the term polyploidy, to describe a cell with more than the expected number of chromosomes.[45] A year later, Öjvind Winge suggested that polyploidy occurred by successive increases in the number of chromosomes. He observed that the number went up by equal increments in Chrysanthemum (18, 36, 54, 72, and 90 chromosomes) and Chenopodium (18 and 36 chromosomes). Polyploid cells have extra sets of chromosomes; cells with an incorrect number of individual chromosomes are called aneuploid.

It is estimated that 30% to 70% of all flowering plants are polyploid.[45] It is difficult to know whether many of the flowering plants are ancient polyploids, or if the chromosome number increased more recently. The broad range for this estimate derives from the fact that the total number of flowering plants is unknown.

Mechanical damage increases cell division and can lead to polyploidy. This suggests one mechanism for the process: errors in cell division, caused by damaged structures within the cell. During division, these structures separate the chromosomes into daughter cells. Failure of a cell to divide successfully can mean that its daughter cells have the wrong numbers of chromosomes.

Polyploidy can also occur during sexual reproduction. If more than one sperm fertilises an egg, the resulting cells could have extra sets of chromosomes. Sperm or egg cells that are malformed, having retained two copies of the chromosomes, might also increase the chromosome number.

It is important to realise that, in plants, the rate of spontaneous chromosome duplication is similar to the gene mutation rate.[46] In plants, polyploidy is associated with high latitudes and areas of recent glaciation.

Plants that self-fertilise or enter into crosses with other species are more likely to have polyploid offspring. There is an extensive, if speculative, literature on the development of polyploidy in plants, but the underlying mechanisms have yet to be clearly established. Although we have been describing polyploidy in plants, it is also a feature of animal cells and, in particular, of cancer.

Changing the number of chromosomes is a powerful evolutionary step. Polyploidy is one of the few recognised ways of achieving *instantaneous* speciation. A new species can be created in a single step, through faulty cell division.[47] For example, a single alteration could create an initial cancer cell. Selection pressures could then hone this first cancer cell and its descendents, to develop the attributes needed for malignancy. Alternatively, after a period of error-prone growth, a descendent cell could be triggered to a form of division where chromosome number was not conserved.[48]

In the following chapters, we consider the development of cancer as the evolution of rapidly dividing cells. We compare cancer cells to a new species, competing with healthy cells in the environment of the body. Ultimately, the success of the cancer is its downfall. By out-competing normal cells, it kills the host and brings about its own extinction.

Cell death

"You do not die all at once. Some tissues live on for minutes, even hours, giving still their little cellular shrieks, molecular echoes of the agony of the whole corpus."
Richard Selzer

Within the body, cells are continually dying and being replaced. In a typical animal, all cells share the same genetic information and cooperate, for the ultimate survival of the organism as a whole. If one cell dies, enabling the organism to survive, the overall evolutionary fitness has increased.

Cell suicide is not accidental, or even unwanted, but is a central part of biology. Tissues in the body are normally held in a state of balance between cell division and cell death. Increased cell division results in growth, as does a decrease in cell deaths. Conversely, decreased proliferation or increased cell death produces tissue shrinkage. The rate of cell death and replacement by division is carefully synchronised and, superficially, results in no difference to the body's internal structure or appearance.

Programmed cell death, known as apoptosis, is an important feature of our normal development and functioning. A classic example is the development of fingers and toes. When a hand forms in the womb, its fingers are initially joined in a single, mitten-like protuberance. Then, cells between the fingers die, in a form of suicide, removing the tissue and separating the fingers. These dying cells are not sick or damaged: their death is essential to produce individual fingers. Throughout development, similar sequences of programmed cell death help shape our tissues.

If the controls on cell death are misapplied or fail, serious health problems can result. An example is Parkinson's disease, which is caused by degeneration of cells in the *substantia nigra*, part of the brain that controls movement. Parkinson's disease is a progressive disorder, characterised by muscle tremors, rigidity, slowed movement and difficulty in balance. It is more common in older people, affecting more than a million people in the United States. An increase in cell death with age is a general feature of body cells.

Cell death as a strategy for survival

The ubiquitous presence of programmed cell death in multicellular animals implies it has a fundamental role in the animals' survival strategy. Not only do cells die to facilitate the survival of the organism, they contain complicated mechanisms to control the process. We have seen that apoptosis helps in the formation of body structures; it is also important for the prevention of cancer.

Long ago, when single-celled organisms first combined to form multicellular creatures, they developed controls to limit cell division and allow intercellular cooperation. Uncontrolled growth would produce a tumour, rather than a biological structure adapted to its environment. Signalling and cooperation are primary activities of cells in multicellular organisms; they limit abnormal growth and induce damaged cells to commit suicide. The ultimate control factor for both cell division and apoptosis may be the relative levels of oxidising to reducing agents, known as the redox state of the cell.[49]

Animal cells, if not controlled, can become like single-celled organisms, which tend to become invasive and spread to new habitats. Cancer cells are the result of such reversion and are a natural consequence of the way animals have evolved. Since the potential for cancer has existed since the first multicellular animals evolved, these creatures have developed ways of preventing their cells from reverting.

One way to stop cells reverting to a primitive or cancerous state is for the body to recognise that damage has occurred and instruct the cells to commit suicide, by apoptosis. If damaged cells are instructed to die, they cannot evolve to generate cancer. However, since cancers are known to grow, we can conclude that some cancer cells must be able to resist these cell suicide instructions. Sometimes, during the error prone cell division phase of cancer development, the cell suicide mechanism becomes damaged. Such cells do not obey apoptosis signals, and therefore have a selective advantage in evolving to a cancerous form.

Mechanisms of cell death

Cell suicide by apoptosis is controlled by the mitochondria, which also supply cells with energy. The mitochondria are the ultimate cellular executioners.[50] Enzymes in mitochondria appear to sense the level of oxidant species,[51] which include molecules or ions formed by the incomplete reduction of oxygen. If the level of oxidants rises above a given threshold level, pores open in the mitochondria. Proteins leak out through the mitochondrial pores and the cell starts out on its pathway to death.[52] This process depends on the concentrations of antioxidants in the cell.[53,54] If the energy in the cell is depleted, it may not be able to follow a controlled pathway to suicide and may be forced to die by necrosis.[55,56,57]

Necrosis

Necrosis is sometimes likened to a car crash: a catastrophic and unregulated event. Cell membranes rupture and the contents are disgorged. However, even necrosis can sometimes be considered as a cell death program.[58] Like apoptosis, necrosis involves free radicals and both can be induced by oxidation damage. However, the physiological effects of apoptosis and necrosis differ greatly. With apoptosis, the nucleus and cytoplasm remain within an undamaged cell membrane. The cell remains are consumed by white blood cells and typically do not cause excessive inflammation or oxidation. By contrast, necrosis releases the cell contents into the surroundings, where they produce inflammation and free radicals.

Autoschizis

Recently, a third form of cell death has been described, called autoschizis.[947] In autoschizis, cells die when the cytoplasm is expelled from the cell, leaving the nucleus apparently intact. This form of cell death was discovered during an investigation of the cytotoxic action of vitamin K on cancer cells.

A new mechanism for cell death in cancer cells should be regarded as a major opportunity. It means that there are additional mechanisms involved in selective killing of cancer cells and implies that there may be new treatments to be explored.

Why cells commit suicide

Programmed cell death, represents a major evolutionary difference between single-celled and multicellular organisms. Staying alive and reproducing are the two most central requirements of living beings. For a single-celled organism, suicide by apoptosis would mean breaking a fundamental requirement for evolutionary survival. Single-celled organisms have survived for millions of years, continually resisting evolutionary selection pressures. To achieve this, they have had to struggle to stay alive, whatever insult they encountered.

However, despite the apparent loss of fitness associated with cell suicide in single-celled organisms, it does occur. Generally, if members of a population of single-celled organisms were to commit suicide, the population would diminish and would risk extinction. However, exceptions to this rule are common. Some single-celled eukaryotes,[a] such as yeast, are known to commit apoptosis.[59,60] During conditions of starvation, such suicide could serve to save related members of a population of cells. An alternative explanation is that apoptosis might be used as a control mechanism, to maintain a stable host-parasite ratio.

Since single-celled organisms have the ability to commit suicide under certain conditions, the development of apoptosis is likely to have occurred at an early stage of evolution.[61] This implies that a form of cell suicide has been available throughout the development of multicellular life. The ability to commit suicide is a fundamental control mechanism in cells, particularly for the development of larger life forms.

Within cells, oxidation levels are used as signals to commit suicide. Low oxidation levels are associated with cellular health, whereas damage to biological structures often involves an increase in oxidising free radicals. Since suicide is an irrevocable step, which ends the existence of the cell and all its descendants, the body requires control mechanisms, to avoid unnecessary apoptosis.

When a biochemical process, such as apoptosis, requires intricate control mechanisms, the operation is usually complicated and involves numerous interactions with the environment. An example is blood clotting, which requires a cascade of chemical reactions, many of which are sensitive to current conditions and signals. If an animal is injured and bleeding, rapid blood clotting can be essential. However, in other cases, such as coronary thrombosis, a small clot could be fatal. The control mechanisms must be able to distinguish between different situations. By analogy, cell suicide in animals requires an extensive control system, with definitive signalling information. If the cell death process went wrong, the result could be catastrophic.

---

[a] The single celled organisms *Dicytostelium discoideum*, *Saccharomyces cerevisiae*, *Schizosaccharomyces pombe*, *Tetrahymena thermophila*, *Trypanosoma cruzi* and *Trypanosoma brucei rhodesiense* are examples that can undergo apoptosis.

Effects of oxidation levels on cell division and death			
Cell State	**Oxidation Level**	**Cell Division**	**Cell Death**
Normal	Low (Reducing)	Inhibited	Inhibited
Slight damage / inflammation	Moderate	Lowered inhibition/ some increase	Inhibited
Moderate damage	Intermediate	Increased	Inhibited / Some apoptosis
Severe damage	High	Increased / Inhibited by damage	Apoptosis
Irreparable damage	Severe free radical damage	Stops	Necrosis

There are various ways for damaged cells to commit apoptosis. A wide range of oxidants has been shown to induce cell suicide,[b] in a variety of cell types.[62,63,64,65,66,67] The response of the cell is complex and depends on the dose of oxidant. Paradoxically, oxidants at moderate levels stimulate cells to increase their growth, whereas, at higher levels, they instruct the cell to die immediately. Generally, the growth of cells in a reducing environment is inhibited. As the level of oxidation increases, the cells are stimulated to divide. At higher oxidant levels, the cells divide faster but start to undergo apoptosis. If levels of free radicals are sufficiently high, the cell is destroyed catastrophically by necrosis, rather than the more controlled method of apoptosis.

When a cell commits suicide, it generates increased levels of oxidants, which have a central role in the process.[68,69,70,71,72] It is possible to induce cell suicide when oxygen levels are low,[73,74] but free radicals are

[b] Examples of oxidants that will induce apoptosis include hydrogen peroxide, nitric oxide, redox-cycling quinones, oxidised low-density lipoproteins, lipid hydroperoxides and diamide.

still involved, since anaerobic cells generate oxidants.[75] Furthermore, cells that are short of oxygen may be more sensitive to some free radicals.[76] Many tumours have low levels of oxygen but retain sensitivity to free radical damage. Delivering free radicals to tumours that are lacking in oxygen could provide an effective treatment for cancer.

Even in single-celled organisms, cell suicide involves and depends upon free radicals.[77] Yeast cells commit suicide if they are dosed with hydrogen peroxide, or if their internal antioxidant, glutathione, is depleted.[78] In animals, redox processes normally interact with more evolutionarily recent mechanisms, to produce an efficient means of programming cell death. However, human cells may have retained the capability to commit suicide by redox mechanisms alone.[79,80,81,82,83] This alternative mechanism has implications for the treatment of cancer. Conventional treatments often depend on inducing diseased cells to commit suicide, by creating oxidising free radicals in the cells.

To explain why tumour cells can be short of oxygen, yet still be subject to oxidative stress, we need to remember that oxidation, despite its name, does not depend on the presence of oxygen; other free radicals can also be damaging to tissues. Indeed, the term oxidative stress is sometimes used to cover free radical damage and hypoxia.[1] Levels of oxidised glutathione, arguably the body's most important, water-soluble antioxidant, can affect the intracellular redox state. Cells stimulated to enter apoptosis can eject glutathione, using molecular carriers in the surrounding cell membrane.[84,85,86,87] Loss of glutathione during apoptosis means the cell does not have sufficient antioxidants to prevent oxidation. Such a shortage of antioxidants explains that cells short of antioxidants can be under oxidative stress, while also being short of oxygen. This suggests a mechanism for induction of redox stress in anaerobic cells.[88] Ejection of glutathione does not necessarily induce apoptosis, sometimes it can promote cell division, while simultaneously making the cell more sensitive to suicide signals.[89,90,91,92,93] Changes in levels of free radicals may drive a cell towards apoptosis, but might not be sufficient to force cell death.

Apoptosis and cancer treatment

The body uses apoptosis to remove damaged cells, which could begin to grow uncontrollably, or even those that show signs of gross genetic change. However, one particular type of cell is reluctant to agree to apoptosis: the malignant cancer cell. Such cells disobey the command to commit suicide. This blocking of apoptosis is an important mechanism for unconstrained growth.

Conventional treatments, such as radiation, generally kill cancer cells by inducing apoptosis. However, treatments that are not completely effective favour the survival of cancer cells that are less able to undergo apoptosis. As a result, follow-up treatments can be ineffective, and the cancer may become multiply resistant to drugs and radiation. The microevolutionary model predicts this resistance mechanism, as cells that do not undergo apoptosis have a selective advantage. The underlying process is similar to bacterial resistance to antibiotics.

Our increasing understanding of the relationship between oxidation levels and cell death offers insights into possible new treatment methods. Many current treatments depend on inducing apoptosis in cancer cells. However, cancer cells often resist signals to commit suicide. Moreover, healthy cells are also subject to the effects of treatment and may be killed.

A cure for cancer is usually considered as complete elimination of all abnormal cancer cells. Ideally, this would be achieved without damaging normal tissues. Such extinction depends on selectively inducing cell death in the abnormal cells. However, cancer cells are not identical, but exist as diverse populations. To suggest a cure, we need to understand which properties are shared by cancer cells but do not occur in healthy cells. These properties will become clear, when we consider the evolutionary development of cancer.

The microevolution of cancer

"One general law, leading to the advancement of all organic beings, namely, multiply, vary, let the strongest live and the weakest die." Charles Darwin

In this chapter, we describe the development of cancer as the microevolution of damaged cells. According to this model, cells that divide rapidly and do not breed true will eventually become cancerous. Many existing hypotheses can be related to this microevolutionary model. Subsequently, we will integrate these different ideas, to provide a more comprehensive explanation.

Evolutionary theory applies to any population of living cells that are subject to selection pressure. According to the microevolutionary model, cancer has no single cause: each individual case evolves along its own unique pathway. Cancers use varied mechanisms to achieve growth.[1] However, there are common features, which arise because of growing cells' shared requirements for evolutionary success.

The model explains the limitations of conventional therapies, such as radiation and cytotoxic chemotherapy, and points the way to consistent methods of treatment. In later chapters, we describe why current therapies often fail, before going on to develop a potentially more productive approach to treatment.

Cancer: an inevitable result of microevolution

The characteristics of cancer cells include uncontrolled growth, failure to breed true and longevity. A more complete list is given below. These features arise in many ways, including some that are not yet fully understood. Nature is more capable than any human of devising ways for life forms to evolve. Microevolutionary principles lead us to expect variation in the biochemistry of cancer cells, even those from the same tumour.[94] Thus, cancer cells might evolve several different biochemical ways to achieve the properties required to become malignant. For example, one cancer might grow rapidly because it has a defect in the gene that inhibits growth, whereas another may over-respond to signals that stimulate cell division.

According to the microevolutionary model, a pre-cancerous cell will generate a varied population of damaged but dividing cells. The requirements for malignancy are similar to those for evolutionary success in a population of single-celled organisms. In the race to leave most offspring, the winning cells will have the properties of rapid division, extended life and high invasiveness.

Single-celled organisms have inhabited the earth far longer than multicellular creatures. Long ago, when individual cells began to group together to form multicellular life forms, they retained the core biochemistry of single-celled organisms. Many such features, important to the success of single-celled organisms, are also used in the development of complex animals.

However, in order to enable cooperation between cells, multicellular organisms added layers of communication, signalling and control mechanisms, on top of the primitive, single-celled biochemistry. Since these controls developed later in the evolutionary process, they have had less time to become robust and resistant to injury. Damage to the relatively recent control mechanisms is likely to occur more frequently than to the older features that multicellular animals share with their single-celled ancestors.

Characteristics of cancer

Throughout the different classes of cancer, malignant tumour cells have a consistent set of properties. These are considered to be:

- Growth, in the absence of "go" signals
- Growth, despite the presence of "stop" signals
- Evasion of cell suicide (apoptosis)
- Ability to promote local blood vessel growth, known as angiogenesis
- The potential for an unlimited lifespan, called immortalisation
- Ability to invade nearby tissues, and to travel and create distant tumours, known as metastases
- An abnormal number of damaged chromosomes

The most visible characteristic of cancer is the uncontrolled multiplication of cells in the body. Cells divide by a process called mitosis, when each cell splits into two daughter cells. In the healthy body, cell division is tightly controlled, so our tissues stay a consistent size and shape. Cells have a natural cycle of growth and cell division. Normal cells act constantly on signals to inhibit this cycle, thus maintaining the current tissue structure, whereas cancer cells keep on dividing. This suggests that the regulation of cell division, essential to multicellular organisms, has failed. Cancer cells have lost their ability to cooperate with the body, instead they evolve and grow like single-celled organisms.[95,96,97,98]

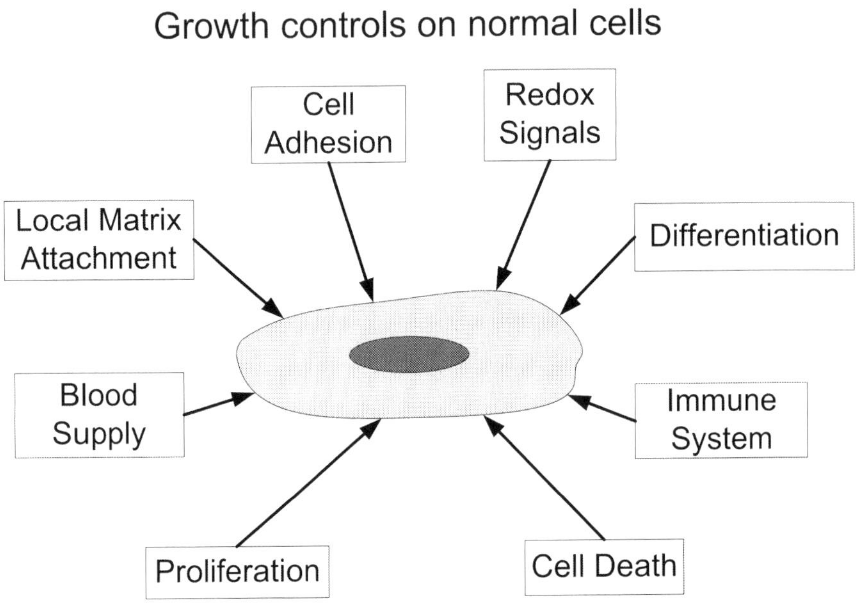

Another characteristic of cancer is that the cells do not breed true.[178] Daughter cells often differ from the parent. The process of cell division involves duplicating the cell's contents. However, errors can occur at any stage of the process: when the DNA is copied, for example, or when the duplicated chromosomes are transported into the daughter cell body.

Cancer cells live a long time and may be immortal. In contrast, normal cells die after a given time or number of cell divisions. Cells from a given species will divide only a specified number of times; this number helps determine aging and longevity. In cancer cells, the cell mortality mechanism is faulty. Indeed, one reason for the mortality of human cells is to prevent them from dividing uncontrollably and producing cancer. A healthy cell that breaks free of the controls on its growth will divide only so many times, before it dies.

Speciation

The term species, despite being a fundamental criterion of biological diversity, is poorly defined. Biologists find this lack of a precise definition useful for describing the large and varied organisation of living things. In sexual creatures, such as many animals and plants, species can be defined as groups that do not interbreed under natural conditions. For asexual creatures, the position is more difficult. When organisms do not engage in any form of sexual activity, species may be inferred from genetic, physiological or structural differences.

The generation of a cancer cell can be compared to development of a new species, which has the properties of a single-celled organism. The characteristics of malignancy are similar to those of bacteria, which are likewise imposed by evolutionary pressure. Although derived from body cells, cancer cells have a different genetic makeup. In particular, malignant cancer cells have an altered number of chromosomes. Such a genetic difference between organisms in a population will generally lead to evolutionary competition. While we do not suggest that a cancerous cell is a new species under the biological definition of the word, the cell is sufficiently different from healthy cells for the description to be apt.

When a cell becomes cancerous, it changes from its original, cooperating multicellular form and starts to act for the replication of its own genome, which has become markedly different from that of the host. A cell with these characteristics competes with its neighbours. We use the analogy of a new species to describe this behaviour; some might prefer the concept of a "selfish cell", analogous to the idea of a "selfish gene." [a]

Cancer cells do not normally acquire the properties described above all at once. The most important feature of cancer is faulty cell division, which is followed by natural selection.[b] If a cell with a faulty

---

[a] Evolution acts on the whole organism (phenotype) at the population level, rather than the level of individual genes. However, it is easier to analyse the process mathematically at the gene level.

[b] By faulty cell division we mean any process that can lead to the daughter cells having a different genetic makeup to the parent. This could be a mutation, chromosome duplication or some other abnormality.

division process produces daughter cells, they will differ from the parent cell. Some of the daughter cells will be less able to survive; such cells may grow slowly, or die. Other cells may be viable enough to survive and occasionally divide: these will continue to produce a population of increasingly varied, abnormal cells. Occasionally, some of the errors will produce cells that, under current conditions, are hardy and divide rapidly. These robust cells will produce more offspring and will be favoured by natural selection.

Healthy human cells generally contain identical genetic information, with a few notable exceptions. In contrast, the genes in cancer cells differ substantially from those in normal cells. Abnormal numbers of chromosomes are a clear sign of potential malignancy. This means the cells will tend to evolve separately, looking after their own genetic makeup, rather than that of the host. With minor genetic variations, the effect will be small. However, large-scale chromosome duplication and deletions are massive alterations, consistent with speciation and competition.

Inheritance

The salient feature of the errors in cancer cells is that the cellular changes pass on to descendent cells: they are inherited. According to the conventional view of cancer, inheritance is assumed to mean that changes result from mutations in the DNA. However, this view is unnecessarily restrictive. Changes in chromosome numbers, or in the machinery that separates chromosomes during cell division, could have a similar effect, as could any inherited structural changes to the cell. A physician reading this latter statement may reject it, because it appears to suggest Lamarckian inheritance, the passing on of acquired characteristics. This is not the case. Most biologists are aware of special cases that do not depend on the genes or DNA. As an example of this, we will describe a strange phenomenon in the single-celled organism, paramecium.

Paramecia are covered in a series of small hairs, or cilia, that beat together to provide directional movement. Most paramecia have an ordered arrangement of such cilia. However, the surface of the cell can be damaged, to produce a small patch in which the direction of the cilia is reversed. Biologists have discovered that such rearrangements can pass to daughter cells, during division, for several generations. This effect was demonstrated in 1965, in a series of experiments by Janine Beisson and Tracy Sonneborn.[99] One explanation for this surprising finding is based on physics or bioengineering. This suggests that the orientation of new cilia is determined locally by the existing structure, in a process called structural guidance.

Another example of structural inheritance occurs when treatment with streptomycin destroys the chloroplasts in euglena, a small, green, single-celled organism, found in fresh water. Chloroplasts are disk-like structures within cells; they contain chlorophyll and are the site of photosynthesis, the process by which plants convert sunlight to chemical energy. The treated cells no longer contain chloroplasts and neither do their descendants.[100]

Normal and Cancer Cells

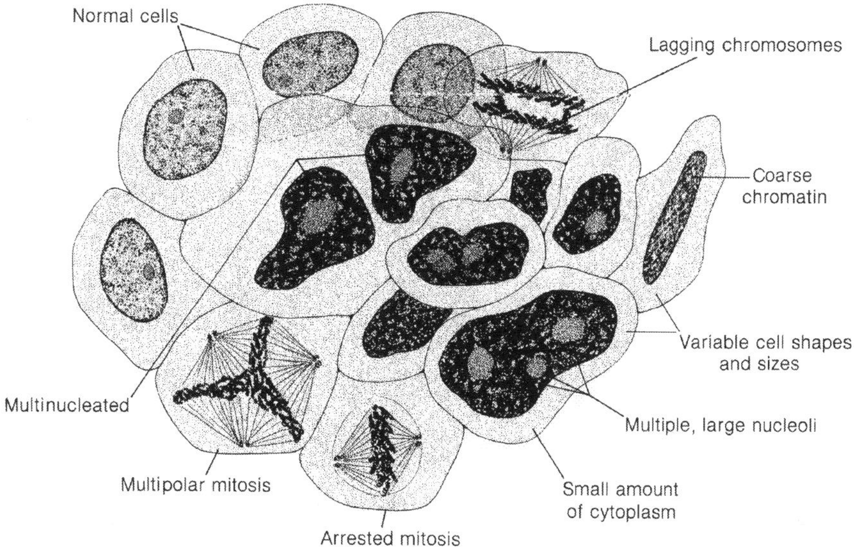

This diagram illustrates aberrant mitotic cell division in cancer cells. This cell division produces cells with abnormal number of chromosomes and varying amounts of DNA, in unusual nuclei. Also evident are cancer cells with multiple nuclei and others with nuclei so enlarged they almost fill the cells.

Cell division in cancer cells can involve a change in the number of chromosomes, which suggests that structural mechanisms could influence the future inheritance of the cell. When a cell divides by mitosis, protein molecules pull the chromosomes into separate daughter cells. Structural alterations to the arrangement of these proteins could result in one daughter cell getting more of the separating chromosomes.

If viable, the resulting daughters would have different numbers of chromosomes from the parent cell, as would their descendants.

Despite cases such as these, genetic mechanisms dominate inheritance and, as far as we are aware, no important structural guidance mechanism has been demonstrated in either healthy or cancerous human cells.

Oxidation and antioxidants

Oxidants cause biological effects, which depend on their concentration. A simple rule of thumb is that the environment of healthy cells is reducing, rather than oxidising. Cells use free radicals and oxidants, such as hydrogen peroxide (H_2O_2), as signalling and control molecules. Reactive oxygen species, such as hydrogen peroxide, superoxide ($^\bullet O_2$-) and the vicious hydroxyl radical ($^\bullet OH$) are widely generated in biological systems.[c] If left unchecked, these could damage the cells. Consequently, cells have evolved complex antioxidant defences, which limit the production of reactive molecules and reduce them to low levels.[101] An important reason for this is to prevent cancer.[102]

Cell division mechanisms depend on redox control and signalling.[103,104] Even small changes in the redox state can be important to regulation of cell growth and differentiation, indicating that oxidation lies at the core of cell signalling mechanisms.[105] A moderate increase in the level of oxidants can promote cell growth and division. Growth factors act through oxidants and, in particular, by increasing hydrogen peroxide.[106,107,108,109] Hydrogen peroxide blocks the primary signalling mechanism that inhibits growth in the cell.[110,111,639] Cancer-causing genes, called oncogenes, can act by increasing the level of oxidants.[112] Thus, cells that are in the process of becoming cancerous generate a more oxidising state, which facilitates cell division.

---

[c] The dot in the chemical formula, eg $^\bullet OH$, indicates an unpaired electron, a characteristic of a free radical.

Effect of oxidation on a population of cancer cells

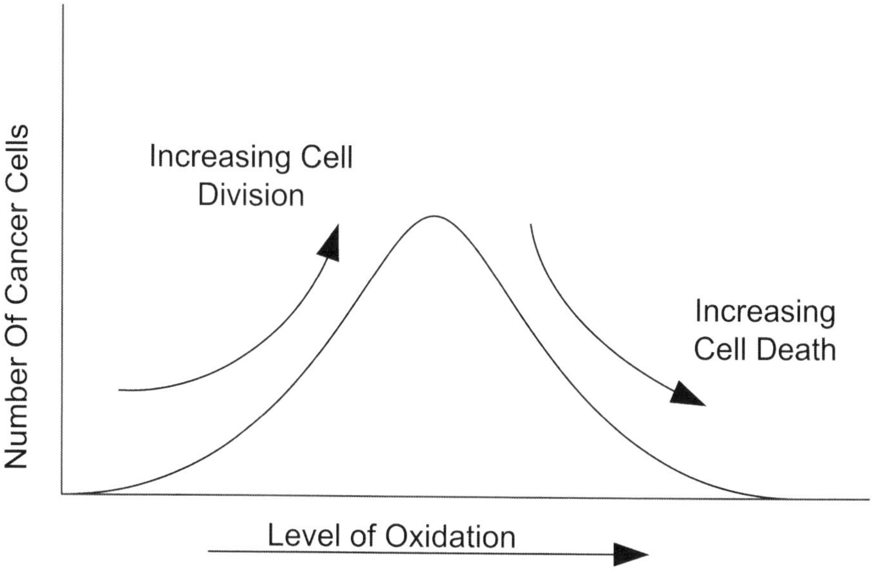

The level of oxidants in a cell has numerous effects. Oxidants influence the cell's ability to divide, can signal the cell to commit suicide, and are intimately involved in the progression of the cell to a malignant condition.[116,113] Although a moderate increase in cellular oxidants signals a command to divide, a further increase could be damaging and cause the cell to die.[114] Intracellular production of higher levels of active oxygen species, such as hydrogen peroxide, is associated with prevention of cell proliferation.[101] Initially, as the hydrogen peroxide increases, the cells grow. Then, as the level becomes too high, they die.

Cancer cells divide and grow, even under crowded conditions, which is unusual. When healthy cells are grown in culture, they increase in number while they are dispersed. However, as the density of cells increases, the growth rate falls and they stop dividing. This is known as density-dependent growth inhibition. When cells crowd together and touch, as often occurs within body tissues, they stop growing because of contact inhibition: the cells sense the crowding and cease dividing.

Contact inhibition is associated with a decrease in the concentration of oxidants within the cells.[115] Removing oxidants from the culture medium produces a reducing environment, which can mimic the effects of the contact inhibition signals. As a result, the cells stop dividing. Control of contact inhibition may depend on the cells' internal redox state, which is directly influenced by the sum of *grow* and *stop growing* signals received.[115] Unlike healthy cells, malignant cells ignore contact inhibition signals and continue to divide.

Oxidation and cancer

Cancer cells generally have an oxidising internal environment, with higher concentrations of free radicals and lower levels of antioxidant enzymes.[101,645] Compared to healthy cells, their antioxidant defences are deficient.[116,117,118,119,120,121,122] Although a few cancers show increased levels of certain antioxidant enzymes, none has a full complement of defences against free radical attack. As far as we know, all but one type of cancer is deficient in two important antioxidant enzymes,[d] called catalase and glutathione peroxidase.[645] The lack of these enzymes is a weakness in cancer cells, offering a potential way to attack the cells with new treatments.

Cancer cells increase their internal oxidation levels as a redox control mechanism, to maximise growth. This method appears to be ubiquitous in cancer cells. While the cancer is growing, the body's levels of oxidants and inflammatory markers increase.[123] Later, we will see that cancer's dependence on redox changes suggests mechanisms for its selective eradication.

The action of antioxidants on cancer cells is far from simple: it may depend upon the cells' redox state and the stage of the cancer. Some cancers may find an external supply of antioxidants beneficial. For example, in healthy mouse cells, the antioxidant supplement, NAC,[e] can decrease growth, by lowering the oxidant level.[124] However, in cancer cells, another antioxidant, the enzyme thioredoxin, stimulates growth and prevents cell suicide.[125] This apparent paradox may be explained if we consider that the levels of oxidants in some malignant cells are so high as to be almost fatal to the cell. Thioredoxin may lower these levels enough to restore the cell to relative health and stimulate growth. Increased levels of thioredoxin occur in many human cancers and are a sign of resistance to therapy.

---

[d] The granular cell variant of human renal adenocarcinoma apparently does not have low catalase and glutathione peroxidase levels, but may have a modified redox biochemistry.

[e] The normal cells were from a mouse embryonic fibroblast culture and N-acetyl-cysteine (NAC) is a sulphur-containing antioxidant supplement, which increases the amount of glutathione within cells.

Oxidation and reduction are central to our microevolutionary model. We have stressed that healthy cells maintain themselves in a reducing state.[15] They use oxidants and free radicals as signals to initiate cell division,[126] responding to external and internal chemical signals to grow and divide. Hydrogen peroxide, nitric oxide (NO) and superoxide ($^{\bullet}O_2^-$) are all used as signals for a cell to divide. These reactive molecules were once thought to be wholly damaging to the cell, because they cause free radical damage. However, at low levels, these chemicals, especially hydrogen peroxide, are vital for cell signalling.[127] The effects of such signalling reveal a fundamental role for the redox balance in the cell. Given the parsimony of nature, this may be no accident, as it also provides a mechanism to prevent cancer.

This is how the mechanism could work. Moderate increases in oxidation levels stimulate cell division. However, if the level of oxidation in normal cells increases too much, the cell initiates a suicide program. This curbs the possibilities for uncontrolled growth, which might otherwise lead to cancer.

Cancer cells are generally under oxidative stress,[128,129,130] which may be an essential feature of cancer growth.[131,132,133] Natural selection favours cancer cells with a high level of oxidants, because they divide more rapidly. Cellular oxidation therefore occurs early in cancer development,[134] and tissue oxidation is associated with development of cancer.[135,136] Cancer cells can flourish when they have developed an increased response to redox growth signalling and an oxidative internal environment, which favours multiplication.[137,138]

The body uses redox signalling for both cell division and cell suicide.[139,140] If the oxidant levels rise greatly, the signal causes it to destroy itself, rather than divide.[141] The body needs to protect itself against the possibility of rogue cells going haywire and causing cancer. By using different levels of the same signal to initiate both cell division and cell suicide, the body has a mechanism to catch and kill cells that have lost their antioxidant controls and are dividing abnormally rapidly.

We have seen that cancers start with error-prone cell division, producing a variety of daughter cells. Some of these precancerous cells might happen to have fewer of the antioxidant enzymes or other controls that normally suppress oxidation and the generation of free radicals. This could result in cancer cells with a more oxidising redox state, which stimulates division and growth, without being so oxidising that it causes a high proportion to commit suicide by apoptosis.

The story does not stop there. One immediate side effect of losing antioxidant controls is that the cell and its DNA will be damaged by free radicals. Thus, by becoming more oxidising, the cell has achieved the two basic features required to break free as a malignant cancer: it divides rapidly, and its daughter cells have modified DNA, together with other oxidation-induced errors.

From this description, it is obvious that any factors that make the healthy cell and its environment more oxidising will increase the risk of cancer. Since almost any insult to a tissue results in oxidation, inflammation or both, the confusing variety of factors that increase cancer risk begin to make sense.

Redox cycling

Redox cycling is an important mechanism, by which cancer cells can be destroyed. Some molecules can be reversibly oxidised and reduced. For example, Vitamin C, as ascorbate, is oxidised by losing two electrons to form dehydroascorbate. Dehydroascorbate can then gain two electrons from the cellular metabolism, to reduce it back to ascorbate. This process of oxidation and reduction can go round and round, forming a redox cycle.[142]

In healthy cells, this cycle is used to regenerate ascorbate from dehydroascorbate, so it can continue to act as an antioxidant. Since ascorbate's antioxidant action requires it to donate electrons, it needs to replace the donated electrons before it can continue to prevent oxidation.[f] The ascorbate-dehydroascorbate cycle is used to maintain a healthy reducing environment in normal cells.

In cancer cells, however, the mechanism fails, causing oxidants to be generated. One mechanism that can cause this production of oxidants is the presence of free iron within the cell. Ascorbate reacts with iron, producing hydrogen peroxide and ascorbyl radicals.[143,144] The hydrogen peroxide then reacts with the iron, releasing highly reactive and damaging hydroxyl radicals. The iron is then available to react with another molecule of ascorbate.[g] Given a supply of vitamin C, this reaction can cycle round, producing large amounts of hydrogen peroxide and hydroxyl ions, which eventually kill the cancer cell.

---

[f] In dynamic flow, with frequent large doses, vitamin C can donate its electrons and be excreted. Therefore, the reducing electrons are available without using cellular energy.
[g] The reactions are Fe(III) + ascorbate → Fe(II) + ascorbate• and Fe(II) + H_2O_2 → Fe(III) + •OH + OH^-

The selective effects of vitamin C

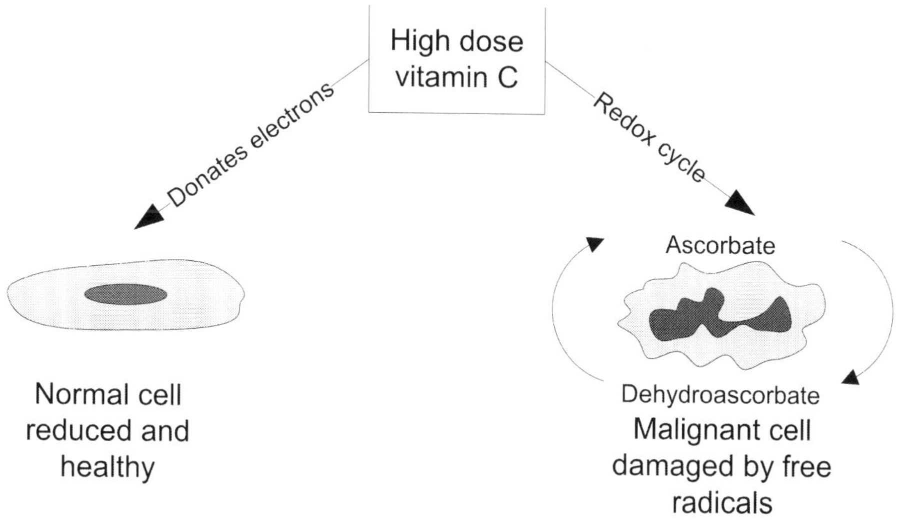

Other dietary antioxidants can also take part in redox cycles. In particular, several forms of vitamin K can react with transition metals, such as iron and copper, to generate oxidants. The antioxidant supplement, alpha-lipoic acid,[h] can act to prevent such oxidant production by vitamin K.[145] However, in cancer cells, alpha-lipoic acid also participates in a free radical generating redox cycle. Alpha-lipoic acid can be reduced to dihydrolipoic acid, which acts as a powerful antioxidant within cells. Dihydrolipoic acid is then oxidised to alpha-lipoic acid. In cancer cells, this redox cycling of alpha-lipoic acid can generate free radicals, particularly in the presence of large amounts of vitamin C. The oxidising redox cycles of these and other dietary supplements occur in cancer cells, but not in normal cells. Therefore, such cycles provide a potential mechanism for selective eradication of the disease.

Immortal cells

Cells that are both damaged and dividing would usually have a reduced life expectancy. Many injured cells would be unable to compete or survive. In some cases, the damage itself would be terminal, providing a form of selection, since cells with damage that shortens their lifespan

[h] References to alpha-lipoic acid should be taken to be r(+)-alpha-lipoic acid unless otherwise indicated.

will leave fewer offspring. However, if the damage resulted in an extended lifespan, the cell would be expected to produce a larger number of offspring. Thus, selection pressure will favour the longer-lived cells in the colony. Given enough time and cell divisions, errors will be selected that result in an extended, and possibly immortal, cell line.

Single-celled organisms, such as bacteria, provide a model for this process: they have evolved an effectively immortal lifespan. Human cells have the biological machinery required for immortality. For example, the germ cells that form sperm and ovum cells are immortal. However, normal body cells have mechanisms to prevent an indefinite lifespan. Once again, we see that a feature of cancer cells is a consequence of cellular microevolution. However to be immortal is not enough; in order for the cancer to grow, it needs to produce offspring.

Limiting cell division

Multicellular animals, such as humans, prevent the growth of cancers by setting a limit to the number of times that cells can divide. One consequence of preventing cancer in this way is that animals age and die. A cell from a given species can divide a maximum number of times, which is called the Hayflick limit, after Leonard Hayflick, who described the phenomenon in the 1960s.[146] Almost a century earlier, in 1881, a German biologist called August Weismann suggested that,

"death takes place because a worn-out tissue cannot forever renew itself, and because a capacity for increase by means of cell division is not everlasting but finite".[147]

The maximum age attainable by an animal species is related to the Hayflick limit. Human fibroblasts - cells that make connective tissue, such as cartilage - will divide for about 60 cell divisions. The maximum age for man is about 120 years. Mice live up to three and a half years, and their fibroblasts divide about 20 times. Chickens, which live up to 30 years, have fibroblasts that divide 30-40 times. The Galapagos tortoise can live for about 175 years, or more, and has fibroblasts that will go through 125 cell divisions.

The mechanism underlying the Hayflick limit is related to the shortening of telomeres. A telomere is a region of highly repetitive DNA, at the end of a chromosome. Each time a cell divides, the telomere is shortened and, unless it is repaired, the chromosome eventually becomes damaged and unstable. The presence of telomeres had been known since 1938, when they were described in a lecture by Hermann Muller. Barbara McClintock confirmed the finding in 1941.[147]

A Russian biologist, Alexey Olovnikov, provided an explanation for Hayflick's limit on cell division. He was thinking about the limit, when a train he was travelling on entered a Moscow subway station. He realised that the engine of a train must cover some of the railway track at the end of the line, so the passenger cars can never reach the true end of the line. By analogy, he realised that a protein (the train) copying DNA (the track) would not reach the telomeres at the end of the molecule.[148] It was later found, as predicted, that telomeres in human fibroblasts shorten as cells divide.[149]

Limiting the number of cell cycles helps to prevent cancer but shortens lifespan. Disturbed cells, which enter a phase of rapid cell division, are unable to grow indefinitely. With each cell division, the telomeres are shortened and the resulting cell line dies. This can be viewed as one of our primary defence mechanisms against cancer and other forms of abnormal cell growth.

As mentioned, certain germ cells are immortal. Such cells use an enzyme, telomerase, to restore telomeres to their full length. Telomerase is sometimes called the cellular immortalizing enzyme. While typical body cells contain the gene for telomerase, it is inactive, preventing unlimited cell division. The use of telomerase by normal cells is highly restricted during early development.

Unfortunately, the cell cycle control provided by the Hayflick limit appears to be inactivated in cancer cells. Like germ cells, cancer cells contain telomerase, which repairs the telomeres and may be a necessary factor for repeated cell division in cancer.[150] Cancer cells have increased levels of telomerase, although they do not reach the levels found in germ cells.[1] When cells become cancerous, telomeres may get shorter, but those below a minimum length are replaced. Telomerase repairs the telomeres, prolonging the life of the cell. Most malignant tumours have active telomerase. Activating the telomerase gene is a mechanism by which cancer can subvert the cell cycle limit and achieve unlimited growth potential.

Anaerobic selection

When a new colony of cancer cells starts growing, it soon reaches a volume that is large, relative to the diffusion of oxygen into the tissue. Such cells experience a reduced supply of oxygen and other nutrients: they are suffocating. An everyday analogy is building a town in a green field site. Simply adding new houses will work for a short time but, unless roads and services are adequate, people will not be able to gain access and the town will not prosper.

As a tumour grows, pre-malignant cells come under selection pressure to develop an anaerobic metabolism, so they can manage with less oxygen. This means changing the controls on the cellular metabolism.[151] The method they use is glycolysis, which is the anaerobic conversion of glucose to lactic acid, for energy.[152] Normal cells also use glycolysis, for example, muscle cells use it when sprinters require a sudden burst of energy.

Even when oxygen is plentiful, cancer cells generally rely on glycolysis to generate energy. The cell does not have to develop new genes to switch to anaerobic metabolism: instead, those damaged cells that over-express existing glycolysis genes have a selective advantage in the environment of the growing tumour. However, this form of metabolism is inefficient, so the colony experiences growth restriction.

New blood vessels

As the tumour grows, its oxygen supply becomes inadequate. A second solution to this problem is for the tumour to grow more blood vessels. Potential cancer cells that can stimulate the development of blood vessels in the tumour have a greater ability to grow. Given time, a colony of dividing cells, which generates damaged offspring, will result in some cells with a mechanism to stimulate blood vessel growth.

Once again, the cells already have the genes to stimulate or inhibit blood vessel growth, as this is a function of normal physiology. They do not need to find a new method, just to activate a gene that promotes blood vessel growth, or inactivate a gene that suppresses this growth. Eventually, given sufficient cell divisions, combined with errors and selection pressure, such cells will arise.

Local invasion

Since cells in the tumour are competing for resources, a cell that can migrate a short distance has an advantage. It can find a new habitat, where growth is free from the constraints of competition with other cancer cells. Given time, microevolution will generate cells that can move and invade the surrounding tissue. The genes for this behaviour already exist within the cells, but are inactive. A cancer cell that expresses these genes gains the opportunity to prosper in a new environment, so is favoured by natural selection.

Spreading to new sites

A single tumour with a reasonable blood supply can grow to a large size. However, as it consists of a number of genetically different cell types, competition is intense. A biological model for this is the evolution of species on an isolated island. The spread of tumours to distant sites is, in some ways, analogous to the dispersion of species between islands. Such dispersion of animals has been studied intensively. Knowledge about biological diversity can be used as a model for the metastatic spread of cancer.[153]

Population dynamics on islands was described, in 1967, by ecologists Edward O. Wilson and Robert McArthur, who proposed an "equilibrium theory of island biogeography."[16] Species move from the source area, which is the nearest mainland or another island, at the immigration rate. This rate is dependent on the distance between the source area and the sink island. The immigration rate is balanced by the sink island's extinction rate. To maintain a species, the sink island needs to have a suitable habitat for the invading animal.

As the area of the island increases, the number of species it can support also increases. If an island has ten times the area of a second similar island, it will contain about twice as many species. If a species can escape the confines of its island, it may find another where it can start a new colony, free from the constraints of competition. A recent investigation of the diversity of beetles and spiders on roundabouts and traffic islands, in the town of Bracknell, England, found that larger areas contained more species.[154]

Most organisms that leave a source site are lost during travel, or do not find a site for a new colony. The conditions in the new colony need to provide a suitable habitat. When they do, a new colony is formed and rapid growth may be possible. A cancer cell forming a tumour at a suitable new site has a similar growth advantage, although most cells that break away from the original tumour site will be lost. The metastatic spread of cancer is a predictable result of microevolution and is a special case of geographic spread and species diversity.[i]

As a tumour grows, not only does it get bigger but also, we predict, its cells will become more diverse. The spread of metastases results in

---

[i] One difference is that cancer spread within the body is more three dimensional than the spread of most island species, involving volumes of tissue rather than areas of habitat.

geographical isolation of cells at the new site, and a correspondingly rapid increase in variation. The diversity helps the cancer to withstand treatment, as it improves the chances of at least some cells being resistant to the therapy. This might explain why larger cancers and those with more metastases are harder to eradicate.

Microevolution

The microevolutionary model can be summarised as a sequence of phases. These phases are analogous to the generation of a new species and its spread. In the case of cancer, redox processes underlie each phase.

Phase One

Damage involving redox mechanisms increases cell division. The cells enter a more oxidising state to facilitate growth. The dividing cell does not breed true because of increased error generation.

Phase Two

Multiple cell divisions occur, with selection pressure for survival and growth. Growth is limited by the supply of nutrients and an anaerobic metabolism is favoured. Cells are under selection pressure to become immortal and resist apoptosis. At some point, the cell becomes "selfish" and reverts to unicellular behaviour, a process associated with alterations in chromosome number.

Phase Three

The disturbed cells come under selection pressure to generate a blood supply, invade other tissues and migrate to other sites. This phase is associated with an increase in cellular diversity and, hence, greater resistance to extinction caused by therapy or other factors.

Consistency

We have described a model of cancer in terms of microevolution. The theory is internally consistent and has the ability to explain core phenomena associated with the disease. Furthermore, it places cancer research in a broad biological context. However, just because an idea is logically reliable, it is not necessarily scientifically correct. We need to show that this approach is consistent with conventional ideas and that it has predictive value. In particular, can this model provide a key to effective treatment?

A malignant cell

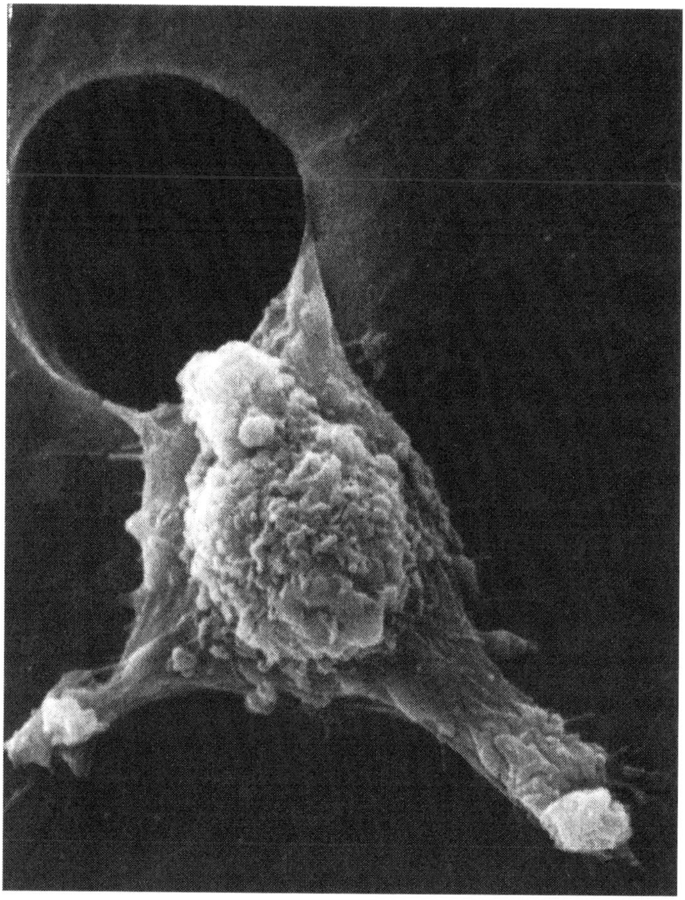

This scanning electron micrograph of a malignant cell shows the cell extending "arms" or pseudopodia to assist its movement through tissues. It illustrates the independent nature of a "selfish" cancer cell.

Conventional theories

"We are usually convinced more easily by reasons we have found ourselves than by those which have occurred to others." Blaise Pascal.

The microevolutionary model is consistent with conventional ideas on the development and growth of cancer. However, current models are generally more limited in their ability to describe the process. Some of these explanations may represent component mechanisms, selected by the evolutionary process. Before we consider these conventional ideas in detail, we will briefly describe the changes that occur in cancer cells.

Cancer cells

Under the microscope, normal cells have a regular appearance, whereas malignant cancer cells look disorganised. Their size and shape becomes variable, with deformed internal structures. The appearance of the cancer cell nucleus has been described by Dr William Hahn, of the Dana-Farber Cancer Institute,

"If you look at most solid tumours in adults, it looks like someone set off a bomb in the nucleus."[155]

This description illustrates the magnitude of the differences we need to explain. Loss or gain of genetic material during malignant cell division often gives rise to abnormal cells, with giant nuclei that contain enormous amounts of DNA. These atypical nuclei usually contain a greater number of chromosomes than do healthy cells.

Dividing cells have fine, tangled, thread-like structures in their nuclei, called mitotic figures, which are often distorted in cancer cells. The mitotic figures help control the separation and division of the chromosomes into the daughter cells. The presence of irregular mitotic figures may be an indication that a tumour is malignant. As the cancer develops, the resultant cells change.

The process of cell specialisation is called differentiation. In normal development, this process leads to varied types of body cell, such as nerve, muscle or blood cells. Malignant tumours contain cells that no longer resemble their tissue of origin; these are said to be undifferentiated or anaplastic. The changes are sufficient for us to consider malignant cells as belonging to new species. Such cells operate to maintain their own genetic identity, rather than that of the organism as a whole.

A sequence of mutations

According to standard theories, cancer development involves a set of mutations on the genes of an isolated cell. Occasionally, this sequence of mutations causes damage to a set of genes that normally prevent cancerous growth. This can happen if radiation or a carcinogenic chemical mutates a cancer-related gene, often by the formation of free radicals.

Mutations in two types of gene are of particular interest. Tumour suppressor genes, which tell the cell not to divide and grow, can become damaged, allowing unrestricted growth. Mutations can also lead to over expression of genes that promote cell growth, called proto-oncogenes. This sequence of mutations allows the cell to break free of normal growth controls and to proliferate. Some time later, additional mutations occur, allowing the cells to invade nearby tissue. Eventually, cancerous cells break out into the bloodstream and form metastases at distant sites in the body.

This theory is consistent with the idea of microevolution, in that it lists the stages of cancer growth and spread in terms of mutations. If we add the concept of selection pressure to this sequence of mutations, we arrive at a description that is similar to the microevolutionary model. However, although such a sequence of mutations is consistent with microevolution, it is inadequate to explain the full story.

A recent study reported that mutations in three specific genes were sufficient to create cancer cells.[156] Other scientists criticised the study for failing to explain why it took more than "60 population doublings" from the introduction of the first of these genes until tumour cells were produced, or why the tumour cells had abnormal numbers of chromosomes.[196] However, the study shows that it is possible, with selected mutations, to initiate a process of rapid microevolution, which results in a cell type with some of the characteristics of a tumour.

According to the conventional mutation model, a sequence of five or six specific mutations leads to cancer. Each additional mutation in the sequence increases the risk of a cell becoming cancerous. The probability of each mutational event is low. While spontaneous mutations can increase the risk of cancer, the rate is too low to generate the necessary genetic changes found during cancer development. Moreover, this theory does not explain adequately why cancers generally occur years after exposure to a mutagen, such as a dose of radiation. The existence of this delay supports the suggestion that faulty cell division initiates a process of microevolution in the damaged cells.

Cancer cells differ from normal cells by more than a few minor mutations. Conversely, a few small mutations would not be expected to cause the changes that have been observed in malignant cancer cells. DNA mutations are generally used to provide an explanation for minor differences in a biological population. Such variation may require geological timescales to lead to a new species.[157] By comparison, the time scale for cancer development is short, perhaps a few years; this is not long enough for a few random DNA mutations to result in the radically different physiology of malignant cancer cells.

The main problem with the mutation theory is that a mutation usually changes a single protein, whereas the changes in cancer are extensive and fundamental to the cells' structure and function. Cancer cells do not look like healthy cells that have suffered a small number of mutations. Such mutations would not usually cause the gross changes necessary to produce a cancer cell.

Despite its limitations, the sequence of mutations model has been a basic dogma for decades. The model is productive, in that it provides a genetic basis for cancerous changes, and points the way to ideas that are more biologically plausible.

Mutation sequence theory extended

To overcome limitations in the mutation sequence theory of cancer development, theorists incorporated the notion of additional cellular damage. This extended theory proposes that a cellular mechanism for the repair of DNA becomes impaired by a mutation. This means that, over time, the cell accumulates mutations. This theory indicates an additional source of variation in the cell colony, which could facilitate microevolution.

Cells have several methods for repairing genetic damage. One such mechanism involves mismatch repair genes. DNA consists of two complementary strands, in a double helix structure. The nucleotide building blocks in the two strands must match up exactly. When a mutation occurs in one of the strands, the two DNA strands no longer match. The function of mismatch repair genes is to identify and repair defective DNA, which occurs from time to time. This process has been compared to the action of a computer spell checker.[158]

If the cellular repair mechanisms cease to function effectively then, over time, mutations can build up. Unrepaired DNA damage will be reproduced and passed on to all future cells. If the repair mechanisms are themselves damaged, the effective mutation rate increases. The driving force in this generation of errors is cell division. Over time, the number of errors a cell accumulates is determined by cellular proliferation.[1] Given enough cell divisions for the processes of natural selection to work, damage to cellular repair mechanisms could indeed result in metastatic cancer. However, there is a problem with this proposal. Unless the variation is substantial, the selection would simply result in genetic drift: the resulting cells would change gradually with time.

For the model to fit the facts, we need to demonstrate a rapid means of generating what is, effectively, a new species of cell. In ecology, it takes millions of years of evolution to generate a new species. In cancer development, we do not have the luxury of geological timescales. The mechanism must cause rapid speciation, and must be consistent and common enough to occur in a third of the population of humans and other mammals, as they age.

To be consistent with our speciation analogy, damaged DNA repair mechanisms must generate enough variation to produce a malignant cancer cell, in a reasonable number of generations. The result of unrepaired DNA damage could simply be a large number of small mutations, which would damage the viability of the cells. There might be insufficient time, in terms of number of cell generations, to produce cancer. In order for this hypothesis to be supported, the researchers would need to demonstrate that DNA repair mutations generate sufficient variation to create cancer cells rapidly and reproducibly. Currently, this remains to be clearly demonstrated. However, this modified theory is a more viable model than that of simple mutations and remains within the microevolutionary paradigm.

Stem cell mutation

Stem cells are unspecialised cells, which give rise to specific specialised tissue, such as blood or skin cells. When a stem cell divides, it produces one dissimilar, specialised, daughter cell and one replacement stem cell. It has been suggested that stem cells may develop cancer more quickly than other cells. The involvement of stem cells overcomes some of the objections to the mutation theory of cancer development, described above.

Most body cells are specialised to perform a particular role. For example, neurons transmit electrochemical signals, muscle cells form contractile fibres, which enable movement, and fat cells store excess energy as lipids. Stem cells, on the other hand, are relatively unspecialised, less differentiated than typical tissue cells and are often similar to the original germ cell from which the individual grew. Most tissues contain a number of stem cells, which are able to divide and produce other cells, to replace those lost in everyday wear and tear. Stem cells can replace cells in damaged tissues; for this reason, there is extensive research to see if they can be used to repair organ damage, perhaps caused by a heart attack or stroke.

Stem cells have many of the characteristics of cancer cells. They are undifferentiated, can divide to produce dissimilar daughter cells, are mobile and may be virtually immortal. Since they are present in most tissues, it is easy to see how they are an attractive target for the mutation theory of cancer. It might take fewer mutations to convert a stem cell to a cancer than it would a specialised cell. Damage to the stem cell DNA repair mechanism could provide a starting point for microevolution to malignancy.

However, the involvement of stem cells does not explain the massive changes in the structure and behaviour of cancer cells, compared to healthy cells. A sequence of small mutations would probably not lead to such changes in the timescales available. A feature of stem cells and tissues that offers a more promising approach is the cell division process. The rapid division of stem cells may be implicated in the microevolution of cancer.

Controlling cell division

Cancers grow by cell division, which is controlled by two types of gene. Some make division more likely, while others inhibit the process. Genes that control cell division, repair genetic errors or influence apoptosis are often called oncogenes, because of their involvement in promoting the growth of cancer cells. This name is misleading, as it suggests these genes cause cancer. A more appropriate name might relate to their specific function, for example, cell division genes. A typical oncogene does not cause cancer but promotes a related behaviour, such as increased cell division.

Cell division and the genes that promote it are essential elements of normal tissues. In healthy tissues, tumour suppressor genes balance the oncogenes that promote growth. These suppressor genes, which are also called repressor genes, act to inhibit cell division and growth. A suitable balance between the expression of growth promotion genes and that of suppressor genes is necessary for the maintenance of healthy tissues. An important influencing factor is the level of oxidants in the cell. A high level of oxidants can lead to growth stimulating genes being switched on and repressor genes being switched off.[101]

While the faulty expression of genes controlling cell growth, division and cell death appears to be essential for cancer, this does not necessarily indicate that mutations in these genes are the cause of cancer. Although such changes could lead to clonal growth of similar cells, they are not sufficient to account for the level of diversity found in cancer cells.

When a tumour suppressor gene malfunctions, the result can be uncontrolled cell growth. A single suppressor gene, called p53, is mutated in about half of all known cancers. P53 is perhaps the most notorious tumour suppressor gene; it was discovered in 1979 by David Lane, in Scotland, and also, independently, by Arnold Levine in Princeton. The p53 gene is the most commonly mutated gene in human cancer. It was named after the molecular weight of the protein it produces.[a] The function of p53 in cellular control illustrates how redox mechanisms signal and regulate cancer gene expression.

The p53 gene inhibits cell division in damaged cells, giving time for the cell to be repaired before dividing again.[159] It regulates cell division and apoptosis, by preventing cells with damaged DNA from dividing and, when damage is too great, promoting cell death. Cell division is a delicate process. Preventing division until the cell has been repaired reduces the likelihood of unnecessary errors in the daughter cells. When a cell is damaged, the p53 protein concentration increases, slowing the cell cycle. DNA repair mechanisms can then repair the DNA. When the DNA is restored, the p53 levels are lowered and cell division can proceed.

With a large amount of DNA damage, p53 triggers the cell suicide program. Normally, the p53 gene causes irreparably damaged cells to commit suicide, by apoptosis. Activation of the p53 gene either inhibits

---

[a] P53 has a molecular weight of 53 kilodaltons.

cell division or signals the cell to commit suicide. Since cell division is controlled by oxidant levels, these actions suggest a redox mechanism.

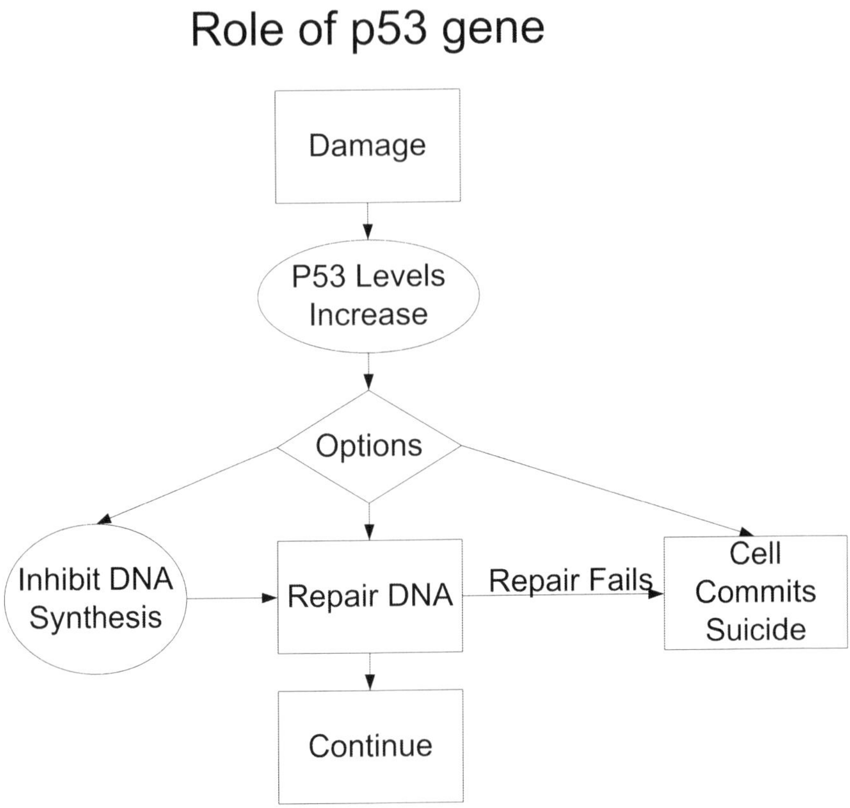

The actions of the p53 gene are regulated by the level of oxidants in the cell.[160,161,162] Expression of the p53 suppressor gene activates a number of additional genes, which make proteins that react to or generate oxidants.[163,164] P53 increases the expression of redox-related genes and leads to the formation of reactive oxygen species, which damage the mitochondria and ultimately result in cell death.[163] Thus, p53 causes cell suicide by inducing the formation of free radicals, which leads to the oxidation of mitochondria. This cell suicide can be inhibited by antioxidants.[165,166] However, some antioxidants promote the expression of p53 in abnormal cells, causing the cells to commit suicide.[167,168,169] Conversely, the p53 gene may be inhibited in cells with a specific antioxidant enzyme deficiencies[b] and more oxidising environment especially in the nucleus.[170,171] Vitamin supplementation is reported to lower the risk of mutation of the p53 gene, consistent with the general role of dietary antioxidants in prevention of cancer.[172]

If the p53 gene is absent, a cell does not commit suicide when damaged and may grow uncontrollably, producing a tumour. If p53 itself is damaged, the cell can accumulate further genetic injury while cell death is inhibited. As we shall see, radiation and chemotherapy treatments act, in part, by stimulating apoptosis in cancer cells. Tumours without a functional p53 gene may be resistant to such therapies. Loss of the p53 gene occurs in about 70 percent of colon carcinomas, 50 percent of lung cancers, and 30 to 50 percent of breast cancers. Hereditary loss of the p53 tumour suppressor gene increases the risk of breast cancer and sarcoma, but does not produce cancer in most other tissues.[1]

Some cancer-related genes can stop cells committing suicide. The first such oncogene to be identified was called bcl-2.[173] This gene was named because of its association with B-Cell Leukaemia. Many more oncogenes that prevent cell death have since been found.[174,175] When the bcl-2 gene is expressed, it generally increases hydrogen peroxide, but at the same time it apparently results in a more reducing environment in the cell, particularly the nucleus.[176,177] The role of oncogenes highlights the involvement of redox reactions in the control of cancer cell death.

Faulty cell division theory

A cell can fail to divide correctly for many reasons, not all of them based on genetic mutation. Environmental damage to the proteins that are responsible for copying and dividing the DNA between the cells could result in each daughter cell having a different genetic makeup. In this case, a physiological or biochemical change in the cell structure results in an inheritable genetic difference between the daughter cells. Errors during cell division provide an alternative to mutation as an explanation for the changes leading to cancer. Faulty cell division offers a neat explanation for cancer, and fits well with the microevolutionary model.

Increased cell division may lead to increased genetic errors.[178] Every time a cell divides, there is a risk of producing cells that are not identical to the parent cell: the process is error prone. Although all the

---

[b] P53 activity may have a specific requirement for an intact thioredoxin system, rather than a general dependence on the intracellular reducing environment.

cells in a tumour grow from a single cell, cell division in cancer cells is corrupted.[179,180]

In faulty cell division, a mutation or other event damages the mechanism for separating the chromosomes.[181,182,183,184,185] A daughter cell might get more or less than its share of chromosomes. Some of the chromosomes may be incomplete or broken. The errors would be compounded with each generation, because of the faulty cell division mechanism. Furthermore, cancer cells have abnormal numbers of chromosomes and faulty cell division may be a mechanism for cells to gain or lose chromosomes.

Of course, many of the daughter cells resulting from each faulty division would die. Although the growth of errors might be exponentially fast, it would be moderated by selection for viable offspring. Such a population could be under even greater selection pressure than cells with damaged DNA repair mechanisms.[186] This process could introduce a high degree of cellular variation, consistent with that required for rapid microevolution to cancer.

Aneuploidy

A specialised version of the faulty cell division model is the aneuploidy theory of cancer. Aneuploidy occurs when a cell has an abnormal number of chromosomes. Human cells are diploid; the 46 chromosomes they contain can be grouped as 23 pairs. However, when cell division is faulty, too few or too many chromosomes may be transferred to the daughter cells. This can happen in many ways. One cell could get a small extra fragment of a chromosome. Alternatively, a cell might receive both sets of chromosomes, ending up with twice the normal amount of genetic information, or 92 chromosomes, in 23 quads of identical chromosomes. It is clear that aneuploidy can result from, and can produce, errors in cell division.

In cancer, cell division does not conserve the chromosome number. Malignant cells have non-standard numbers of chromosomes. The aneuploidy theory of cancer originated over a century ago.[196] The theory states that all cancer cells have abnormal numbers of chromosomes,[187] and the evidence seems to confirm this suggestion. For example, in one study of approximately 27,000 cancers, all had chromosomal aberrations.[188] Another survey of 2,400 cancers found chromosomal variations consistently.[189] Typical malignant cancers contain cells with multiple copies of damaged chromosomes.[190] There may be some forms of malignant cancer with the correct number of whole chromosomes, but such cancers appear to be rare.

An abnormal number of chromosomes may be a defining factor in human malignant cancer.[191,192,193,194] Irregular numbers of chromosomes are also found in some non-cancerous conditions, such as Downs syndrome, and in some aged cells.[195] The presence of an atypical chromosome complement does not necessarily mean a cell is cancerous, but cancer cells without aneuploidy are exceptional.

Peter Duesberg, the scientist who suggested HIV does not cause AIDS, is a notable champion of the aneuploidy theory.[196,197,198,199,200,201] Duesberg argues that the most critical feature of a cancer cell is that it has the wrong number of damaged chromosomes. This model is an extreme version of the faulty cell division theory. However, it is supported by experimental observations on the number of chromosomes in cancer cells. Furthermore, the model is consistent with the extent of damage observed in the nuclei of advanced cancer cells.

This theory is important to our story, because polyploid changes in chromosome number are one of the few, known mechanisms for instantaneous formation of a new species. Variation in the numbers of chromosomes in daughter cells provides a way of explaining the rapid and all too frequent formation of malignancy.

Evolutionary perspective

If cellular microevolution is the driving force for the development of cancer, then a varied population of cells, dividing with errors and subject to natural selection, will eventually become cancerous. However, the mechanism for generating a cancer cell may vary from one case to another. One cancer could arise from multiple mutations, another from stem cell damage, while error prone cell division might be the cause of others. These different ideas are alternative pathways to the final destination of malignancy. Different cell lines take different paths, depending on the particular conditions of their microevolution. These processes are the cells' response to selection pressure. Microevolution does not specify the mechanism used to achieve the goal: it merely selects for cells with the most effective response to the conditions.

Each of the above models for cancer development could occur. However, evolution in cancer is rapid. Limited numbers of mutations, as in the standard theory, are unlikely to introduce sufficient genetic information for evolutionary pressure to produce the new species of cells that we call cancer.

Faulty cell replication, by contrast, could result from mutations in the controlling genes. In some cases, faulty cell division may be a natural extension to the standard theory. Modifications that involve damage to cell division mechanisms and, specifically, alterations to the chromosome number, are likely to provide the injection of information that is required, together with a high cell death rate through natural selection.

Microevolution provides a scientific framework for explaining the development of cancer. As we have seen, the various conventional models for cancer development can be viewed in a microevolutionary context, within which a coherent picture emerges. In the following chapter, we will examine a number of less conventional views, in the same way.

Unconventional theories

"For a successful technology, reality must take precedence over public relations, for Nature cannot be fooled." Richard Feynman

A quick internet search reveals so many alternative cancer "cures", that an uninformed person could suppose curing cancer is easy. Although many of the remedies have little scientific support, this mass of misinformation contains several interesting ideas. It is worth considering some of the more promising proposals, to see whether they have any biological basis.

In this chapter, we describe several unconventional theories from a microevolutionary perspective. Cancer specialists tend to ignore these approaches, or consider them obviously incorrect. We will show that the medical establishment sometimes dismisses ideas too readily. For example, modern stem cell theory is a variant on a theory of cancer as embryonic tissue, proposed by John Beard, about a century ago.

Beard's trophoblastic theory

At the start of the twentieth century, John Beard, an embryologist from Scotland, suggested that cancer originates from leftover embryonic cells. The placenta is the organ that channels nourishment from a mother to her foetus. Beard noted that cancerous tissue is similar to the trophoblast,[202] a layer of cells that produce the placenta. Following fertilisation, the egg and sperm grow into a small, hollow ball of cells. Trophoblast cells arise from the wall of this ball, and burrow into the lining of the womb, in order to form the placenta.

Beard described this burrowing characteristic as similar to the invasive nature of metastatic cancer cells. Recent research supports this suggestion.[203] Many, though not all, cancers express antibodies from trophoblast cells.[204,205] Invasive cancer of the uterus can arise directly from trophoblast cells, and may be highly sensitive to conventional chemotherapy.[a] Beard considered that cancer was a result of unrestrained growth of trophoblast cells. He viewed cancer as a form of false placenta, growing from cells that had moved to an unexpected position in the body during development.

---

[a] Choriocarcinoma.

Pancreatic enzymes

In 1906, Beard proposed that enzymes from the pancreas provide the body's primary anticancer defence mechanism. The pancreas is best known for secretion of the hormone insulin, which controls the levels of sugar in the blood. Failure of the pancreas to produce insulin leads to diabetes. The pancreas is a large gland near the stomach; it secretes digestive fluid, containing enzymes that break down proteins, into the intestine.

The foetal pancreas begins producing enzymes early in development. Beard hypothesised that enzymes from the pancreas are used to control placental growth and cause the trophoblast cells to differentiate. We now know that genes associated with differentiation of the trophoblast are involved in pancreatic development,[203] providing a biochemical link between these two organs.

Beard suggested that pancreatic enzymes could be an effective treatment. In the early 20th century, some physicians in Europe and the United States tried treating advanced cancer using injections of pancreatic enzymes. They reported considerable success, with variations that could be explained by the quality of preparations used or by differences in cancer type. Following Beard's death, in 1924, the use of this treatment declined.

Beard's theory and stem cells

At first impression, Beard's ideas seem to have little in common with conventional theories, but this is not the case. Beard's theory predates modern research, but shares many features with recent ideas that cancers develop from stem cells.[206] We can speculate that, over time, Beard would have modified his concepts to accommodate stem cells. Given that Beard's idea precedes current models by about a century, it is justifiable to suggest that the currently fashionable stem cell mutation theory is an extension of Beard's original ideas.

The basic premise of Beard's ideas is compatible with the microevolutionary model. Although the conversion of a healthy cell to a malignant one involves microevolution, there is insufficient time to generate new and effective proteins to support malignancy. Our protein molecules have evolved over millions of years: much longer than the few years available for cancer development. Thus, a cancer cell is largely restricted to modifying the existing cell machinery. It can switch off genes that inhibit cell proliferation, or switch on others that promote growth. However, it does not have time to evolve a new gene to provide a novel enzyme or other protein, with functions not found within the capabilities of a normal cell.

Most cells in our bodies have the complete complement of genes, including those used by trophoblast cells. A liver cell, for example, has the same genes as the fertilised egg from which it was derived. In the process of specialisation, the cell becomes differentiated, by selectively switching on some genes, while deactivating others. Oxidants can provide the stimulation for differentiation and cell specialisation.[207,208,209,217] The formation of cancer is often described as de-differentiation, or reversion, to the characteristics of an earlier phase. While this simplification is not born out by detailed analysis, it is descriptive of some processes of cancer development.

As Beard recognised, the invasive features of malignant cells are also seen in trophoblast cells, during development of the placenta. Whole groups of genes can be switched on or off together, so the genes normally expressed in trophoblast cells, if activated, could give an evolving cancer cell many of the features of malignancy. All the genes for trophoblast behaviour might be activated together, providing the cell's offspring with an increased ability to migrate through and invade normal tissues. This would confer a large selective advantage, in a single mutation.

As a secondary consequence of the genes being switched on collectively, the cells might gain the same proteins as are found on the surface of trophoblast cells.[b] This more modern interpretation of Beard's ideas could justify his assertion that the hormones or enzymes that control the growth of the trophoblast during pregnancy might have value as a cancer treatment.

If researchers class Beard's theory as a discounted historical idea, they risk missing fruitful ideas for investigation and treatment. Beard deserves the credit for being a biological innovator and scientist ahead of his time. However, we should not place too much emphasis on the specific details of Beard's hypothesis. Had he been aware of the increases in scientific knowledge since his death in 1924, it is likely that he would have modified his ideas. If he had survived to modern times, we like to think that John Beard would have been a champion of stem cell research.

---

[b] Human chorionic gonadtropin (HCG) is found on the surface of trophoblast cells and on some cancer cells.

Anaerobic theory

There are two major forms of cellular respiration: aerobic, which uses oxygen to generate energy, and anaerobic, which does not. For many anaerobic cells, oxygen is a poison. In human physiology, anaerobic metabolism occurs when we exercise heavily. In the absence of oxygen, our muscles use glucose to produce energy, generating lactic acid. This allows athletes to achieve short bursts of effort, above those possible by increasing oxygen intake. We all have experience of feeling breathless after exercise, when we take in oxygen to recover from such exertion.

The idea that the metabolism of cancer cells is deranged originated early in the 20th century. In 1931, Otto Warburg won the Nobel Prize for research into cell metabolism.[210] Later, he showed that cancer cells produce lactic acid from glucose, even in the presence of oxygen. The mechanism is similar to that used by muscle cells, when oxygen is in short supply. Cancer cells, however, are more dependent on glucose as a source of energy.[211,212,213,217] Warburg concluded that cancer cells have defective energy metabolism.[214] A second Nobel Prize winner, Albert Szent-Gyorgyi, came to a similar conclusion,[215] and recent reviews confirm the importance of this approach.[216,217]

According to the anaerobic theory, cancer cells are thought to live without oxygen. Towards the end of the 20th century, elements of the medical establishment took an extreme view of this theory: assuming it implied that cancer cells need no oxygen, and might fail to thrive in its presence.[218] It has since been shown that cancer cell lines can grow in the presence of oxygen. Furthermore, research suggests that cancer cells can use oxygen, in addition to their anaerobic metabolism.[219] Once cancer cells had been shown to tolerate oxygen, the anaerobic theory of cancer fell into disfavour. Despite this, researchers who disregarded the implications of the anaerobic theory may have placed unnecessary constraints on cancer research.

Reginald Holman, a medical bacteriologist, suggested we reconsider the anaerobic theory of cancer. He pointed out that claims that cancer cells are anaerobic have not been refuted, as widely believed. Indeed, the presence of anaerobic metabolic mechanisms in cancer cells has been confirmed.[220,221,222] In addition, high levels of oxygen and the resulting oxidation can slow cancer cell growth.[223,224] Most tumours are deficient in oxygen and contain cells that use anaerobic metabolism; this appears to be a common feature of the growth of tumours.

According to the microevolutionary model, shortage of oxygen is an important selective feature for cancer. One of the first problems a growing tumour faces is a lack of oxygen and nutrients. Unless a tumour finds a way to acquire a local blood supply, it generally does not grow larger than a millimetre in diameter. Even when blood vessels invade the tumour, they are not as well formed as normal vessels. Blood vessels in tumours often have sluggish blood flow and leaky walls, providing an inadequate supply.[225,226,227]

A growing tumour needs a continually increasing supply of oxygen and nutrients. Once the tumour consists of more than about ten cells, the cells start to come under selection pressure, due to lack of oxygen. Some of the proliferating cells may rely more on anaerobic metabolism, which does not need oxygen.[228] These cells have a selective advantage and will tend to flourish. Thus, during growth, cancer cells tend to become less aerobic. They use a primitive and simple form of metabolism, getting their energy mainly from glucose and other sugars. Compared to healthy tissues, they are less dependent on oxygen.

The periphery of a tumour is an active region, where host and cancer cells interact.[229] This area may have a good blood supply and be well supplied with nutrients. Conversely, the centre of a large tumour is short of oxygen and nutrients, and is often described as the dead, or necrotic, centre. If a cancer cell is more than about a fifth of a millimetre away from a blood vessel, growth is prevented and necrosis can set in.[1]

The lack of oxygen in solid tumours gives them a unique physiology and may lead to development of more aggressive tumour cells.[226,230,235] Lack of oxygen also makes radiation and chemotherapy less effective.[231] At the same time, it provides a feature that may be used to develop new therapies. One suggestion has been to use anaerobic bacteria to help control cancer. These can grow under the low oxygen conditions in tumours but not in the rest of the body, where the oxygen supply is higher.[232]

Lack of oxygen can be associated with spread to other sites. The more anaerobic the tumour, the greater is the selection pressure on the population of cells. Treatments such as radiation, which kill cells locally, shrinking the tumour, can also increase the local supply of oxygen, reviving cancer cells that would otherwise die.[233] Ironically, a treatment that is effective in shrinking a tumour may ultimately change the local microenvironment and promote cancer spread. Treatments that kill sensitive but less aggressive cells could reduce the competition, allowing the more virulent cells to spread quickly.

The anaerobic theory of cancer is consistent with microevolution. Because of the prevailing shortage of oxygen, selection pressure will tend to favour cells with an anaerobic metabolism, even if it is not an essential feature of tumour growth.

Acid accumulation

A derivative of the anaerobic theory is the idea that cancers are more acidic than healthy tissue. In 1933, Dr William Hay introduced the idea that disease is caused by accumulation of acid in the body. Glucose can make tumours more acidic, as lactic acid is produced and local carbon dioxide is increased.[234,235,236] More recently, this hypothesis has been taken up by Robert Barefoot, who claims that a calcium-based alkalising diet will destroy cancer.[237,238] However, there is little evidence that tumours are short of calcium, or that the acidity in some tumours is more than a side effect of metabolism. Some cancers deposit calcium, which can be used for tumour detection: in breast cancer mammograms, for example. Such deposits are likely with high concentrations of calcium, in a relatively alkaline environment. The idea that an alkali diet, with additional calcium, will redress this balance has little scientific support.

Varied mechanisms

The current conventional theories of cancer development are consistent with microevolution, as are some unconventional theories. They also provide models that are similar in various essential features. Current stem cell ideas were clearly predated by Beard's trophoblast theory. The fashionable idea of angiogenesis (new blood vessel growth), is related to tumour blood supply limitations, which result in a growth-limiting, anaerobic environment for cancer development. Both old and new ideas have a place among the diverse mechanisms that a cell can use to break free from normal physiological controls on growth.

Conventional medicine uses several descriptive approaches to track the developmental progress of cancers. We will now consider how the different stages of tumour growth and development relate to the microevolutionary model.

Stages in cancer growth

"The older a cancer is, the worse it is. And the more it is involved with muscles, veins and nutrifying arteries, the worse it is, and the more difficult to treat. For in such places incisions, cauteries and sharp medications are to be feared."
Theodoric, Bishop of Cervia (1267)

In this chapter, we introduce some of the terminology used to describe cancer.[a] The profusion of medical names and jargon associated with cancer can act as a barrier to understanding. Frequently, the names are Latin or Greek versions of simple English terms. For example, a leukocyte is a white blood cell and its name comes from the Greek for white cell. Leukaemia is a cancer of these white blood cells.

Physicians describe cancers in terms of the tissue of origin and the stage of growth. Cancers from different tissues often have characteristic growth rates. Growth stages provide a way of classifying successive phases of the microevolutionary process. For example, a patient might have a "stage II bladder cancer". In this case, stage II indicates that the cancer has invaded the bladder's muscular wall.

In this chapter, we explain how the clinical stages relate to the underlying biological processes of tumour growth.

Benign or malignant

Physicians divide tumours or growths into two categories: those that expand locally, known as benign, and those that spread aggressively and invasively, which are called malignant. Benign tumours are relatively safe and do not invade the surrounding tissues or spread to distant sites. In some cases, benign tumours function like normal cells of the tissue from which they were formed. The body can encapsulate such tumours with connective tissue, to limit their growth. Large benign tumours can be surgically removed, restoring the body to health. In many cases, complete surgical removal provides a cure. Although benign tumours are abnormal, they are less dangerous than malignant growths, because they have not fully escaped the growth controls that keep normal cells in check.

---

[a] Additional details can be found in the appendix and the glossary.

Cancers are often first noticed as lumps or masses in a tissue. Benign tumours are frequently encased in a well-formed capsule. Malignant tumours, without such a capsule, tend to infiltrate surrounding tissues. This difference helps doctors to separate malignant from benign tumours. A distinct, self-contained lump is more likely to be benign, whereas a mass that is inseparable from surrounding tissues, irregular in shape and lacking a well-defined border, is more likely to be malignant.

Location

Cancer can occur in any tissue of the body. In the year 2000, the most common cancers for females in the United States were those arising in the breast, lung and colon. The most frequent sites for males were the prostate, lung and colon. Such cancers as these are formed from epithelial, or lining, tissue. They are called carcinomas, and are the most common type of cancer.

The site of origin of a tumour may influence its symptoms and determine the treatment. The organ or tissue from which a cancer is derived is important, both for its name and for medical classification. The origin may give a general indication of how the cancer will grow and when it will show symptoms. This influences how early a diagnosis is likely to be made. For example, a visible skin tumour will generally be detected early. By contrast, a sarcoma,[b] developing from connective tissue in the abdomen, may grow to several pounds before it produces symptoms.

Grades and stages

Physicians grade the cellular features of tumours, in order to select the most appropriate therapy. The grade classifies cancers in terms of their microscopic appearance, structure and composition. The visible characteristics of cancer cells are often correlated with how the cells behave and how aggressively they will grow. Grade one cells are well-differentiated and are the least malignant, grade two are intermediate cells, which are moderately differentiated, and grade three cells are poorly differentiated and are the most malignant.

The stage refers to the degree to which a cancer has grown and spread; an early stage may be more treatable than a later one. TNM, a widely used international cancer staging system, describes the cancer using the size of the Tumour, extent of lymph Node involvement and

---

[b] Tumour arising from connective tissue, such as muscle, bone or cartilage.

presence of distant Metastases or sites. However, there are other classifications in common use. Colon cancer uses the Dukes' system, named after its originator, an English doctor called Cuthbert Esquire Dukes (1890 -1977). Some female reproductive cancers are described using the FIGO staging system.[c]

In an older system, cancers are grouped into four main stages, denoted by Roman numerals, I to IV. Stage I refers to localized cancers that can usually be cured by surgery. Stage IV usually represents inoperable or metastatic cancer. A recurrent cancer is one that has returned after a period of treatment. The stages are defined precisely, but differently, for each kind of cancer. To be specific, we will outline the stages for colon and rectal cancer, but the general features apply to other tissues.

- **Stage 0**: The earliest stage of colon cancer. The cancer involves only the lining of the colon, called the mucosa, and is confined to one or more polyps bulging from the surface.

- **Stage I**: Involves more than just the inner lining of the colon. The polyp has progressed, extending into the colon wall.

- **Stage II**: The cancer has spread beyond the colon or rectum to the surrounding tissue, but has not yet spread to lymph nodes.

- **Stage III**: The cancer has spread outside the colon or rectum and into the local lymph nodes, but not to other organs.

- **Stage IV**: The cancer has spread to other organs in the body, such as the lungs or liver.

---

[c] Fédération Internationale de Gynécologie et d'Obstétrique.

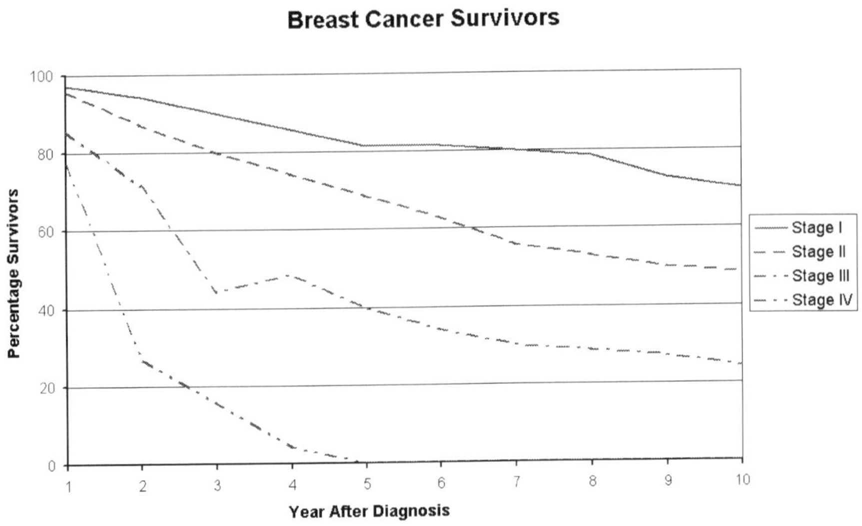

This graph illustrates the relationship between stage of cancer and survival time, in patients with breast cancer.[1]

Cancer growth

By the time a tumour is detected, it may have a mass of about a gram. This means the original tumour cell and its descendants have divided at least 30 times, to produce around one billion cells. At this stage, the tumour mass contains more than just cancer cells. Normal host cells, such as those in blood vessels, nourish the tumour. In many cases, stimulated immune cells have also invaded the tumour, producing a low-grade inflammation.

Precancerous stage

Most tumours grow for years before the person notices them. Studies have shown that benign tumours of the inner lining of the large intestine take from three to five years for a new polyp to form. It usually takes a similar period for the polyp to progress to a malignant carcinoma. By the time it is discovered, a malignant tumour is often at a late stage of its life cycle.

The appearance of a tissue can change long before it becomes cancerous; such abnormal cellular changes often precede cancer. These alterations are known collectively as precancerous lesions. The changes are described by a number of terms, including hyperplasia (an increased number of cells), dysplasia (alteration in the size, shape and organization of cells), and neoplasia (new cell growth). For example, increased cell growth in the inner lining of the uterus (endometrium) is called endometrial hyperplasia, which often precedes cancer in this tissue.

Non-invasive stage

Before tumours spread, or metastasise, to other tissues, they spend a period as non-invasive lesions. This is the earliest stage of true cancer. The tumour remains in its site of origin, without spreading. An example of this form is a non-invasive, or *in situ*, cancer of the cells lining the cervix. Such a cancer is on the surface or mucosal lining of the cervix, and has not penetrated deeper into the tissue.

Typically, a non-invasive tumour lacks the network of blood vessels necessary for growth. This lack of an adequate blood supply means that the cancer is unlikely to have sent cells into the blood stream to seed tumours in new locations. For this reason, non-invasive cancers can be surgically removed, resulting in a complete cure. Unfortunately, cancers at the non-invasive stage generally do not cause clinical symptoms, so are frequently undetected.

Angiogenesis

Smaller tumours can obtain nourishment by simple diffusion. A tumour can reach a diameter of one or two millimetres before its inner cells are too far from essential nutrients and oxygen for adequate provision. For the tumour to continue growing after this, small blood vessels (capillaries) must form to create a blood supply. This process is called vascularisation, or angiogenesis. Vascularisation is a normal, physiological, growth process, by which the body can repair damage.

Small tumours may exist without blood vessels for many years, as a harmless cluster of cells. Growth of such clusters will be limited by their lack of oxygen and nutrients. Under these conditions, cells that can thrive without much oxygen will have an advantage. However, if an error in cell division means cells are able to instruct surrounding tissues to provide a blood supply, the microenvironment changes. Catastrophically, the tumour begins to generate blood vessels and uncontrolled growth becomes possible.

Tumour cells do not usually start by being able to stimulate capillary development, but those that develop this capability have an enormous selective advantage. Years and many generations of cancer cells may be required for the micro-evolutionary process to produce cells capable of angiogenesis. Gaining a blood supply does more than simply provide the tumour with nutrients: it gives the cancer cells a means by which to spread.

Since the growth of a tumour requires angiogenesis, restriction of blood vessel development is a promising therapeutic approach. In animals, antiangiogenic drugs can shrink tumours by destroying the local blood vessels and by preventing their production.[239] This finding has not yet been translated into an effective treatment for humans.

Thalidomide

Thalidomide, a notorious drug, is a powerful antiangiogenic treatment that inhibits the growth of new blood vessels.[239] It was introduced as a sedative, in 1957, with apparently few side effects. In the early 1960s, it was given to expectant mothers, to prevent morning sickness. Unfortunately, inhibition of blood vessel growth during foetal limb development caused birth defects. The "thalidomide babies" had deformed flipper-like limbs (pharcomelia). The effects were not limited to limb malformations; some brain damage was also reported.

Thalidomide illustrates many of the difficulties involved in evaluating drug safety and utility. Drugs typically have a number of biological effects. Thalidomide's use as a sedative depends on a separate pharmacological action to the mechanism by which it produces deformity. Furthermore, its side effects are species dependent and require a specific metabolism.[240] Such a species-specific response was unlikely to be picked up during animal testing.

Thalidomide comes in two forms, known as optical isomers. These have the same molecular structure but are mirror images, like the left and right hands of a glove. Both keep your hands warm, but only the left glove fits the left hand. The isomer of thalidomide that acts as a sedative is called R-thalidomide. This does not cause birth defects, unlike the other isomer, S-thalidomide. Interestingly, the problem cannot be solved by giving only R-thalidomide, as some of this is converted to S-thalidomide within the body.

As long as thalidomide is not given to pregnant females, it is an effective sedative with few side effects. More importantly, it has a powerful anti-inflammatory effect, when used in leprosy patients. The drug has been used for treatment of several skin diseases,[241] and for inflammatory disease of the gut.[242] It has also been used in AIDS patients, to prevent body wasting.[243,244]

Because it restricts the growth of blood vessels, thalidomide may be a useful drug in the treatment of some forms of cancer.[245,246] Obviously, pregnant women should not be subjected to its toxic effects, which also include nerve damage. It is worth noting that the side-effects of thalidomide may be related to production of free radicals and oxidation.

Blood vessel growth and free radicals

Thalidomide prevents blood vessel growth by stimulating local oxidation.[247,248,249,250] Since a primary purpose of blood vessels is to supply oxygen to the tissues, it is unsurprising that free radicals and oxidants should be involved in signalling to produce new vessels,[251,252,253,634] or that vascularisation can be inhibited by antioxidants.[254] In particular, vitamin C is a potent inhibitor of new blood vessel growth,[255] and may protect against cancer in this way. Lack of oxygen promotes blood vessel growth, but vitamin C can inhibit this action.[256, 257]

There are several possible ways in which capillaries may be stimulated to grow and the mechanism may differ from one cancer to another. Cancer cells may release or stimulate the release of proteins, oxidants or other factors, which encourage the growth of blood vessels. One such protein is called vascular endothelial growth factor (VEGF). VEGF induces endothelial cells, the building blocks of capillaries, to enter a tumour nodule and start developing small blood vessels. As the endothelial cells divide, they secrete additional chemicals that stimulate the growth and movement of the cancer cells. Both normal and cancer cells respond to such signals. At this stage, healthy body cells may be collaborating with the cancer to enable its growth and spread.

Tissues also produce factors that inhibit the growth of blood vessels. Instead of promoting blood vessel growth, microevolution can have the same effect by selecting cells that fail to produce the chemicals that normally inhibit vessel growth. There is evidence to suggest that tumour cells can lower their levels of inhibiting factors, with the result that angiogenesis can start. The effect of preventing inhibition is similar to that of stimulating angiogenesis: the tumour gains a blood supply and overcomes the growth limiting effects of oxygen and nutrient starvation. The balance between stimulators and inhibitors determines whether a tumour is able to develop a blood supply.

Microinvasion

Invasion of local tissues begins when a cancer cell separates from the other tumour cells. Ordinary cells stick to each other using adhesive molecules, which keep the cells in an organised structure. Most cells have several types of proteins on their surface, allowing them to bind to similar cells or to the surrounding extracellular matrix. The matrix is a network of proteins and cell secretions, which provides structural support for tissues. If cells are unable to attach to the extracellular matrix, they may commit suicide by apoptosis. Cancer cells that can inhibit apoptosis therefore have a selective advantage, as they may be able to break free of the matrix and move to other tissues.

If a mixture of chick embryo cells is placed in a culture dish, the cells tend to group together with others from the same original tissue. The cell adhesion molecules that produce this clumping are involved in many aspects of human biology. Early in development, they help in embryo formation. Later, they act as traffic signals, directing the action in wound healing, inflammation and cancer. However, the characteristic lack of oxygen in tumours can change the expression of adhesion molecules and facilitate the spread of cancer.[258,259,260]

Invasion and dissemination

A solid cancer invades nearby tissues by breaking through the basement membranes, which form barriers between tissues. A cancer can use different mechanisms to achieve this invasion, including secreting proteins to dissolve the connective tissue. After breaching the membranes, cancerous cells can invade surrounding tissue and enter the bloodstream.

The body has two main fluid transport mechanisms, the blood stream and the lymphatic system. Lymph is a clear, straw-coloured fluid, similar to blood plasma. It drains tissues and discharges its cargo into the blood. Lymph carries white blood cells and is important in fighting infections. A cancer cell can use either blood or lymph vessels to travel around the body.

In order to spread throughout the body, the cancer cells of a solid tumour must be able to accomplish several difficult tasks. They need to separate from their neighbouring cells, break through supporting membranes, and tunnel through tissues until they reach a lymphatic gland or blood vessel that will transport them. The cells are then transported throughout the body. To break out of a blood vessel, they must migrate through its inner lining, which forms yet another barrier. Providing the habitat is hospitable, the cells can then start to form a new tumour.

Lymph nodes sometimes trap cancer cells that have entered a lymphatic vessel, producing local enlargements. Lumps in the lymph glands of the armpit can indicate the spread of breast cancer, for example. Cells not trapped in this way can enter the blood vessels and spread to distant parts of the body, such as the liver, lungs, and brain. The cells that escape to these distant sites still have the properties of rapid growth, and may form secondary tumours, or metastases.

Metastasis

Once in the bloodstream, cancer cells spread throughout the body. Some cells might lodge in the capillaries of another organ, before entering it and growing into a new tumour. Fortunately, not all the cells within a tumour are able to spread. Initially, only a few of the tumour's varied cells possess the genetic alterations for becoming invasive. Cancer cells can leave a tumour as individuals or in small groups. Such small groups of cancer cells join with lymphocytes and platelets, which may provide additional protection in the hostile environment of the bloodstream.[1] These cellular clumps may help the cells attach to the blood vessel wall at potential sites for a new colony.

When they enter the blood, cancer cells are subject to mechanical stress, protein-consuming enzymes and the direct action of the immune system. Only a small fraction of the cells that leave the tumour and enter the bloodstream, or lymph, will survive to form a secondary tumour. Melanoma cells injected into the bloodstream die rapidly, with only one percent remaining the following day. It is claimed that a breast tumour sheds a million cells per day, but less than 0.1% of these cells are detected in the blood, as they are soon destroyed.[1]

A new metastatic colony might not have the ability to form blood vessels, since it descended from a small sample of cells or even a single cell. The cells that form the colony are not necessarily the same as those that stimulated blood vessels in the original tumour. Roger King estimates that only about one in ten colonies will be able to generate a blood supply in the short term.[1] These colonies will generally grow more rapidly than the original tumour. Colonies that do not stimulate blood vessel growth may die or remain the size of a grain of rice, for years. Lack of oxygen and nutrients holds back proliferation, until some cells evolve the ability to stimulate a blood supply. Once this has happened, the secondary tumour can engage in explosive growth.

The spread of cancer to different sites in the body is claimed to be characteristic of the disease in humans, though rare in experimental animals.[1] However, this suggestion is a result of poor animal modelling. Cancer rates increase with age, whereas laboratory experiments are usually of short duration and involve young animals.[261] Metastases are found in normal veterinary practice, in cats, dogs, horses and cattle, for example.[262,263] The pattern of spread found in humans is not a distinguishing feature, it is found in many other mammals, as they age.

The location of a primary tumour influences how it spreads. Taking the analogy of animals migrating to an island, the "distance" for cancer cells to migrate can be measured in terms of the connectivity between the population source (original tumour) and location of the metastases (the sink). The architecture of the circulatory system provides one explanation for the location of metastases. Cancer cells often establish new tumours "downstream" of the original tissue, in what is described as the "first-pass" organ.[1] This means that tumours spread preferentially to certain organs of the body. An example is the lungs, which are a common site for metastases, as they are the first tissues with small blood vessels that are reached by blood leaving many organs.

However, not all cancer spread can be explained by the nature of the circulation. Clinical observation suggests that cancers have preferred sites for developing new tumours. At the end of the 19th century, Paget suggested that the spreading cancer cell (the seed) required a welcoming environment (the soil) in which to grow.[1] Some tissues provide a more favourable habitat for the invading cancer cells. Breast and prostate cancers often spread to the bones, whereas lung cancer often forms metastases in the adrenal glands.

Clinical effects

As we have mentioned, tumours can take years to develop before they produce clinical manifestations. The symptoms of malignant tumours often arise from the local effects of abnormal growth. A tumour can compress, invade or destroy normal tissues. In many cases, the primary tumour may become apparent before the metastases cause symptoms. Surgical removal of the primary tumour may then be only a temporary measure, as metastases could continue to grow and spread for several years before being noticed.

In addition to local symptoms, malignant cancers can cause generalised effects. They produce substances such as tumour necrosis factor, which circulate in the bloodstream, causing loss of appetite and body wasting (cachexia). Tumour necrosis factor is released by white blood cells and, occasionally, by tumour cells.

A tumour within an organ can interfere with its functioning. Cancer of the adrenals, for example, can destroy the gland and cause hormone insufficiency (Addison's disease). With hormone-producing tissues, the opposite may be true. Benign tumours of the parathyroid gland can release of large amounts of parathormone, causing high levels of calcium in the blood. This leads to fatigue, muscle weakness, nausea, anorexia and constipation.

Occasionally, the growth of a tumour can lead to rapid and dramatic symptoms. Benign ovarian cysts can rupture, producing abdominal discomfort. A freely growing tumour may sometimes twist at its connection to the body, cutting off its own blood supply. Stopping the flow of blood and oxygen will cause cell death, but could also cause internal bleeding, toxicity and severe pain.

Some cancer sufferers report symptoms that are not related to the location of their tumour. This type of symptom can be the initial effect of a small tumour and is useful for early detection. Some dramatic syndromes are caused by production and release of large amounts of hormones. For example, small-cell cancers of the lung may release a hormone that stimulates the adrenal glands. The adrenals then produce excessive amounts of hormones, causing hypertension, muscle weakness and high blood glucose.

The cancer classification system is huge and depends on the cell characteristics and the tissues of origin. There are, however, enough consistent features of cancer cells to group them as a single disease. Tumours progress through a series of stages, consistent with the evolution of a population of cells under selection pressure. Describing this progression is important, as it can suggest possible mechanisms for treatment. However, to prevent cancers getting started in the first place, we need to understand the causes of the disease.

How cancer starts

"Since, both in importance and in time, health precedes disease, so we ought to consider first how health may be preserved, and then how one may best cure disease."
Galen (130-200)

Cancer has been linked with many possible causes. As a result, prevention advice generally depends on avoidance of factors associated with development of the disease. The microevolutionary model takes an integrative approach, which brings together the different causes of cancer. According to this model, anything that promotes oxidation, inflammation and error-prone cell division has the potential to cause cancer.

Prevention of disease is more effective than treatment. As we learn more about the causes of cancer, we may be able to suggest ways to prevent it. One well-known action a person can take to avoid cancer is not to smoke tobacco. Another measure is to eat good food: a diet high in antioxidants can help prevent cancer.

Unfortunately, the prevention of cancer is a difficult topic for study. Even under the best conditions, making scientific advances is a challenging process. It took many years to show that smoking is a cause of lung cancer. Although there was substantial evidence that separate components of tobacco smoke were carcinogenic, this was not enough to establish smoking as a cause. Scientists had to build up evidence for the association between cigarette smoking and lung cancer from population studies.

Science is based on the collection and organisation of refutable knowledge. The best scientific evidence comes from small, simple experiments that give a definite result and are easy to repeat. Such data were not available to test the hypothesis that cigarette smoking causes cancer in humans. By comparison, population studies are a weak form of science, as the evidence they provide can be interpreted in several ways. For example, an alternative suggestion for the data on smoking is that the genes that cause cancer also make people more likely to smoke.

In this chapter, we give a brief overview of some of the mechanisms that researchers have linked with the initiation of cancer. This list is diverse and apparently unrelated. However, the underlying theme is that all the mechanisms lead to increased cell division, oxidation or genetic damage.

Oxidation damage

Antioxidants generally prevent the development of cancer. This is because oxidative free radicals promote cell growth and division, so factors that prevent oxidation will hinder proliferation. In healthy cells, antioxidants establish a reducing environment and tend to restrain growth. They also neutralise free radicals and inhibit several stages of cancer development, since, as we have explained, oxidising free radicals are involved in many carcinogenic processes.[264]

Many cancer causing agents cause mutations, either directly or indirectly, by free radical mechanisms. Examples include X-rays and the ultraviolet radiation in sunlight, which generate free radicals in the body. Asbestos fibres and other irritants in the lung cause free radical damage and promote cancer. Smoking releases free radicals into the bloodstream, initiating and encouraging cancer in tissues.

One of the most effective ways of avoiding cancer is to ensure the body has a good antioxidant intake. A second approach is to eliminate sources of inflammation, which also generate free radicals.

Inflammation

Occasionally, the body recognises cancer cells as foreign and mounts an immune response that destroys the tumour.[265] In such cases, the immune system produces antibodies to destroy the cancer cells. This mechanism is a recurrent explanation for rare cases of spontaneous remission. Tumour necrosis factor is a local hormone, released by white blood cells during inflammation, which can kill cancer cells.[266] It was identified and named because of its ability to destroy tumours. At first, researchers considered it an anticancer agent; however, its action is not specific to tumours and it is highly toxic.

Paradoxically, tumour necrosis factor sometimes promotes cancer growth.[267,268,269,270] Recently, substances that inhibit its action have been suggested as anticancer treatments.[273] This paradoxical action is not restricted to tumour necrosis factor: other inflammatory mechanisms can both promote and destroy cancer.[271] Inflammation itself is involved in the immune destruction of tumours, yet is also associated with malignancy and cancer growth. Interestingly, we can resolve these contradictions by taking an evolutionary view.

Tumours have been described as wounds that do not heal.[272] However, this idea is not new. In 1863, Rudolf Virchow proposed an association between cancer and inflammation.[273] Physicians explain such inflammation as the body's response to the tumour. Inflammation is the body's way of fighting disease, so we might expect to find an association with cancer. However, recent research has confirmed that chronic inflammation is a cause of cancer.

Any prolonged insult to cells can result in cancer. The precipitating factors are increased cell division, damage and time. Chronic inflammation has these characteristics; it is therefore likely to increase the risk. Carcinogenic chemicals increase cell division and cause damage, leading to inflammation. Chemically damaged cells can release inflammation-stimulating hormones. Alternatively, the cells may die, activating immune system cells to clear them away, which also causes inflammation.

Inflammation occurs in response to tissue damage. It stimulates repair mechanisms, including an increased blood supply. The tissues become red and swollen, as blood is encouraged into the area. Chronic inflammation causes tissues to change and stimulates physical alterations known as remodelling, including the generation of new blood vessels. A common example of a remodelling process is the formation of calluses on the hands or feet, caused by repeated mechanical stimulation.

One explanation for the association between inflammation and cancer is that cancer cells themselves actively stimulate inflammation. This is an excellent evolutionary strategy, because it helps the cancer cells gain a blood supply, allowing them to invade local tissues and migrate to other parts of the body. In the early stages of development, tumours lack oxygen and nutrients. By stimulating an inflammatory response, a tumour can take advantage of the resulting blood supply, to assist its growth and spread.[274]

Prevention of inflammation may be an effective strategy for avoiding cancer and slowing its progress. Anti-inflammatory drugs such as aspirin can protect against cancer.[275,276,277] Research on treatment with anti-inflammatory drugs is in progress.[278,279,273] However, once the disease is established, many cancers have a mechanism for reducing inflammation throughout the body.

At this late stage, inflammation at the site of a tumour can release chemicals into the blood, promoting the tissue wasting often seen in cancer patients. These chemicals lessen the body's ability to produce an inflammatory response, suppressing inflammation throughout the body. This feature may provide a selective advantage by conditioning tissues to accept metastatic cells in locations distant from the tumour. Such tissues may provide a more hospitable microenvironment for the invading cells.

The cell cycle

The cell cycle, a series of events involving growth, replication and division of cells, depends on gene activity. Each stage of the cell cycle is under genetic control. In the first stage, the cell increases in size. In the second, the cell copies its DNA. In the third stage, it prepares to divide. Finally, in the fourth stage, it splits into two daughter cells.

Typically, genes control cell growth and division by generating proteins that form a cascade of signals. These signals relay messages from the outer cell membrane to the nucleus. Growth factors are chemicals that signal cells to divide: a growth factor binds to a receptor on the outer cell membrane, activating it to send a signal to the cell nucleus. Within the nucleus, proteins start copying the genes involved in cell division.

Correspondingly, tumour suppressor genes produce a cascade of signals that inhibit cell growth. Both stimulatory and inhibitory signals act on a "clock" in the nucleus, which determines when cell division will occur. Errors in the signalling cascade can cause the clock to enter the cell cycle more frequently. A cell that develops such errors has taken an initial step towards becoming cancerous.

An oncogene associated with cancer is called ras, as it was originally isolated from rat sarcoma virus. The protein produced by the ras gene affects cell cycle signalling. Ras genes are widely distributed among animal species, as we might expect from a gene acting on such a fundamental part of cell function. Mutant proteins, encoded by a defective ras gene, can result in continuous cell cycle activation signals, which increase the rate of cell division. Perhaps a quarter of all human cancers contain cells with overactive ras genes.

Mutation

To form a large tumour, a cell must divide and produce billions of cells. This means the cancer cell has to break free of normal controls that prevent unwanted division. Damage to the genetics of the cell is at least partly responsible for this loss of control. As the cell divides, the genetic changes and abnormal growth patterns are passed on to the daughter cells.

Healthy cells maintain a balance between pressures to grow and divide, remain the same, change to another cell type, or die. A small number of genes are thought to control this process. Proto-oncogenes encourage cell growth and division and, in mutated forms, they are over-expressed in cancer cells. The name proto-oncogene is misleading, because such genes have important functions in healthy cells. However, the defective forms, or oncogenes, can interfere with any of the signalling processes mentioned. They thus affect the cell cycle and can initiate inappropriate cell division, decrease cell repair mechanisms or inhibit cell suicide. Mutations that increase the expression of proto-oncogenes may stimulate cell growth.

Tumour suppressor, or repressor, genes are fewer in number and normally act to prevent proliferation. These genes are part of the cell's fundamental control of cell division. If released from the effects of repressor genes, cells will engage in more frequent division. Mutations that interfere with the correct functioning of tumour suppressor genes can therefore result in over-stimulation of cell growth and division, which could ultimately result in cancer.

Chromosomal translocation

The term translocation refers to what happens when a section of a chromosome breaks off and moves to another part of the chromosome, or even to a different chromosome. Movement of a section of a chromosome is a large mutational change, which may have substantial consequences for the cell. Such chromosomal translocation is found in some forms of human leukaemia and lymphoma.

During translocation, two separate genes can fuse together and, occasionally, the newly combined gene can stimulate tumour development. The Philadelphia chromosome, named after the city in which it was found, was the first such translocation linked to human cancer. Its discovery in 1960, by Peter Nowell and David Hungerford, was a landmark. It was the first consistent chromosomal abnormality found in human cancer. More than 90% of patients with chronic myelogenous leukaemia have the Philadelphia chromosome.

Cancer promoting genes

Studies of genetic modifications induced by cancer-causing viruses have led to the discovery of genes that are integral to the control of cell growth. In some cases, a common virus or bacterial infection can lead to cancer. Viruses can alter the sensitive balance of promoter and suppressor genes, which control cell division. The first viruses associated with cancer were discovered in the 1960's. One of these was the Epstein-Barr virus, followed in 1970 by the hepatitis B virus. Hepatitis B or C can cause liver cancer. *Helicobacter pylorus* is a bacterium known for causing stomach ulcers, which can also cause cancer. The papiloma virus can cause cancers of the genitals, anus and cervix.

A large number of cancer-promoting oncogenes have been found in retroviruses. Retroviruses are relatively small, with many varieties. In animals, retroviruses have been particularly implicated in cancer. The biology of these viruses is unusual, which makes them useful in cancer research. Whereas most organisms store their genetic information as DNA, retroviruses contain protein and use RNA as their genetic material. They also contain an enzyme called reverse transcriptase, which can copy RNA and produce DNA, enabling them to add their own genetic material to the host cell. This is the opposite, or reverse, of the normal mechanism, in which RNA is made from DNA, hence the name retrovirus. As explained earlier, the human immunodeficiency virus (HIV), thought to be implicated in AIDS, is usually described as a retrovirus.

Retroviruses can steal genes from animal cells. Sometimes, a retrovirus will incorporate some of the host cell's genetic material into its own RNA. When the retrovirus infects a cell, it erroneously copies these host DNA genes into its viral RNA where they become part of the viral genome. This feature of retroviruses led to the discovery of oncogenes. Genes found in apparently normal human DNA and implicated in cancer were almost identical to retrovirus genes. Although retroviruses can induce tumours in animals, this rarely occurs in humans.

The first "cancer gene" to be identified was called the Src gene.[a] In the early 20th century, it was found that the Rous sarcoma virus could convert healthy cells into abnormally dividing cells. This virus appeared to contain only four genes, but infection could lead to cancer. In 1970, it was shown that the ability of Rous virus to transform the cell depended on its src gene.

The src gene is similar to a gene found in the host cell. Indeed, the gene is widespread in animals, which suggests it arose early in evolutionary development. In animal cells, src is a regulator of cell growth. Src may have originally been an animal gene that a retrovirus picked up accidentally, during infection. However, the version of src in

[a] The src gene is pronounced "sark" as it originated with a sarcoma virus.

animal cells is not identical to the viral gene. The viral form has slight alterations and is a cancer-promoting gene, as it instructs the cell to divide more often than normal.

Other retroviruses produce tumours by inserting their own genes into critical sites in the animal DNA. When this insert is next to or within a proto-oncogene, which promotes cell division, the viral insert may cause the animal gene to become overactive. By switching on genes that increase cell division, the virus has moved the cell towards development of cancer. Alternatively, it may inactivate a tumour suppressor gene, so cells are no longer inhibited from dividing.

Tumour suppressor genes

A person who inherits a mutated tumour suppressor gene will have a higher risk of developing cancer. We have previously described the most common tumour suppressor gene, p53, but there are many more. Early evidence for the existence of tumour suppressor genes came from studies of human heredity.

In 1971, Alfred Knudson studied retinoblastoma, a cancer of the back of the eye. This cancer may arise occasionally in adults, but an inherited form occurs in children. By separating the childhood and adult cancers, Knudson was able to suggest a two-hit mutation hypothesis for this disease.

In children, a tumour suppressor gene is linked both with eye cancer and with bone tumours. In adults, the same gene is associated with cancers of the lung, cervix, breast, prostate and bladder. In the inherited disease, a child gets a mutated gene from one parent. This single mutation is not enough to stimulate a cancer, as the second copy (allele) of each gene can function normally. However, if a mutation occurs in the remaining healthy gene of a retinal cell, it may produce cancer.

With the non-inherited form of the disease, it is claimed that both copies of the gene need to be mutated in the same retinal cell. Two such mutations, happening in both copies of the same gene, in the same cell, are much less likely. The non-inherited form of the disease is therefore far less frequent.

Maintaining tissue health

The triggering events for cancer are those that lead to the growth of a colony of cells, which undergo evolutionary selection. Factors that increase the rate of cell division and generate damage will promote the development of cancer. X-rays, carcinogens, viral infection and mutagenic chemicals can cause free radicals and genetic damage.

There are numerous individual rules by which people can lower their risk of cancer. The primary mechanisms for such prevention are avoidance of tissue inflammation and oxidants. However, many of these risk factors are part of our environment and cannot be avoided within a modern lifestyle. Conversely, however, advances in our knowledge of nutritional science provide practical methods for reducing the risk.

Nutrition

"Don't give the same food to the sick and the healthy." Croatian proverb

Poor nutrition is arguably the leading cause of cancer.[1] A substantial proportion of cancers, perhaps forty percent, could be prevented by changes in diet and lifestyle.[280,281] Researchers claim that changes in diet could lead to a decrease of about two-thirds in breast, colorectal, and prostate cancers.[286,282] and half the number of lung and other cancers.[286,283] This may be true, although an orthomolecular approach to nutrition, providing an optimal intake of nutrients, would be more likely to produce such results.

Orthomolecular medicine uses natural substances, found in a healthy diet, for the prevention and treatment of disease. According to this approach, nutrition comes first in disease prevention; drug treatments and other therapeutic interventions should only be used when specifically indicated.

The relationship between diet and cancer is complex; it could take decades to gain even a modest understanding of this subject. Currently, the evidence from population studies is unclear. This is not surprising, because of the difficulty of extracting solid information from poor quality data. Population studies are inaccurate, difficult to perform and do not produce consistent results.[284,285] Many of these studies are questionnaire-based and hence unreliable, because people have poor memories and are not always totally honest. Researchers obtain measures of nutrient and vitamin intake by estimating the levels in the foods reported: another potential source of error. Furthermore, interactions between nutrients are difficult to separate in such studies.

Despite these methodological difficulties, scientists have suggested many possible links between diet and cancer. For example, a high dietary fat intake is reported to be associated with breast, colon, rectal and prostate cancer. Changes in the diet, such as eating less red meat and more fruit, vegetables and legumes (peas and beans), are frequently recommended to reduce the risk of cancer.

Multicellular plants and animals have had hundreds of millions of years of evolution to produce non-toxic substances that prevent cancer. The survival of multicellular organisms depends on their ability to restrain cancerous growth. Hence, basic biological principles predict that vegetables should contain numerous antioxidants and other anticancer agents. As expected, dietary chemopreventative substances have been found, which influence the pre-cancerous stages of tumour development. Fruit and vegetables contain high concentrations of such substances. Examples include cauliflower, broccoli and cabbage, which contain anticancer nutrients, such as isothiocyanates and dithiolthiones.

We stress that prevention and cure are two different things. It should not be assumed that a diet to help prevent cancer would also aid recovery from the illness; optimal dietary requirements could change during the development of the disease. Thus, a diet that prevents cancer occurring might accelerate the growth and development of an established tumour, or shorten the lifespan of a terminally ill patient.

General reviews of cancer and nutrition can be found elsewhere.[286,287] In this chapter, we describe the potential use of nutrition to reduce the incidence of cancer. In later chapters, nutrients of particular interest for cancer therapy will be examined in detail.

Vegetables

The evidence linking consumption of vegetables to prevention of cancer is contradictory. For many years, doctors and governments have encouraged the idea that vegetables and fibre are good for health and reduce the risk of cancer. However, low dietary fibre may not increase the risk of breast cancer.[288] More surprisingly, a large study of American health professionals has concluded that frequent consumption of fruit and vegetables does not appear to confer protection from cancer of the colon or rectum.[289]

In other studies, a greater consumption of whole grains, fruit and vegetables is associated with a reduction in colon cancer incidence,[290] although the studies do not explain the claimed protective effects. Fruits and vegetables contain many substances besides dietary fibre, including antioxidants and vitamin C. Notably, high doses of vitamin C have many of the advantages of fibre, including the production of large, soft, bulky stools, with a short transit time.

The anticancer effect of dietary fibre may be caused by the presence of inositol hexaphosphate (IP6), which is abundant in animal and plant cells.[291,292] This ingredient appears to have many beneficial effects. IP6 appears to be an effective anticancer ingredient in vegetables high in dietary fibre. It prevents cancer growth, rather than killing the cells.[293,294,295] IP6 is absorbed from the stomach and gut,[296] and is taken up by cancer cells.[297] It inhibits the growth of many cancer cells lines, but not of normal tissues.[298,299,300,301] IP6 causes cancer cells to revert to a more normal appearance and behaviour, a process known as redifferentiation.[302,303,304]

In 2002, a review cited 28 animal studies, showing that IP6 has anticancer effects in breast, colon, liver, leukaemia, prostate, sarcomas and skin cancer.[305] No human studies were reported, which is consistent with the lack of financial incentive associated with this substance. However, one group has reported positive findings from a preliminary clinical study on six patients.[295] Overall, they found a reduced tumour growth rate in the treated patients, together with some indications of regression. After the initial treatment, one patient with liver metastases refused chemotherapy. Her doctors treated her with dietary supplements: IP6 and inositol. The study claims that, after 14 months, her cancer showed a reduced growth rate.

The anticancer mechanism of IP6 is not fully understood. However, IP6 binds iron and thus acts as a potent antioxidant.[306,295] In early stage cancer cells, this action will tend to place the cell in a more reducing state, causing redifferentiation and inhibiting growth. By binding iron, IP6 may prevent ascorbate-induced apoptosis of cancer cells.[142] For this reason, IP6 and redox cycling mechanisms, which are described later, may each provide mutually exclusive approaches to cancer treatment.

IP6 can inhibit cancer in rats and mice. In rats, the effect is greater when the IP6 is given in drinking water, rather than as a component associated with dietary fibre.[307,295] Rats given IP6 showed a reduction in breast tumour number and incidence, whereas controls did not. Pure IP6 was more effective than a high fibre diet for preventing experimental cancer. In the words of the researchers:

"Thus, for cancer prevention, prophylactic intake of IP6 may be not only more effective, but also more practical than gorging on large quantities of fiber."[291]

The US government's five-a-day programme claims that people whose diets are rich in fruits and vegetables (five to nine servings a day) have a lower risk of cancers of the lung, mouth, pharynx, oesophagus, stomach, colon, and rectum.[308] Such people are also less likely to get cancers of the breast, pancreas, ovaries, larynx, and bladder. The effect is stronger among older subjects.[311] Salad vegetables may be protective against breast cancer.[309] However, the results are not fully consistent and some studies indicate little or no protective value from vegetable intake on cancer.[310]

Vegetarians avoid red meat, which does not supply dietary fibre and is associated with colon cancer.[311,312,313,314,315,316] As might be expected, vegetarians suffer less from many diseases, including colon cancer. It is unclear whether this finding relates to diet specifically,[317,318,319] as vegetarians differ in other aspects of how they live. Generally, the western diet is cancer promoting, as it contains a high proportion of sugar-based carbohydrates, together with red and processed meats.[320, 321]

In 1992, epidemiologist Gladys Block reviewed a large number of studies, which suggested that fruit and vegetables protect against cancer.[322] People who ate the least fruits and vegetables had double the risk of cancer, compared to those who ate most. These claims included results for cancer of the mouth, stomach, oesophagus, lung, pharynx, uterus, pancreas, and colon.[323] Vegetables, particularly raw ones, appeared to be most protective. Cruciferous vegetables such as cauliflower, broccoli and cabbage[324,325] may be especially beneficial.[326,327,328,329,330,331] Garlic, onion and leeks may be protective for stomach, rectal and prostate cancer.[332,333,334]

Numerous substances found in vegetables could contribute to their observed anticancer activity. Many of these substances are antioxidants. Research has already shown some of these, including vitamin C, vitamin K and alpha-lipoic acid, to be outstanding anticancer agents. Unfortunately, the profusion of components in vegetables can sometimes act as a smokescreen.

The aim of scientists in this area should be to separate the components, investigate their properties, and discover any potential for synergy. This reductionist approach has been the basis of our advancing scientific knowledge for centuries. It forms the basic philosophy behind the use of nutritional supplements. Such supplements provide the isolated substances in higher quantities, offering potentially greater beneficial effects with low toxicity. Popular advice, to consume five helpings of fruit and vegetables per day, may ignore important aspects of the nutrients we consume. Surprisingly, a rigorous, reductionist approach to nutritional science contradicts the conventional dietary approach and supports the isolation and use of specific supplements.

Iron

Taking iron as an example, we can see the potential for identifying individual components of the diet. In cancer cells, unbound iron is a source of oxidative free radicals.[335] During their development, cancer cells with increased levels of iron are likely to have a growth advantage, as the resulting internal oxidation level stimulates growth. The microevolutionary model therefore predicts that increased iron consumption would promote cancer.[336,337,338] People with genetic iron-handling disorders, such as haemochromatosis, would similarly be at increased risk of cancer.[339]

The literature supports these predictions. Cancer and precancerous cells accumulate iron, along with related metals, such as copper and gadolinium.[340,341] Cancer cells often have enhanced mechanisms for absorbing these metals.[342] Oncogenes can alter the biochemistry of iron, disrupting the close control of levels found in normal cells.[343] Since cancer cells may require free iron to promote oxidation, increased dietary iron promotes cancer growth,[344] while restriction reduces proliferation.[345,346] As tumours accumulate iron preferentially, patients might suffer anaemia. Unfortunately, treating this deficiency could help the cancer to grow.[347]

On the positive side, cancer patients have physiological mechanisms to restrict the availability of iron to tumours.[348,349,350] Furthermore, dietary factors that remove iron, such as IP6, slow the growth of tumours.[351,352,353] These characteristics of iron in cancer cells are a result of cancer's microevolutionary requirement for high levels of growth-promoting oxidation.

Dietary antioxidants

Just as increased levels of oxidation favour the development of cancer, so antioxidants can work to prevent it, by helping to establish a healthier redox environment. Antioxidants can cause early cancer cells to redifferentiate,[1,172] slowing proliferation and inhibiting the evolution of a tumour. Food contains a profusion of antioxidants with the potential to prevent or even treat cancer.[1] The large number of these substances makes identification and analysis difficult and means that a full understanding of nutrition will take many decades of scientific research. Notable examples include grape seed extract,[a] and green tea.[b,354,355,356]

---

[a] Grape seed extract contains proanthocyanidins.

[b] Green tea contains epigallocatechin-3-gallate.

From a cancer prevention viewpoint, perhaps the single most important factor in the diet is the antioxidant density, which is the amount and variety of antioxidants relative to the number of calories.

Information on antioxidants in food is inadequate. Furthermore, findings concerning the effects of dietary antioxidants on the development and progression of cancer are inconsistent. This variation could be a result of the different pathways an evolving group of cells can take to achieve malignancy. In addition, different compartments of tissues, or even different components within a cell, can have distinct redox environments.[357,358] The action of an antioxidant depends on which parts of the cell it enters and the nature of the cellular components it can act upon. Despite these factors, antioxidants derived from food offer the potential for prevention of a large proportion of cancers. As we explain later, certain dietary antioxidants may also provide a means of effective treatment for patients currently considered hopeless.

Fish oil

Fish contains omega 3 oils, which may offer protection from cancer. Omega 3 oils[c] have been shown in animal studies to protect against cancer, while high levels of omega 6 oils[d] are associated with breast cancer.[359,360,361,362,363,364,365]

Flax seed oil and some algae provide vegetable sources of omega 3 oils. Flax seed is also a good source of dietary fibre and phytoestrogens, which may be beneficial to menopausal women.[366] Ground flax seeds have been shown to reduce tumour size and spread in mice.[367,368,369,370] Feeding flax seed to nursing mother mice lowered the size and number of breast cancer tumours in their offspring.[371,372] Similar results have been reported for prostate cancer in mice.[373] In humans with prostate cancer, preliminary results with flax seed suggest beneficial effects.[374]

Chlorophyll

Chlorophyll is the molecule in green plants that is responsible for photosynthesis. It is of fundamental biological importance and forms part of the ultimate power source for life on earth. However, the evidence for it being a powerful anticancer agent is insubstantial.

---

[c] Omega 3 oils, derived from alpha-linolenic acid (ALA), include eicosapentaenoic acid (EPA) and docosahexaenoic acid (DHA).

[d] Omega 6 oils include linoleic acid, gamma-linolenic acid and arachidonic acid.

Chlorophyll may be able to bind to cancer-forming chemicals that sometimes occur as a result of cooking our food.[e] It has been suggested that such bound molecules are less able to be absorbed into the body. Animal and laboratory experiments,[375,376] together with preliminary clinical trials,[377,378] indicate a beneficial effect.

Selenium

The mineral selenium is a dietary antioxidant that might protect against cancer. Dietary sources of selenium include Brazil nuts, brewer's yeast and sunflower seeds. It is possible that much of the population is deficient in selenium, as dietary intake can be poor, especially in areas where soil levels are low. For men, low selenium intakes are a cancer risk. The incidence of cancer in men may be reduced by a selenium intake of 200 micrograms per day.[379,380,381,382,383,384,385,386] Females appear less sensitive to low intakes of selenium and, in particular, breast cancer may not be greatly influenced by low selenium levels.[387,388,389,390,391] However, selenium may protect both men and women from colon and lung cancer.[387,392,393]

Vitamin B

The microevolutionary model predicts that a lack of B vitamins will increase the risk of cancer. Deficiency can cause DNA damage, which promotes variation in dividing cells. These errors may initiate or accelerate the development of cancer. The model also predicts that this increased risk of cancer would be found preferentially in cells that divide rapidly, such as those in the lining of the gut.

Scientists have proposed vitamin B_{12} as a protective agent against cancer. In mice, this vitamin inhibits tumour growth, kills cancer cells and extends lifespan.[394,395,396,397] Deficiency of vitamin B_{12} can increase the rates of genetic damage,[398] providing more variation for cells evolving towards cancer. There is evidence linking B_{12} deficiency with breast cancer.[399,400] Vitamin B_6 works alongside B_{12}, folic acid and the amino acid methionine, in preventing cancer.[401]

Folic acid is usually grouped with the B vitamins; it is found in dark green, leafy vegetables. Together with vitamins B_{12} and B_6, it is involved in DNA synthesis and repair. Folic acid deficiency leads to increased DNA damage. In the United States, about one person in ten does not take enough folate, implying that such damage could be a common problem.[402]

---

[e] An example is the protective effect of chlorophyll on aflatoxin, a potent carcinogen.

The cancer risk associated with folate deficiency appears to be higher for colon cancer than for breast cancer. One explanation may be that the gut contains more rapidly dividing cells than the breast. There is evidence that folic acid plays a role in preventing cancer of the gut,[403,404,405,406,407,408,409,410,411,412] with differing results in men[413,414] and women.[415,416,417,418] Folic acid deficiency is also associated with an increased risk of breast cancer,[419,420,421,422,423] but this seems a lesser effect, related to alcohol consumption. Alcohol interferes with the body's use of folic acid and can increase the risks associated with deficiency.[424,425,426] However, some authors have suggested that increased blood levels of folate could be preventative against breast cancer in women with high alcohol consumption.[427] High doses of folic acid may reverse some of the DNA damage that can lead to cancer.[428,429,430,431,432]

Vitamin D

When skin is exposed to the sun, it produces vitamin D. Since vitamin D is protective against cancer, people who are avoiding the sun for anti-aging purposes, or for reduction of skin cancer risk, should supplement their diet with vitamin D. Otherwise, the attempt to avoid melanoma may backfire, producing a greater overall cancer risk.

This sunshine and cancer paradox illustrates the difficulties associated with simplistic health advice. In temperate climates, advice to avoid the sun could increase the overall rates of cancer. Northerly areas of the United States have higher rates of cancer of the breast, colon and ovary than the south. Population studies suggest that sunlight has a protective effect against cancer of the prostate,[433] ovary,[434] breast[435] and other forms of malignancy.[436,437] Prospective studies support these findings.[438,439,440,441,442,443,444,445,446]

Even gentle exposure to the summer sun can generate large amounts of vitamin D.[447] A little sunshine generates much higher values than the current recommended dietary intakes.[448] Active vitamin D in the body has anti-cancer properties. Both normal and cancerous cells convert circulating vitamin D to a form that is more active against cancer.[449,450,451,452,453] This active form of vitamin D inhibits the growth of prostate,[454] colon,[455] and pancreatic[452] cancer cells.

Carotenoids and vitamin A

Vitamin A and its derivatives, such as beta-carotene, can reduce the risk of cancer.[172] Laboratory animals that are deprived of vitamin A have an increased cancer risk. Cells deprived of vitamin A become dedifferentiated, having an appearance similar to some cancer cells. Derivatives of vitamin A, such as retinoic acid, promote cell differentiation.[1] Vitamin A may be effective against leukoplakia, which can precede cancer and is characterised by warty patches in the mouth.[456] In 1971, Raymond Shamberger showed that vitamin A could decrease the number of chemically induced experimental skin cancers. This has been supported by numerous experimental[457] and population studies.[458]

Carotenoids are substances that give colour to vegetables. Hundreds of carotenoids exist in nature and several are common in normal foods. A well-known example is beta-carotene, familiar in carrots. Over time, these minor nutrients have become popular subjects in studies of diet and cancer.[459] One of the lesser known carotenoids, lycopene, may be particularly protective. Tomatoes contain relatively high levels of lycopene. Cooked tomatoes may provide more lycopene, although they have less vitamin C.

Studies have indicated that high intakes of lycopene are associated with a lower risk of prostate cancer.[460,461,462,463,464] Lycopene has also been used to treat prostate cancer. In one study, 26 men took 15mg per day of lycopene for three weeks, before surgery. Control subjects consumed five servings of fruit and vegetables. The results indicated the superiority of the supplement, showing reduced cancer spread in the supplemented group.[465] Although the results were statistically significant, the researchers suggested the findings offered no firm conclusions, because of the small sample size. This is a strange interpretation of statistics. In a small sample, statistically significant results indicate either a large effect, or a relatively rare fluke result. It is conventional to accept such results as valid, since they can be checked by replicating the experiment.[15]

A second study appears to confirm the benefits of lycopene against prostate cancer. Thirty-two men took 30 mg of lycopene per day, in food and supplements, for three weeks before surgery. The results showed increased cancer cell death and other indications of improvement.[466] These small, independent studies were conducted within a short time interval. Since the results of the two agree, this suggests a large effect or, less likely, fluke results. On this evidence, prostate cancer sufferers might choose to supplement with high levels of lycopene. Even if the positive results are eventually refuted, current findings suggest potential benefits.

According to early studies, beta-carotene might reduce the risk of cancer. Paradoxically, later studies suggest it might increase the risk, in smokers.[467,468,469] These researchers appear to have misunderstand the nature of antioxidant protection. Smoking increases the level of oxidation. In the lungs of smokers, beta-carotene and similar dietary antioxidants would give up their electrons and cease to act as antioxidants. To remain effective under oxidative conditions, dietary antioxidants need a continuous supply of electrons, such as that made available by a dynamic flow of vitamin C through the body.[15,470] In dynamic flow, vitamin C provides electrons for other antioxidants, which otherwise become ineffective or even act as pro-oxidants.

Despite the limitations of carotenoids as primary antioxidants, the evidence suggests that dietary carotenoids can prevent cancer.[471,472,473,474,475,476,477,478,479,480,481] These studies imply that carotenoids may work together with other antioxidants to provide protection, justifying the suggestion that a variety of vegetables in the diet is a good anti-cancer strategy.

Detection and diagnosis

"The greater the ignorance, the greater the dogmatism."
Sir William Osler

The earlier a cancer is discovered, the longer is the subject's life expectancy. This has nothing to do with the effectiveness of treatment: it is a consequence of the fact that the total time required for a cancer to kill a subject does not depend on when it is found. Earlier detection means that the time from initial diagnosis to final outcome is increased.

Progress in medical investigations, such as magnetic resonance imaging, means it has become easier to obtain an early diagnosis. Because doctors can find cancers more quickly, the apparent success of treatments, judged by the five-year survival rate, may increase with time. This error is sometimes called lead-time bias.

Early detection

There are widespread claims for the advantages of early detection.[482,483] Both doctors and patients believe that the sooner a cancer is discovered, the easier is it to treat. This is certainly true in some cases, as the cancer is localised and less developed.[484] Finding a tumour at an early stage is advantageous to the patient. For example, if a breast tumour is found before it spreads to other parts of the body, it might be cured with surgery.

Earlier detection implies increased life expectancy, by definition. However, it does not necessarily mean the person lives longer than they would have done if the cancer had remained undiscovered. For example, assume a woman develops a cancer at 65, which is discovered when she is 70. She has a life expectancy of two years and dies at age 72. Let us next suppose that new diagnostic methods find the cancer when she is 66. Her life expectancy on diagnosis is six years, but she still dies aged 72. Despite the improved life expectancy at the time of diagnosis, her overall lifespan is not increased. Whether or not patients would prefer an early or late diagnosis in this case would depend on the individual.

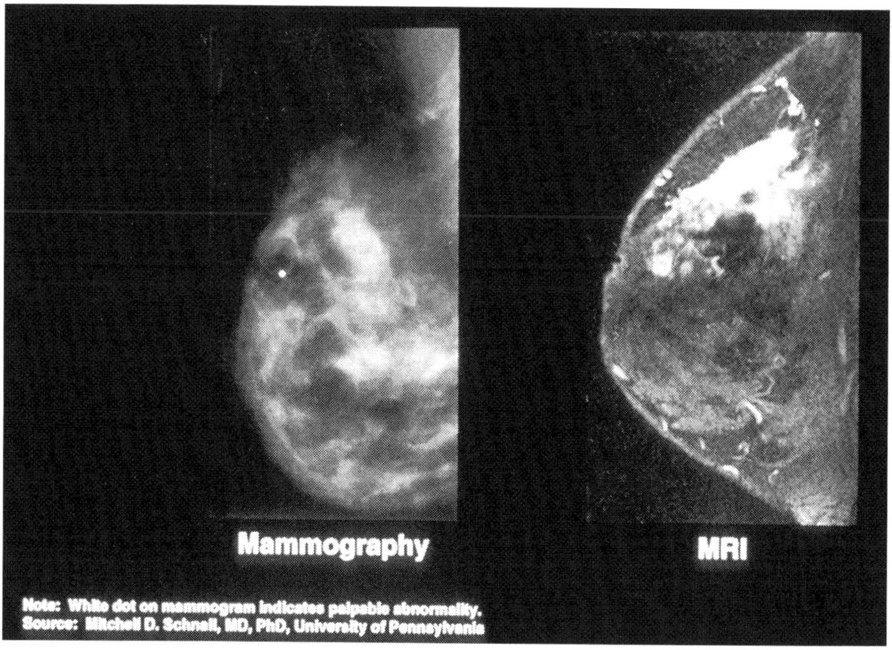

This image shows how developing technology allows improved staging of tumours. The X-ray image, on the left, has less soft tissue detail than the more recent technique of magnetic resonance imaging, on the right.

While championing early detection, the medical profession also claims that modern treatments improve lifespan. However, many small cancers remain dormant and do not become clinically significant. Better detection of such cancers could lead to apparently increased rates of "cure". However, some of these cures could relate to tumours that, had they remained undetected, would not have caused clinical problems. In the past, such tumours may have gone unnoticed, saving patients the stress of treatment for what they believe is a deadly disease.

From these examples, we can see that it is hard to separate apparent improvements in life expectancy, due to improved detection, from the real improvements that are due to beneficial treatments. Thus, taken alongside better screening and earlier detection, claims for the benefits of newer treatments are difficult to substantiate scientifically.

One way to approach this problem is to see what happens to patients who are diagnosed at an early stage and then elect to go untreated or to be treated conservatively. A study in Sweden followed 223 such patients, with prostate cancer, over a ten-year period. The mean age at diagnosis was 72 years. The authors report that only 8.5% (19 out of 223) patients died of prostate cancer within 10 years.[485] However, in total, 124 patients died during the follow up period, of which the large majority (105) died of other causes.[a] The disease-specific survival rate was equally high in a subgroup of patients that met current indications for radical prostate surgery (87.9%; 51 out of 58). A follow-up study by the same researchers describes the survival rates over 15 years, and concludes that, without reliable prognostic indicators, an aggressive approach to all patients with early disease would entail substantial overtreatment.[486]

The results of this study indicate that a man in late middle age, diagnosed with early prostate cancer, will probably die from some other cause before succumbing to the disease.[487] Since the side effects of radical prostate surgery include impotence and incontinence, people promoting this treatment need to be certain that the benefits outweigh the potential risks for the individual concerned.

The authors of the Swedish study point out that patients treated for this slow-growing cancer need to be compared to untreated controls. Conservatively treated patients with early prostate cancer have a favourable outlook, and the number of deaths saved by initial radical treatment is small. However, if the cancer is advanced, or has spread to other sites, some patients could benefit from aggressive therapy.[486]

When researchers make claims that modern treatments extend life, they must compare the outcome for the new treatment with that from the equivalent stage of cancer in the past. Unless they can be sure of this, the comparison is not valid. Even if the stages are the same, scientists criticise such comparison of results across time. Other factors, such as standard of living, may be responsible for any differences between the two groups. Claims for minor treatment improvements over the years, based on historical evidence, should be viewed with suspicion, as there are too many unknown variables.

For example, the decline in mortality rates for colon cancer, over the period 1950-1984, could have resulted from earlier detection and lead-time bias, as described above.[488] Furthermore, treatment methods are only one of the things that might have changed. Health improvements could relate to improved diet, reduced levels of pollution,

---

[a] Patients who had tumour progression were treated with the hormone oestrogen or orchidectomy (castration).

or other factors that have varied with time. Clearly, it is in the interests of clinicians, medical institutions and charities to claim the credit for improvements; however, such claims are rarely subject to independent challenge and validation.

Richard Gordon, author of the classic comedy *Doctor in the House*,[489] relates a breakthrough in preventative medicine, called the Beachy Head principle.[490] Beachy Head, a cliff on the south coast of England, has been a suicide site for many years. For several years, suicides occurred at an increasing rate of about four a year. Then, from 1975 onwards, the reported number of suicides suddenly halved. In the period 1965 to 1969, 65% of deaths on Beachy Head were suicides. From 1975 to 1979, only 32% of deaths were suicides. Surprisingly, this remarkable reduction did not involve fencing or other safety apparatus. Rather, it was due to the appointment of a new coroner, who was reluctant to classify deaths as suicide. Such changes in attribution can have large effects on mortality statistics, making historical comparisons hard to interpret.[491]

Diagnosis

An accurate diagnosis is the first step towards treatment. People are often the first to notice their own cancer, so everyone should be aware of the basic symptoms of the disease.

The American Cancer Society provides seven warning signs for cancer. These are:

- Unusual bleeding or discharge
- Changed appearance of a mole or wart
- Persistent lump or tissue thickening
- Persistent cough or hoarseness
- Changes in excretion from the bowel or bladder
- A sore that does not heal within two weeks
- Indigestion or trouble swallowing

These and similar warning signs are not very specific. Clearly, many people who experience indigestion do not have cancer.

A standard medical investigation may begin with the clinical history of the patient. The doctor then carries out a physical examination and, perhaps, laboratory tests. Diagnostic imaging may be required, using ultrasound, X-rays (including CT scans), or magnetic resonance. Occasionally, these methods may involve radioisotopes, to make an organ or abnormality visible on a medical image. Endoscopy, using a flexible, lighted tube, allows the doctor to inspect parts of the body or take tissue samples for microscopic examination.

Biopsies are often the most authoritative test for diagnosis of cancer. A biopsy is a small sample of tissue that is removed for examination. In some cases, an entire small tumour may be removed. When the tumour is large, a section may be used for the biopsy. Biopsies taken during endoscopic examination are usually small (<5mm in length). With a needle biopsy, the doctor removes a small core from suitable tissues, such as breast or brain lesions, using a hollow needle. In some cases, fine-needle aspiration is used, to take samples of isolated cells.

During an operation, a surgeon can take a biopsy, which is then frozen for rapid microscopic examination. Technicians slice the sample into thin sections, staining them to highlight cell features. Experts in pathology then check to see whether the sampled tissue is healthy, or contains malignant cells, indicating that remaining parts need removing.

Screening

It is possible to screen healthy people for various types of cancer. Such tests may detect cancer at an early stage and many people opt for regular tests for cancer and other diseases. However, the tests are difficult to interpret and require sophisticated analysis.

We can take smear testing for cervical cancer as an example of screening.[492] To save one life, about 1000 women must be tested. From 1976 to 1996, a study in the South West of England indicated that for every 10,000 women screened, 1,564 had a positive test. Of these subjects, 818 had a follow up investigation and 543 were found to have abnormal cells. Without screening, 80 women were expected to develop cervical cancer by 2011, of whom 25 would die. Screening might prevent 10 of these deaths. One way of viewing such results is as insurance,[493] in which the cost of regular checkups is weighed against the catastrophic risk of developing cancer.

To help explain the difficulties associated with screening, let us suppose an apparently healthy woman has a test for a rare cancer. Only one in a thousand people get this particular form of cancer. The test is 99% accurate. If you have cancer, the test will find it 99 times out of 100. If you do not have cancer then, 99 times out of 100, the test will indicate you are clear. The woman's test comes back positive, indicating she has the disease. How worried should she be? How great is the chance she has the disease?

Strangely, the chance of this woman having cancer is less than 10%.[b,494] Most non-mathematicians, including physicians, have difficulty understanding this problem. Let us consider some concrete figures. Suppose a million people take the test, of whom 1000 have the disease and 999,000 are healthy. The test is 99% accurate, so 9,990 healthy people test positive, compared to only 990 sick people. This means that nine times out of ten, people testing positive are healthy and do not have the cancer. This is known as a false positive. In cases like this, where the disease is rare, the chance of a person having the disease (1 in 1000) is lower than the probability that the test is wrong (1 in 100).

The consequence of such a high proportion of false positives is that many healthy people suffer the anxiety of thinking they have cancer. They are likely to undergo further, more invasive tests, or may even be treated for a disease they do not have.

Tumour markers

Cancers can be recognised by their altered gene activity. Tumours produce substances that can be measured, thus acting as cancer markers. These markers can help physicians monitor the effects of treatment, by indicating whether a tumour has returned, for example. Increased blood concentrations of a tumour marker may indicate re-growth. Such markers may provide an estimate of the proportion of cancer cells that are growing actively and are likely to be aggressive.

One such marker is called prostate-specific antigen (PSA), which is a protein produced by the cells of the prostate gland. Both benign enlargement of the prostate and prostate cancer become more frequent with age. Most prostate cancer occurs in men over the age of 65 years. It is normal for men to have low levels of PSA in their blood, but levels increase with prostatic enlargement and cancer. PSA levels alone do not give sufficient information to distinguish between benign enlargement and cancer. Furthermore, prostate cancer can occur even when PSA

---

[b] The chance this woman has the disease is the probability of having cancer given that the test result is positive (9%). Most people who have a positive test will be disease free.

levels are low.[495] However, doctors take the results of PSA tests into account when deciding whether to check for signs of prostate cancer.

PSA screening for prostate cancer is controversial, since no-one has established that it saves lives.[496,494] Some advocates recommend screening for all men over 50. Such screening would cost an estimated 12 to 28 billion dollars, in the first year alone.[494] However, while small cancers may be detected, there is no evidence that finding a small tumour will increase lifespan. Most men who show an elevated PSA level on screening do not have cancer. When such men are biopsied, only 25%-30% are found to have prostate cancer.[497]

The frequent false positives associated with this test lead to further medical procedures, with associated risks and financial costs. The process creates anxiety for patients and their families. Small cancers, detected by the PSA test, might never become a threat to life. However, even the follow-up biopsy may cause side effects, such as bleeding and infection. Over-investigation also increases the risk of complications from unnecessary surgery or radiation treatment.

Understanding the effects of screening requires a high degree of numeracy. Psychologist Gerd Gigerenzer illustrates this in his discussion on mammography screening for breast cancer. He describes four ways of looking at the same data, from four clinical trials, involving 280,000 women in Sweden. The percentage risk reduction was 25%. This large figure is quoted frequently, and is likely to mislead. In this case, four people in 1000 who did not participate in screening died, compared with three in 1000 people who did participate. The absolute risk reduction was the difference between these two figures, i.e. 4-3 = 1 in 1000 people.

The absolute risk reduction, expressed as 1/1000, seems much less persuasive, as an argument for screening, than the equivalent relative risk reduction of 25%. We stress that these two figures both refer to exactly the same data. The number of mammographies needed to save a single life was 1000. Finally, the increase in life expectancy, as a result of screening from the age of 50 to 69 years, was only 12 days. Gigerenzer calculates that a person can gain a similar increase by reducing their car travelling mileage by 300 miles each year.

Costs of screening

In 2004, a study evaluated the medical and non-medical costs associated with false-positive prostate, lung, colorectal, and ovarian cancer screens.[498] Unsurprisingly, the authors concluded that false-positive findings among those undergoing cancer screenings are relatively common. The vast majority of positive findings were false-positive results, which often lead to follow-up investigations that do not result in a cancer diagnosis. Such investigations can be expensive and provide little health benefit for an apparently healthy individual. Screening the population for cancer is more difficult to justify than most people realise.

Measuring treatment success

"We have not lost faith, but we have transferred it from God to the medical profession." George Bernard Shaw

Once doctors have diagnosed cancer, the next task is to begin treatment. Three common approaches are surgery, radiation therapy and chemotherapy. Typically, a treatment or combination of therapies is selected, depending on the patient's current condition, the type of cancer, its extent and rate of progression. The patient's response to the treatment is monitored continuously.

Most common solid tissue cancers do not respond well to treatment. With the exception of surgery, conventional treatments are often ineffective and can produce severe side effects,[499] especially when the disease is advanced. It is hard for physicians to have nothing to offer desperate patients, even when no effective treatment is available. Consequently, chemotherapy may be given for common, late-stage cancers, even though the therapy is unlikely to provide a cure and has unpleasant side effects.[499]

In carrying out the research for this book, we found little evidence that the selection of treatment involves quantitative cost-benefit analysis. New treatment methods are tested in clinical trials before they can be adopted. Despite this, the true cost of such treatments in terms of quality of life may greatly outweigh the benefits, even in cases where life is extended. If it is acceptable to subject patients to toxic and debilitating treatments, even if the chance of success is low, then surely it should be equally acceptable to administer promising, non-toxic treatments, even when the supporting evidence is incomplete.[15]

Cost-benefit analysis

All cancer treatments should be supported by a cost-benefit analysis. Potential benefits include increased lifespan, reduction in pain or symptoms, and an improved quality of life. These factors can be weighed against the costs of treatment, both financial and health-related. Such analyses would allow physicians to calculate the financial cost of each day of extended life. In addition, they could estimate the number of days of incapacity or severe side effects associated with each additional day.

Patients also need to know the number of symptom-free days they might expect, following treatment. This information would make it easier for both patients and doctors to make informed decisions about treatment choices.

A person undergoing treatment for cancer might hope for an improved quality of life, an extended life expectancy, or both. However, some cancer therapies, particularly cytotoxic chemotherapy, have side effects that reduce the quality of life during the period of treatment. Unless such treatment provides a greatly extended life expectancy or an overall improvement in the quality of life, it may be unjustified.

Currently, many treatment studies use tumour shrinkage as a measure of success: if the treatment shrinks the tumour, then the response is considered positive. According to this criterion, a complete response to treatment is for the tumour to shrink to undetectable levels. If the tumour gets smaller, then the treatment is thought to be partially effective. Although such regression might seem an obvious indicator of treatment success, the evidence for these claims in terms of benefits to the patient is unclear. There is little scientific justification for the suggestion that tumour shrinkage is associated with an increased lifespan.

Tumours regress when the balance between cell proliferation and apoptosis shifts towards increased cell deaths. However, tumour shrinkage can be transient, as the cancer adapts to the treatment and becomes resistant to the therapy. The microevolutionary nature of the disease means there will be selection pressure in favour of therapy-resistant cells, which may be more aggressive. Patients whose tumours have been made smaller through chemotherapy, for example, do not necessarily live longer, or even have an increased quality of life, than they would otherwise have done.[696]

Five-year survival and cure

A conventional way of measuring response to treatment is the five-year survival time: the proportion of patients who are still alive five years after diagnosis or treatment. For some cancers, shorter intervals are used. However, since cancer can return if only a few isolated cells remain in the body, it is difficult to know whether a patient is fully clear of the disease.

If a treated patient is checked regularly, the disease is often found to have recurred within the first few years. When a patient has been symptom-free for five years, and no sign of the disease is present, the patient may provisionally be described as cured. However, since such cures are not absolute, a relapsing patient may yet die of a cancer from which they were supposed to have been "cured".

Although the use of five-year survival may be useful for comparative controlled clinical trials, it is not suitable for general use as a guide to successful treatment. The only reliable indicator of a cancer cure is long-term survival. The five-year survival figure is a poor measure for general application, since it can be dominated by improvements in diagnosis.[500]

In order to illustrate the difficulties of measuring improvements in treatments, we will consider prostate cancer. Prostate cancer is one of the most common male cancers and is a leading cause of death. Epidemiologists at the American Cancer Society estimated that, in 2001, about 198,100 men would be diagnosed with cancer of the prostate, and approximately 31,500 would die from the disease.[501]

Since the 1980s, there has been a dramatic increase in the incidence of prostate cancer.[502] However, this apparent increase may be a result of the introduction of the prostate specific antigen (PSA) test, or other factors that enable earlier detection of the cancer. From 1980 to 1992, the incidence of prostate cancer more than doubled. The rates then fell a little between 1992 and 1995. Since around 1995, rates have been more stable. Mortality rates for prostate cancer increased slightly during the mid 1990s, before falling back. The mean rate in the period 1997-2001 was 31.5 people per 100,000, compared with 31 people per 100,000 in 1975.[502] This means that, despite an apparent doubling of incidence, morality rates have hardly changed over the same period. These figures may be taken to suggest that the war on prostate cancer is being lost. It is worth noting that the incidence of symptom-free prostate cancer, found on microscopic examination of the tissue, increases with age.

Men whose prostate cancer is diagnosed at an early stage, without spread into other organs, have a five-year survival rate close on 100%. However if the cancer has spread to distant sites, only about a third of patients survive five years or more (32.6%).[502] The cancer's stage affects its microscopic appearance. If the cells appear relatively normal and differentiated at the time of diagnosis, the survival rate approaches 100%. However, if the cells are undifferentiated, having lost the characteristics of healthy prostate cells, the survival rate drops to 73.5%. Men who are younger than 50 years when diagnosed are less likely to survive for five years, as their disease is often more advanced when found.[503]

There are many possible measures of response to treatment, such as the size of a solid tumour or a reduction in the number of cells, in the case of leukaemia. However, patients may show an apparently favourable response to treatment, without their lifespan being improved. The following table is modified from information provided by Dr Ralph Moss, based on published data on response to treatment and proportion of disease-free survivors.[499] The table illustrates the relationship between response to treatment and survival in cancer patients.

Treatment response and actual survival		
Cancer	Response %	Survival %
Breast (stageIII/IV)	75	Rare
Small Cell Lung	90	10
Stomach	50	Rare
Ovary	75	10-20
Multiple Myeloma	75	Rare
Leukaemia (acute)	75	20
Leukaemia (chronic)	75	Rare
Prostate	75	Rare
Bladder	60	Rare

Will Rogers phenomenon

Previously, we described how earlier detection can lead to extended five-year survival statistics and make newer treatments seem more effective. This is not the only source of bias that causes new therapies to appear more beneficial. With time and better diagnostics, the criteria for staging cancer can change. This process is called the Will Rogers phenomenon, known medically as stage migration.[504,505] The attribution to Will Rogers, a cowboy-humorist, came from his joke,

'When the Okies left Oklahoma and moved to California, they raised the average intelligence level in both states."

The implication, that the less intelligent Okies moved to California, where they were above the average intelligence, can be applied to groups of cancer patients. If the least healthy patients currently classed as stage 1 are re-classified as stage 2 then, on average, both groups will be healthier. This is stage migration.

Improvements in cancer detection mean that local spread and metastases are discovered earlier. Thus, a person previously thought to have a stage I cancer may be allocated to stage II. Alternatively, a cancer previously classified as stage III might be re-classed as stage IV, if improved techniques mean metastases are found more quickly. We should notice that, with improved diagnosis, the movement is from the less to the more advanced stage. Such movement will improve the average prospects of both groups.

The category the patients move from will lose those members with the most advanced cancers; therefore, the average for this category will be improved. However, the category these patients move into will gain members with less advanced cancers, so the average health of this group will also improve. Stage migration causes an apparent improvement in response to treatment in all cancer stages. It is important to note that this progress is an artefact of the reclassification: no patient's health has actually improved.

Publication bias

The medical literature tends to publish only positive studies.[506] Journals may reject negative studies or studies with low significance. Thus, negative studies and those showing only slight significance tend to end up in the filing cabinet, and are never sent for publication.[507] Where papers are accepted, the more significant and positive results are published in higher impact journals.[508] The literature therefore gives a rose-tinted view of progress in cancer treatment.

So far, we have ignored the commercial bias that can be introduced by pharmaceutical companies and other medical industries. A drug company has a financial incentive either not to publish negative results found in trials of its products, or else to present the data in the most positive light. Furthermore, in describing publication bias, we need to include ghost-written articles. Pharmaceutical companies have a massive financial interest in clinical trials and reports of drugs. Recently, pharmaceutical companies have engaged in swamping the medical literature with ghost-written articles, which are little more than advertising.[509,510] Doctors are paid for agreeing to be the "author" of such a paper. In some journals, half of all medical articles dealing with drugs may be ghost-written. Diana Zukerman has described the result as "checkbook science".[511] The production of scientific papers as a branch of marketing distorts the reporting of medical research,[512] to everyone's detriment.

Surgery and radiation

"A chirurgien should have three dyvers properties in his person. That is to say, a harte as the harte of a lyin, his eyes like the eyes of a hawke, and his handes the handes of a woman." John Halle (1565)

One of the oldest treatments for cancer is surgery. Surgery is probably the treatment of choice for early cancers, as it is more likely to produce a cure. The main problem is that by the time the cancer is diagnosed, many patients have tumours that have spread to distant sites.

It may not be possible to remove the whole of a tumour. In this case, operations should have an expectation of direct and identifiable benefit to the patient before they are performed. For example, surgery may be used to reduce pain or to return function to a part of the body. Such palliative surgery includes the removal of tumours that are obstructing the oesophagus or colon.

A number of factors influence the decision to use surgery. The first is whether the patient can be cured by removal of the tumour. If not, surgery may be used in combination with other forms of treatment, such as chemotherapy. This can be a successful strategy; for example, survival rates in childhood muscle tumour (rhabdomyosarcoma) have increased substantially when surgery is combined with chemotherapy. Such a large response to combined therapy, although found with some childhood cancers, is rare. In the case of non-small-cell lung cancer, surgery and chemotherapy have been shown to provide only marginal extra benefit, compared with surgery alone.[513]

In people with a predisposition to develop a particular cancer, surgery can be used to reduce the risk. A controversial example is mastectomy, for cases where development of breast cancer is hereditable and highly likely. Another example is the surgical correction of undescended testicle, a condition that occurs when one of the testes remains inside the body during development, instead of descending into the scrotum as normal. Since this can increase the risk of testicular cancer and has other undesirable effects, the defect may be corrected surgically.

Radiotherapy

Radiotherapy is the use of high energy ionising radiation to kill cancer cells. A large proportion of people with cancer receive radiation therapy.[499] Typically, the radiation is in the form of X-rays or gamma rays, although occasionally subatomic particles, such as neutrons, may be employed. This ionising radiation causes oxidation, generating free radicals in the tissue. These free radicals damage the cancer cells directly.

When exposed to radiation, some cancer cells die because they sustain so much damage that they cannot replicate. Others recognise the damage and commit suicide by apoptosis. Radiotherapy injures rapidly dividing cells more than slowly growing tissues. Actively dividing cancer cells are thus more sensitive to radiation than most healthy cell types.

With internal radiation therapy, sometimes called brachytherapy, a radiation source is placed in or near to the tumour. Typical sources include isotopes of caesium, iridium, gold and iodine. By contrast, external radiation therapy delivers radiation from a source outside the patient. The location and shape of the tumour can be identified by clinical imaging, using a magnetic resonance body scanner, for example. Then a suitable wavelength or energy for the radiation beam is chosen. The beam's energy, position, shape and dose are carefully determined to give a maximum dose to the tumour, while minimizing damage to the surrounding healthy tissues. Internal radiation generally results in fewer acute secondary effects than therapy using an external source.

Radiation is often administered in repeated small doses, which helps minimize complications. Repeated doses can also be more effective in killing the cancer. The maximum dose to healthy tissue surrounding the tumour can be reduced by moving the energy beam, so it takes several paths though the body, crossing at the tumour.

It is claimed that radiation can be curative in a few cancers, such as carcinoma of cervix,[514,515] or Hodgkin's disease.[516] However, radiation is less effective against most common, slow-growing cancers. In Hodgkin's disease, treatment is associated with a small risk of secondary tumours, cardiac events and infection-related mortality. This suggests that improvements to reduce side effects might be possible.[517,518]

Radiation therapy can have acute side effects, which affect rapidly renewing tissues, such as those in the gut and the lining of the bladder. Cells in these tissues divide frequently and, like cancer cells, are more susceptible to damage than slower growing cells. Side effects take time to build up and include scaring (fibrosis), loss of normal tissue, and formation of abnormal openings (fistulae). These effects are related to the total dose of radiation. Although radiation can cause cancer in the long term, this risk is considered negligible compared to the presence of an existing and potentially fatal disease.

Radiotherapy works by causing oxidation, which produces cancer-damaging free radicals. Because of this, the effectiveness of radiation therapy can be limited by a shortage of oxygen in the tissues. If the tumour has an anoxic environment and an inadequate blood supply, it is harder for radiotherapy to generate free radicals by oxidation. Since the development of solid tumours involves anaerobic conditions, radiation therapy may be less effective for such tumours.

The internal structure of tumours is variable, with some areas of cells existing under anaerobic conditions. Such cells divide slowly and are therefore likely to be unresponsive to radiation therapy, which must be given at doses that are not harmful to healthy body cells. Cells that are not destroyed by radiation will remain and multiply within the treated tumour, thus the microenvironment will tend to select cells that are resistant to the effects of radiation, including free radical damage. In such cases, radiation may shrink the tumour but not drive the cancer cells to extinction. As cellular resistance builds, repeated radiation treatments become ineffective.

One way to approach the problem of resistance in solid tumours is to use radiotherapy and surgery in combination. The properties of radiotherapy are somewhat complementary to those of surgery. Radiation is relatively ineffective at destroying cells in the centre of a cancerous lump, which tend to be anaerobic and dividing less rapidly. Luckily, these are precisely the cells that are more accessible to the surgeon. On the other hand, radiation is more effective when it comes to destroying small clusters of undetected cancer cells in surrounding tissues. Surgical removal of a tumour in the breast, for example, may thus be followed by radiation therapy to the local area and lymph nodes, in the hope of destroying any remaining cancer cells.

Chemotherapy

"It amazes me how much of what passes for knowledge in cancer therapy turns out to be incomplete, inadequate and anecdotal." Ralph Moss

Chemotherapy is the administration of drugs that selectively destroy or slow the growth of cancer cells. Treatment of cancer with drugs has been practiced for at least 500 years. The early physician, Paracelsus (1493-1541), is sometimes described as the father of chemotherapy.[499] The advantage of chemotherapy over surgery and radiation is that it is not limited to known sites: it can attack cancer cells throughout the body. Anticancer drugs travel through the bloodstream, attacking tumour cells wherever they occur, with the exception of a few locations that the drug does not reach in high enough concentration.

Cytotoxic drugs kill cancer cells directly, whereas cytostatic drugs stop their proliferation. Tumours grow if their rate of cell division exceeds the rate of cell death. Hence, theoretically, both forms of drug have the potential to cure. However, in practice, a cure requires a long period of sustained exposure to the agent, as the population of cancer cells is driven to extinction.

Chemotherapy does not usually result in a cure for cancer. This treatment method has been subject to substantial criticism, which is described in *Questioning Chemotherapy*, a book by Ralph Moss.[499] Commentators have even asked whether chemotherapy provides any advantage, in the majority of slow growing and lethal cancers.[499] In the case of melanoma, for example, chemotherapy may be ineffective and may lead to a reduced quality of life.[519]

Ulrich Abel, a statistician, has provided a comprehensive analysis of published studies of chemotherapy for carcinoma. He found no evidence for increased survival with this form of treatment. The exception was a slight benefit in lung cancer.[520] According to Abel,

"Apart from lung cancer, in particular small-cell lung cancer, there is no direct evidence that chemotherapy prolongs survival in patients with advanced carcinoma."

Strangely, while the treatment has continued to be used, this study has not been followed up since Abel published it, in 1992. The benefit mentioned seems to be specific to a particular form of (small-cell) lung cancer and is confirmed by multiple studies.[521,522,523,524] A similar small advantage may occur with colorectal cancer.[525] However, a more recent study confirms the lack of positive results from chemotherapy in breast cancer.[526] The demonstrated benefits of chemotherapy are not as clear as a patient undergoing such treatment might reasonably expect.

Why chemotherapy fails

Driving a population of cancer cells to extinction may be beyond current anticancer drugs. Imagine a highly effective anticancer drug, which kills 99% of all cancer cells. Suppose this drug is administered to a person with a small tumour, detected at an early stage and having, perhaps, a mass of one gram. Despite having been detected early, this tumour will have been developing for several years and may already have existed for three-quarters of its natural lifespan.[1] A tumour of this size may contain a billion cancer cells. Killing 99% of these cells still leaves ten million cells, which have been selected for drug resistance.

The toxicity of chemotherapeutic drugs limits the dose that can be employed, although side effects can be reduced by giving the drug in divided doses, over a period of weeks. Delivering the drug in this way allows both patient and cancer time to recover from the toxic effects. Suppose this method of increasing the dose allows us to kill 99.999% of the cells in the tumour, and any higher dose is toxic to the patient. Such a dose will still leave ten thousand, highly selected cancer cells alive. These cells may be more aggressive than those that were eliminated. Furthermore, these surviving cells are now free from competition with the other cancer cells. The result could be a tumour that shrinks in size to become undetectable, followed by a period of increased growth by the aggressive cells. For this reason, the drug therapy may be apparently successful, but might not increase the patient's lifespan.

A combination of cytotoxic drugs can be more effective than a single agent, because of variation between cancer cells. Some cancer cells in a population will be resistant to the action of a single drug. However, these resistant cells may be sensitive to the effects of a second drug. By giving drugs with different mechanisms of action, more cancer cells can be killed for an equivalent level of toxicity to the patient. Combinations of cancer drugs can be more effective, but the difficulty of driving every cancer cell to extinction remains.

Variation in the behaviour of cancer cells also limits the action of chemotherapy. For example, cytotoxic drugs often act when the cell is dividing, but not all cells in a tumour are dividing at any given time. During the period between divisions, many cancer cells may be resistant to the treatment. The duration of the treatment is limited by the toxicity of the drug, so it is possible to miss the sensitive cell division period. Thus, cells that are not dividing during the treatment would survive.

Another problem with chemotherapy is that tumours have a varied internal structure. Some cancer cells are located close to blood vessels and receive a high concentration of the drug, leading to their death. Other cancer cells are located in sanctuary sites, some distance from the nearest blood vessel. These are less likely to be actively dividing, because they lack sufficient oxygen and nutrients: they are therefore resistant to the drug. In addition, they experience only a low dose, which may be too small to be effective. A large protein molecule, for example, may take more than a month to diffuse to the centre of a tumour.[1]

To be effective, chemotherapy depends on sustaining selection pressure on the cancer for long enough to wipe out cells that are sequestered away from easy drug access. This requirement implies the use of substances of low or negligible toxicity to normal cells, which is not a feature of current cytotoxic drugs.

Origins of chemotherapy

Chemotherapy began as a by-product of research on chemical warfare, a fact that should warn us of the potential for severe side effects. The first modern chemotherapeutic agents were derived from mustard gas. In 1942, Luis Goodman and Al Gilman, pharmacologists from Yale University, were testing possible chemical warfare agents. They found that some of these agents decreased the number of white cells in the blood, from which they suggested that the substances might be used to treat leukaemia.

The same year, a Yale anatomist called Thomas Dougherty injected nitrogen mustard, a derivative of mustard gas, into mice with tumours. The first mouse had a dramatic response: the tumour softened, shrank and regressed. However, it returned a few weeks later and subsequent treatments were less effective. Although the effect could not easily be replicated in mice, the treatment was tried in a human. It has thus been claimed that chemotherapy was based on experimental results from a single mouse.[499]

Coincidentally, in 1943, an American ship called the SS John E. Harvey was carrying 100 tons of the chemical warfare agent, mustard gas.

The ship was hit by German bombs and exploded in Bari harbour, Italy. Following this exposure to mustard gas, the crew were found to have suffered suppression of their bone marrow and white blood cells. A US army doctor sent tissue samples to America, where it was realised that such changes could be relevant to the treatment of cancer.

From its first clinical use, the problems with chemotherapy became obvious. The first patient, a man with advanced lymphosarcoma, showed a remarkable response to the treatment. As with the first mouse, the man's tumour softened within two days. Within ten days, the patient's swollen lymph nodes had diminished and he appeared to be recovering. Shortly after, he showed a huge loss of white blood cells. However, as the cells recovered from the insult, the tumour grew back, providing only a brief reprieve from the cancer. Repeated treatments gave progressively smaller beneficial responses.

Side effects

The first patient illustrates the difficulties with this treatment. Conventional chemotherapy damages patients' normal tissues, especially rapidly growing cells in the blood, bone marrow and gut. Cancer growth is slowed temporarily but diseased cells become resistant to repeated treatments. Alternatively, the treatment may completely fail to kill the cancer, while still damaging normal tissues.

There are parallels between antibiotics and chemotherapy, as explained previously. Some antibiotics kill bacterial cells, while leaving the host cells undamaged. Such drugs are effective because they exploit differences between bacterial and human cells. In the same way, anticancer treatment with cytotoxic drugs should ideally kill cancer cells, while leaving healthy cells undamaged. Unfortunately, there are few biochemical differences between cancer cells and the cells of the host animal.

Malignant cells are modifications of standard cells and retain many of their characteristics. This similarity means that anticancer drugs must exploit minor differences between cancer cells and healthy cells. Many anticancer drugs act either by preventing cell division or by introducing free radical damage. These drugs often induce cancer cells and other rapidly dividing cells to commit suicide by apoptosis. There is often little margin between the dose that kills cancer cells and the dose that destroys sensitive healthy cells. The resulting side effects, such as loss of cells in the intestine, can be debilitating.

Drugs that impair cell division can prevent the growth of both cancer cells and other rapidly dividing tissues, such as blood cells, produced by the bone marrow. Anticancer drugs often damage the production of red and white blood cells, leading to anaemia, tiredness, bleeding and an inability to fight infection. Hair follicle cells also divide rapidly, which is why cytotoxic drugs can reduce hair growth and cause baldness. One way to minimise these side effects is to give doses of cytotoxic drugs at intervals, to allow rapidly dividing normal cells to recover. If the tumour does not recover as quickly as the healthy cells, intermittent dosing can produce improved results.

Chemotherapy often involves severe side effects, sickness and debilitation. Patients may believe this suffering is justified by the hope of a cure or substantial increase in lifespan. However, in many cases, chemotherapy does not extend the lifespan but merely reduces the quality of life in the period remaining.[499] Hence, cytotoxic chemotherapy for cancer should always be justified by substantial supporting data on quality of life and extended survival. Unless such treatment is expected to provide a large life extension, with improved quality of life, its use may be counterproductive. Without solid evidence for benefit, patients' optimal strategy could well be to decline the treatment and avoid the side effects.

Magic bullets

Nobel Prize winner and early pharmacologist, Paul Ehrlich (1854-1915), coined the term *magic bullet* in 1907. He used it to describe a chemical that, when introduced into the body, will seek out the disease and destroy it, leaving normal tissues unharmed. Ehrlich considered that toxic molecules could be found, which would selectively target a disease-causing organism. The result would be an ideal chemotherapeutic agent, which acted only on the diseased cells.

Although fine in principle, production of a magic bullet is easier said than done. The action of drugs is usually oversimplified. Only a few drugs have a precise mode of action, interacting with a single receptor that recognises the shape of the drug. Such receptors respond to normal chemical signals in our bodies. A corresponding drug typically approximates to the shape of the naturally occurring substance, mimicking its highly specific action. In principle, such drugs can have the characteristics of a magic bullet. However, even when a class of drug acts on a specific receptor, the drugs can have different effects, depending on their distribution in the body. For example, a tranquilliser drug might be highly receptor specific, but if it does not cross the blood brain barrier, it has little clinical utility.

Many drugs do not act on specific receptors but have varying actions on different tissues. Drugs that kill cancer cells are often unspecific in this way. When we describe how they work, we should remember that these drugs affect many systems in the body. Notably, cytotoxic anticancer drugs often involve oxidation and widespread free radical damage.[a,527,528] By focusing on the drug's initial interaction with the cell, which eventually leads to apoptosis, these unfortunate effects might be overlooked.

Many different drugs are used in the treatment of cancer, including cytotoxic drugs, which kill cancer cells. In rare cases, these drugs may provide the primary treatment, even for a localised cancer. More often, they are the last line of attack against advanced cancers, for which other therapies cannot be successfully applied. They are often used as a secondary therapy, following surgery or radiation.

In rare cases, chemotherapy can eradicate cancer. We have already described some of these, which include treatments for choriocarcinoma (a cancer originating in cells from the placenta) and Burkitt's lymphoma (a tumour of the lymph glands). Combinations of drugs were initially used to treat leukaemia and lymphoma, with some success. This early success of chemotherapy soon led to its use in solid tumours. However, solid tumours are often slow growing and resistant to the effects of chemotherapy.

DNA building blocks

Many current cytotoxic drugs interfere with protein synthesis or DNA replication. The building blocks of DNA and RNA are nucleic acids. Some drugs, called antimetabolites, interfere with the production of these nucleic acids. In this way, they prevent the correct use of normal DNA and RNA components, or metabolites, in the body. When these drugs block the enzymes that synthesise nucleic acids, production of DNA and RNA is prevented and cell division does not occur. Some of these drugs, for example, cytarabine, are so similar in structure to the original building blocks that they can be incorporated directly into a gene, disturbing its usual function and causing damage. However, since all dividing cells manufacture DNA, these drugs have corresponding side-effects in normal body tissues.

---

[a] Examples include doxorubicin, mitomycin C, etoposide and cisplatin.

Cross-linking agents

Cross-linking drugs, as the name suggests, bind two molecules together. These drugs are also called alkylating agents. The drug cisplatin is a widely used alkylating agent that is used to treat cancer. In 1979, cisplatin was discovered by chance, in a series of experiments by Barnet Rosenberg, of Michigan State University.

Rosenburg thought the appearance of dividing body cells was similar to that of iron filings in a magnetic field. He became interested in the action of electricity on bacteria, and in whether an electromagnetic field would change the observed patterns. He noticed that bacterial cells grew longer in the electric field, but did not divide. Bacterial cells normally grow by elongation prior to splitting in two during cell division: initially, therefore, he suspected that the current prevented cell division.

However, the cells continued to grow in length, even after he switched the current off, and did not divide until several hours later. Rosenberg realised that an electrochemical reaction was disrupting the cell division, rather than the electric current itself. The observed effect was caused by a reaction between the platinum electrodes and nutrients in the fluid that contained the cells. The chemical produced became known as cisplatin, although its full name is *cis*-diamminedichloroplatinum(II), or DDP. Since cisplatin inhibited cell division, Rosenberg thought it might make a good anti-cancer drug.

Cisplatin, like other cross-linking agents, prevents cell division by binding to the two strands of the DNA double helix. This binding prevents the double helix from separating into two single stands, which is necessary for producing new DNA, RNA and protein molecules. Without new DNA and protein synthesis, the cell cannot divide and growth is impeded. However, the effect of cisplatin is not restricted to cancer cells, it has a general action on dividing cells in the body.

Antibiotics

Several antibiotics prevent cells from producing new RNA and protein. The first of these, actinomycin, was discovered in 1940 at Rutgers University, by Selman Waxman and Boyd Woodruff. Its name indicates its origin: it was found in cultures of the yeast *Actinomyces*. Actinomycin inserts itself between the two chains of the DNA double helix, distorting the spiral shape and blocking production of RNA and proteins. Several other antibiotics also work as anticancer agents.[b]

In addition, some anticancer antibiotics inhibit the enzyme topoisomerase, which influences the shape of DNA and is important in protein synthesis. Regulation of DNA shape and super-coiling is important as, during division, the DNA must unwind for its sequence to be read. Altering the shape of the DNA coiling can prevent cell division. Another mechanism of action involves direct cell oxidation. Doxorubicin, for example, acts on a protein in the cell membrane, producing free radical damage that leads to cell death.[529,530]

Some anticancer antibiotics have more than one mechanism of action. Indeed, it is unusual for an anticancer drug to have a single physiological effect. The drug procarbazine, for example, has several actions against cancer cells. It acts in a similar way to the cross-linking drugs, causing breaks in the DNA and injury to the cell. In addition, it produces free radicals, causing widespread cell damage, like doxorubicin.

An example: a drug study with docetaxel

We will consider the example of a recent drug, docetaxel, which was regarded as a breakthrough for the treatment of stage IV breast cancer. Like the better-known drug, taxol, docetaxel was derived from the yew tree. It appears to have been approved on the basis of uncontrolled trials.[531,499] This example concerns a study on the use of docetaxel in non-small-cell lung carcinoma.[532] The trial was conduced by several collaborating university hospitals, in the United States.

According to the National Cancer Institute's website, about 172,570 people in the US are diagnosed with lung cancer each year and 163,510 die. Small-cell carcinoma of the lung is a distinctive form of lung cancer, in which metastases are often established by the time of diagnosis. Non-small-cell cancer covers a variety of disease types.

The study in question included 80 patients, with advanced cancer that had previously been treated with platinum compounds. They ranged in age from 34 to 79 years. Most of these people developed docetaxel-induced blood diseases. Twelve developed anaemia; 77 suffered a reduction in white blood cells (leucopoenia); and 71 subjects suffered a reduction of neutrophil white blood cells (neutropenia), which was severe

---

[b] Bleomycin, mithramycin, mitomycin and doxorubicin are examples of antibiotics with anticancer effects.

in 11 patients. With the resulting loss of immunological competence, six patients endured severe infections, although none was fatal. One patient died of an unrelated problem. Severe fluid accumulation, described as "usually" responsive to diuretic therapy, was seen in three patients.

There were other problems, in addition to these blood-related side effects. Seven patients suffered damage to the nerves (neuropathy). Three subjects had sores on the lining of the mouth (stomatitis). Three suffered fatigue and one a hypersensitivity reaction. Despite these reactions, the researchers concluded, "Docetaxel was relatively well tolerated in this pretreated population." Readers are free to come to their own conclusions as to the accuracy of this statement.

Given the reported side effects, we might expect large benefits for this treatment to be considered clinically appropriate. However, no complete response to the treatment was observed, as the tumours did not shrink to undetectable levels. Thirteen of the 80 patients had a partial response. The authors claim an increased survival time, but the study had no controls and we cannot see how this suggestion was derived from the experimental data. Only 25% of the patients survived for one year, compared with published figures suggesting up to 40% survival.[533,534]

The primary question in such cases should be how people treated with the drug compare with those who receive no treatment or "best supportive care". Although these kinds of study are rare, we can examine one such report.[534] This followed 103 non-small-cell lung cancer patients who were treated with docetaxel, and 100 who were given only best supportive care. The patients were selected randomly into two groups. The treatment group had extensive side effects, as might be expected. In the treatment group, 29 survived for one year, compared to 19 who survived in the untreated group. However, the survival curves converged and, after 18 months, there was no difference in survival rates. This short-term difference in treatment response would not be convincing to a rigorous scientist. The best supportive treatment group, together with their physicians and carers, knew they were not receiving therapy. The difference in short-term life expectancy may simply have reflected patient expectations.

A favourable response in a small proportion of treated subjects might be used as justification to treat the majority of patients. However, the majority of patients may suffer the side-effects and reduced quality of life associated with the therapy, without any benefit. It is not enough to show that, say four people out of a 100 will have their lifespan extended by a few weeks, if both they and the 96 others endure months of debilitating chemotherapy to achieve this result.[499]

Inhibiting folic acid

Folic acid, one of the B vitamins, is essential for cell division and reproduction. It is used in the synthesis of the nucleotide building blocks for DNA and RNA. One of the early anti-cancer drugs, methotrexate, inhibits an enzyme that produces folic acid, thus preventing synthesis of these essential nucleic acids. Without these DNA building blocks, genes are not copied and cell division is prevented. Unfortunately, cell division in normal cells is also inhibited, leading to severe side effects.

Drugs from plants

During cell division, large protein structures called microtubules pull the two sets of chromosomes into opposite ends of the cell, before separation into daughter cells. Two drugs, vincristine and vinblastine, obtained from the periwinkle plant (*Vinca rosea*), damage the microtubules and thereby prevent cell division. These vinca alkaloids are used in the treatment of Hodgkin's disease and leukaemia.

The drug podophylotoxin, from the tree *Podophylium*, works in a similar way to the vinca alkaloids. American Indians used extracts of the tree to cure warts. The drug company Sandoz used podophylotoxin as a model to develop a synthetic lung cancer drug.[c] However, the derived drug may have a different mechanism of action, inhibiting an enzyme involved in the repair of nuclear DNA.

Extracts from the bark of the yew tree (*Taxus brevifolia*) are reported to have beneficial effects in cancer. This led to the isolation of taxol, a potent anti-cancer agent. Taxol is effective against a range of malignancies, such as breast, lung and prostate cancer. Like the vinca alkaloids, it interferes with the microtubules in separating the chromosomes, thus preventing cell division. However, taxol could not be synthesised commercially, as the molecule was complex and the particular yew tree was an endangered species. The solution was to find a related chemical in the needles of the western yew and to convert this to taxol chemically. The taxol story provides a strong argument for biological conservation.

---

[c] Etoposide.

Hormones

Steroid hormones, such as prednisolone, can inhibit cell division and are used to treat inflammation. However, several forms of cancer are stimulated by naturally occurring hormones. Breast cancer may depend on oestrogen for growth, so drugs that block the action of this hormone, such as tamoxifen and its derivative toremifene, may prevent cancer growth.[535] Conversely, drugs that have an action similar to the hormone progesterone can cause some tumours to shrink. Progesterone often has a physiological action opposite to that of oestrogen.

In males, the growth of prostate cancer can be stimulated by male hormones, such as testosterone. Removal of the testes can slow the growth of such tumours, but a less radical approach is the use of drugs, such as stilboestrol, to counteract the effects of male hormones. The ultimate effectiveness of this drug treatment is questioned and side effects include loss of libido and breast enlargement.[536]

Drug resistance

Cancer cells have the remarkable characteristic of multidrug resistance. During chemotherapy, they rapidly become tolerant to the effects of drugs. This tolerance can be widespread and the cancer cells can become resistant to a number of different and unrelated chemicals at the same time.[537]

Cancer cells become resistant because of microevolution. Given time, the most robust and resistant cells are selected and make up an increasing proportion of the population. The mechanisms underlying multidrug resistance are not always clear, but may involve selection for abnormal chromosomes,[537,538,539] redox mechanisms[540] or failure of cells to commit suicide. Cytotoxic drugs can produce resistance in many ways, for example:

- The blood supply to the centre of large tumours can be limited, so the drug does not penetrate to the central cells.

- Cancer cells can develop modified proteins, which do not interact with the cytotoxic drug.

- Cancer cells may produce larger amounts of the affected proteins, so some protein molecules remain available to perform the targeted functions.

- Cancer cells can incorporate pumps in their outer cell membranes, to remove anticancer drugs.

- Cancer cells can stop manufacturing the proteins that carry the drug into the cell body.

- The cancer can produce extra maintenance proteins, to repair damage faster than the drug can inflict it.

The processes by which cancer cells become resistant should not be considered intelligent. Rather, the treatment eliminates cells that are responsive to it, so unresponsive cells have a selective advantage. The methods by which the tumour responds to, and may overcome, the effects of treatment are simply a reflection of microevolution. Any varied population of cells or organisms will respond to selection pressure by exploiting the available evolutionary options.

Cytotoxic chemotherapy and radiation kill cells by inducing apoptosis. The cell triggers its cell-death program, rather than submitting to the damage produced by the treatment. However, some cancer cells may be highly resistant to committing suicide in this way. Current therapies for cancer are limited by their toxic nature and the biological response of the cancer. Most chemotherapy drugs were discovered years ago and progress has been largely limited to improving methods of delivery.[1] It is time to take a wider biological approach to destroying cancer.

Biological therapies

"The art of medicine consists in amusing the patient while nature cures the disease."
Voltaire

The US National Library of Medicine gives two definitions for biological therapy. According to the first, it is the treatment of disease by the administration of substances that produce a biological reaction in the organism. This definition covers the whole of pharmacology, unless we limit it explicitly to treatment with sera, antitoxins, vaccines, cells, tissues, and organs. The second definition covers treatments to stimulate or restore the ability of the immune system to fight infection and disease. When used to lessen the side effects caused by some cancer treatments, such immunological support is known as immunotherapy, biotherapy, or biological response modifier (BRM) therapy. Conventional cancer treatments incorporate several forms of biological therapy.

Biological therapies aim to enhance the body's inbuilt ways of fighting cancer. Spontaneous remissions, where people recover miraculously from terminal cancer, are more frequent than might be expected.[15] These cases indicate that, occasionally, the body's defence mechanisms can eliminate even advanced cancer. Unfortunately, we only know of such remissions after the event; there is no way to predict which patients will recover.

The study of spontaneous remission is in its infancy. The individual nature of such remission makes it unsuitable for controlled scientific investigation. However, researchers have suggested that cytokines, signalling substances produced by the body, could be implicated in spontaneous remission. Several have been tested as treatments, including interferon, tumour necrosis factor and interleukins. Tumour necrosis factor can kill cancer cells, as its name suggests, but has side effects when employed as a drug.

Medical researchers have spent huge amounts of effort to develop vaccines against cancer. The idea behind vaccine development is to stimulate the immune system, so it will recognise the cancer cells as foreign. White blood cells may then attack and destroy the cells. The actions of a vaccine for cancer are, in principle, no different from those used against infections. For example, the HPV vaccine for cervical cancer immunises against the human papillomavirus, which is a known cause of the disease.[541]

Medical scientists have tried using tumour antigens or inactivated cancer cells to develop vaccines. Unfortunately, most of these attempts have been unsuccessful. Cancer cells contain mainly the same genes and proteins as any other cell in the body. It is therefore difficult for antibodies to distinguish cancer cells from healthy body cells. Moreover, even within a single tumour, cancer cells are not identical, but consist of a varied population. Unless its spectrum was wide enough, a tumour vaccine would simply select for those cancer cells that did not express the antigens targeted by it, which would develop into a resistant population.

Gene therapy has attracted a great deal of interest and one of the promises made to gain funding for the human genome project involved fighting cancer. This project mapped the sequence of bases in human DNA. Theoretically, gaining information about genetic defects that lead to cancer could allow us to repair these errors in the future. The incidence of cancer could be reduced by correcting the oncogenes that increase the risk of the disease. With this in mind, scientists are exploring techniques to allow the insertion of genes into cancer cells.

However, those making claims for gene therapy often neglect to mention the massive chromosome alterations that typify cancer cells. It seems exceedingly unlikely that simple insertion, activation or deactivation of a few genes would cause cancer cells to revert to a normal form. The cellular changes and selection pressures are just too large.

Immune response

One theory of resistance to cancer is called immune surveillance. Our immune systems have evolved to destroy substances they recognise as foreign. They use antibodies, proteins produced by the lymphocyte white blood cells, to bind and neutralise antigens. An antigen is any molecule against which an antibody will react.

The immune system can occasionally identify and destroy emerging cancer cells. Antibodies recognise and react to abnormal antigens on the surface of the cancer cell as foreign or non-self. However, as mentioned previously, tumour cells are derived from normal body cells and may share many similar antigens. These shared antigens do not trigger an immune reaction. Unfortunately, few human cancers generate a strong immune response.[1]

According to the immune surveillance theory, cancer cells often emerge in our bodies but our immune systems destroy them. White blood cells act as predators against cancer cells. Those cancers that do develop have evaded the body's immune surveillance. It has been suggested that some degree of immune activation could occur in people with a form of colon cancer, called hereditary nonpolyposis or Lynch syndrome. People with this disease inherit a faulty DNA mismatch repair system, which causes them to produce mutant proteins. Some of these appear on the cancer cell surface and may be sufficiently different from normal molecules for the immune system to distinguish the cells as foreign.[542] Changes in proteins on cell surfaces provoke an immune response, which can eliminate the cell.

Further support for the immune surveillance theory comes from people with immunological deficiencies. Such deficiencies may result from several causes, including genetic inheritance, infections such as AIDS, anti-rejection drugs used during organ transplants, and conventional treatments for cancer. People with defective immune systems have an increased risk of developing cancer, particularly malignant lymphoma. Such lymphomas may be a result of infection with the Epstein-Barr or human T-cell leukaemia viruses. However, the forms of cancer that kill most people, such as lung, breast, and colon cancer, are generally not increased in patients with immune deficiency conditions.

Sometimes the immune system can attack an existing tumour. Antigens occur on the surface of healthy cells, but the body's immune system ignores them. However, some cancer cells have unexpected antigens on their surfaces. In the laboratory, monoclonal antibodies can be produced to react to a specific antigen. Monoclonal antibodies are specific antibodies, which attack a particular tumour antigen. Tumour-specific antigens are unique to tumour cells, for example, mutated proteins on the tumour cell surface. White blood cells (T lymphocytes or T cells) recognise and eliminate cells that possess these antigens.

Identification of a tumour antigen that is unique to a cancer might provide the key to producing such a tailored antibody. Herceptin, a new breast cancer drug, is a monoclonal antibody.[543,544] Monoclonal antibodies could allow chemotherapy and radiation to be targeted more accurately. For example, a toxic drug may kill cancer cells by triggering cell suicide. If the drug is bound to a monoclonal antibody, the antibody could deliver the therapeutic agent directly to the tumour cell. Although this sounds like a powerful way to attack cancer, it is limited by the fact that many healthy cells possess the same antibodies as cancer cells.

Tumour-associated antigens, found on the surface of cancer cells, can also appear on healthy cells, although they may be expressed in larger levels on the surface of cancer cells. These antigens can serve as targets for antibody attack. Occasionally, the body's immune system can attack the cancer via the tumour-associated antigens, but this can produce unwanted side effects. The antibodies react with the antigen and damage or kill the cancer cells. However, if healthy cells have the same antigens, they also can be attacked and damaged by the antibodies.

One cancer that often carries a tumour-specific antigen is malignant melanoma, which occasionally undergoes spontaneous regression. This suggests that, in some circumstances, an immune response may cure such cancers. The most frequent explanation for this is that the immune system has been stimulated to destroy the cancer. However, an immune response is not the only explanation for spontaneous remission. Many other mechanisms exist for killing cancer cells. Indeed, as we shall see, a relatively small rise in vitamin C intake might destroy some susceptible cancers. Unfortunately, while this could explain rare spontaneous remissions, a small increase in vitamin C intake would be an ineffective treatment for the majority of cancers.

Abnormal antigens, if released into the blood stream, can be used as a diagnostic test. Certain tumour-associated antigens are produced by developing embryos. Later in development, these proteins cease to be generated. However, cancer cells can reactivate the proteins, which then distinguish the cancer cells from healthy cells. Carcinoembryonic antigen (CEA) may be elevated in people with cancer of the gut, and sometimes in lung, ovarian, breast, pancreatic or stomach cancer.

Previously, we discussed John Beard's trophoblast theory, which suggested that cancer cells originate from cells used in the development of the placenta. Consistent with Beard's ideas, the hormone human chorionic gonadotrophin (HCG) is produced by both trophoblast and cancer cells.[545] HCG is made by common cancers; it may play a central role in some features that effect the progression of cancer.[546] In the rat, cancer cells vary in the amount of HCG they express and this correlates with the aggressiveness of the tumour.[547] In mice, immunisation against HCG can slow the development of tumours.[548]

Some studies suggest that cancer is a problem of development and differentiation. Synthesis of HCG is a common biochemical feature of cancer.[549] There may be a relationship between human cancer cells that metastasise spontaneously and expression of the HCG gene. This relationship may indicate that the HCG molecule is an indicator of metastatic cancer.[550,551] Vaccination against HCG has been proposed as a treatment,[552,546] and initial experiments have shown benefits.[553,554,555,556]

Bone marrow transplants

Bone marrow damage is a side effect of chemotherapy and radiation treatments. Marrow is found in the cavities of long bones and is rich in stem cells, which produce red and white blood cells, along with platelets for blood clotting. Both chemotherapy and radiation can decrease the synthesis of blood cells and platelets, causing increased susceptibility to infection, abnormal bleeding and anaemia. Bone marrow transplants, also known as stem cell therapy, can be used to restore the production of blood cells. Complete destruction of the bone marrow, followed by transplantation, is occasionally used as a therapy in patients with cancer of the blood cells (leukaemia and lymphoma).

There are two common approaches to bone marrow transplantation. In an autologous transplant, the patient's own stem cells are sampled before therapy. After high levels of chemotherapy or radiation have destroyed the bone marrow cells, the harvested cells are injected into the blood, to help the bone marrow recover. If there is a cancer of the marrow, as in leukaemia or lymphoma, this approach may be compromised because the sample may contain diseased cells. In these cases, the marrow sample must be purged of cancer before it is returned.

The second form of transplant is known as allogenic. This uses donated stem cells, from a person with a matching tissue type. This form of transplant brings the risk of mismatch between tissues. Because of the danger of this complication, allogenic transplants may be less favoured.

Enzyme therapy

Once a person is born, pancreatic enzymes, in addition to their normal digestive function, could theoretically provide an endogenous cancer surveillance system. As described previously, Beard claimed that cancer arose from germ cells that had strayed into other tissues and become trophoblastic during their migration. Cancer, Beard claimed, was "a natural phenomenon, not a disease". According to this model, it becomes a disease when the surveillance system fails, through a deficiency of pancreatic enzymes.

Beard claimed that pancreatic enzymes, derived from the baby and mother, control the growth of trophoblast cells.[557] Reports from the

1960s appear to show that animals fed pancreatic enzymes show an improved response to cancer. However, a recent study of the effect of oral pancreatic enzymes in mice suggested that the improved survival was a result of better nutrition.[558] Clearly, such nutrition would need to be more beneficial to the mice than to the cancer, in order to be effective.

Oral enzyme treatment is questionable, as the human digestive tract is excellent at destroying and denaturing proteins. Orally administered enzymes might simply be digested. The use of insulin by diabetics illustrates the general physiological ineffectiveness of oral proteins. Insulin is a protein, secreted by the pancreas, which helps control the level of glucose in the blood. In early onset diabetes, the cells that produce insulin are missing or inactive, so insulin is not produced. In patients with this condition, a degree of control of blood glucose can be obtained by injecting insulin. Eating pancreas or taking insulin orally, even in large amounts, is not an effective treatment.

To date, there is no theoretical foundation or solid mechanism for oral enzyme therapy. Beard believed it was necessary to inject the enzymes, to prevent destruction by digestion. It may be that some orally ingested pancreatic enzymes are absorbed into the blood stream. However, this mechanism has yet to be confirmed.

In 1981, Nicolas Gonzalez began studying the use of oral pancreatic proteolytic enzyme therapy, as a cancer treatment. Gonzalez describes 25 cases of people with cancer, who were believed to be unlikely to survive for an extended period. The patients following Gonzalez's treatment program experienced long-term survival or tumour regression. As a result, Michael Friedman, an Associate Director at the US National Cancer Institute, suggested a pilot study, for patients with pancreatic cancer. Survival from pancreatic cancer is relatively rare, so a small study over a short interval was appropriate.

The study began in 1994 and involved 11 patients.[559] The patients had late stage cancer. Nine patients survived one year, five survived two years, and four were still alive three years later. These results greatly exceeded the 25% expectation of survival at one year and 10% survival at two years. This limited study suggests that pancreatic enzymes, together with aggressive nutritional therapy, increased survival over the normal expectation for pancreatic carcinoma. Rigorous experiments are required to replicate or refute these suggestions.

Shark cartilage

In 1992, William Lane and Linda Comac published a book called *Sharks don't get cancer*, which claimed a breakthrough in preventing and treating cancer.[560,561] Lane claimed that sharks almost never get cancer. The reason given was that sharks do not have a bony skeleton but are cartilaginous. In the body, cartilage has few blood vessels. Furthermore, they claimed, cartilage in sharks can inhibit the growth of blood vessels and thus prevent the growth of large tumours. This effect was shown when blocks of cartilage, implanted into chicken eggs, eyes,[562] or near tumours, appeared to inhibit blood vessel growth.[563] According to Lane, extracts of shark cartilage taken orally or by injection would cure cancer.

The idea that sharks never get cancer is a strong hypothesis, in that it is easy to test, but it is wrong. Some sharks do get cancer.[564] Despite the observation that cartilage can inhibit growth of new blood vessels,[565] aggressive cancers can invade connective tissues such as cartilage. Sarcomas are cancers that commonly arise from connective tissues. Sarcomas are not rare in humans and can occur in sharks.

It may be true that sharks get relatively few cancers, but not necessarily for the reasons given by Lane. Most wild animals have a relatively short lifespan and die before they have time to develop cancer, a disease associated with aging. Sharks are fish and have a different metabolism and metabolic rate to mammals. It is therefore not appropriate to compare the rate of cancer development in sharks with that in humans. A more appropriate question is whether sharks get fewer cancers than other cartilaginous fish; however, there is insufficient evidence to make such an assertion with confidence.

Lane does not identify the components in shark cartilage that allegedly make it special. The agents appear to be proteins.[566,567,568] Indeed, beef cartilage has a similar anti-angiogenic effect, as popularised by John Prudden.[569] Lane argues that sharks have a higher proportion of inhibitory substances, which prevent blood vessel growth. Shark cartilage is expensive to obtain and many sharks are endangered species. Lane's interest in sharks seems to arise from his background in marine resources: he was looking for a reason to exploit sharks commercially. The research on cartilage and cancer did not originate with Lane, but with scientists studying basic mechanisms of blood vessel growth.

There is a large gap between the demonstrated fact that cartilage will inhibit blood vessel growth locally and administration of the cartilage systemically, orally or by injection. Lane's proposal would be more convincing had the substances in cartilage been isolated and identified, a process that he appears to discourage.[560] Animal experiments on oral shark cartilage give mixed results.[570,571] Preliminary clinical trials of oral shark cartilage in humans have also had mixed success[572,573] and a recent clinical trial reported no benefit.[574]

Anticancer factors in cartilage may work by mechanisms other than prevention of blood vessel growth, such as impeding cell migration or inhibition of collagen breakdown.[575] Water-soluble extracts of cartilage contain several molecules that prevent blood vessel growth,[576] which may act during wound healing when given orally in humans.[577]

Sharks are a good research animal for studying the inhibition of blood vessel growth. This research may lead to effective treatments for cancer.[578,579] Although the idea that use of shark cartilage is pseudoscience may be putting it a bit strongly,[580] the suggestion that oral intakes of purified cartilage will prevent or cure most forms of solid cancer is currently unjustified. Any hypothesis should be tested, rather than simply discounted. However, based on current evidence, the authors would not take shark cartilage tablets if they had cancer.

Laetrile

Laetrile is a derivative of apricot seeds, which was first described as a chemical substance in 1837. In the 1920s, Ernst Krebs theorised that laetrile could kill cancer cells. Later his son, Ernst Krebs Jr., suggested it was a missing vitamin, vitamin B_{17}. However, it does not meet the minimum nutritional requirements for vitamin status.

In fact, the substance that has been used in most treatments is a related compound, called amygdalin, which is extracted from apricot seeds or almonds. Both substances contain a cyanide group and, under specific circumstances, this can be released. It has been proposed that the cyanide is released into cancer cells rather than normal cells, producing a selective anticancer effect. However, there is little supporting evidence for this.[581,601] The use of laetrile is strongly discouraged by the US Food and Drug Administration and other medical authorities.

Evidence for the use of laetrile is weak. It has failed to produce significant cancer inhibition in cell culture,[582] and animal experiments.[583,584,585] Amygdalin can kill some leukaemia cells in culture.[586] In other experiments, using a rodent model, amygdalin failed to inhibit cancer growth in four types of tumour.[587] A similar study, of three tumour types, also reported no anticancer effect.[588] Another study, which used melanoma and leukaemia cell lines in mice, failed to give positive results.[589] The animal experiments were reported as not providing sufficient evidence to merit going forward with human clinical trials.[590]

A comparative review of 68 cases treated with laetrile found two complete and four partial responses to therapy, an inconclusive result.[591] This was followed by preliminary clinical trials. These trials were undertaken by Charles Moertel of the Mayo Clinic. Moertel also carried out a series of flawed clinical trials on vitamin C and cancer, suggesting, incorrectly, that the vitamin was ineffective.[15] Moertel performed an uncontrolled study on 178 patients with cancer. He stated that laetrile was ineffective in extending life or reducing symptoms, but that it showed toxicity. He concluded that laetrile was a toxic drug and ineffective as a cancer treatment.[592]

Laetrile has been linked to serious side effects. In one study, six of 10 dogs fed laetrile and almonds died.[593] In humans, death from cyanide poisoning is a concern with this treatment.[594,595,596,597,598] Although laetrile preparations contain little free cyanide,[599] laetrile can be broken down to hydrogen cyanide and other substances, by bacteria in the gut.[600,601] In cancer patients, oral doses result in higher levels of cyanide in the urine than do injected doses.[602] By contrast, injected doses of laetrile can be excreted unchanged.[603] A further reported side effect is that oral laetrile causes congenital malformations in pregnant hamsters.[604]

Laetrile has been denounced as unscientific quackery,[605,606,607] although such an assertion is unjustified. A recent investigation showed that specific delivery of amygdalin and the enzyme that breaks it down to produce cyanide might be accomplished, using monoclonal antibodies.[608] If this could be achieved, research on laetrile might ultimately lead to a practical method for selective elimination of cancer cells.

Coffee enemas

Coffee enemas are, perhaps, the most ridiculed of alternative cancer therapies. However, from around 1890 to 1977, this therapy was covered in the Merck Manual,[609] a respected source of conventional medical information. Coffee enemas may have been removed from the manual for reason of space rather than merit.[a] In addition, nursing texts recommended coffee enemas for about a century.

---

[a] Dr Nicholas Gonzalez reports a correspondence with the editor, confirming this reason for removal.

Patients report an improved feeling of health and well-being following a coffee enema. This feeling is thought to be associated with relaxation, rather than being a result of caffeine intake. Such enemas may stimulate improvements in liver and gall bladder function.[610,611] Cancer patients could find them beneficial to help remove the toxic products of cancer metabolism. Ridiculing coffee enemas may simply reflect prejudice. However, there does not appear to be evidence for this therapy being more than a minor benefit to cancer patients.

Diet-based therapies

The Gerson therapy is a well-known and controversial dietary cancer treatment, which also involves coffee enemas. In one study of melanoma, the five-year survival rates for Gerson compared well with those reported for conventional therapy.[612] However, this method of comparison is fraught with difficulties, as factors other than the treatment could explain the results. For example, conventional therapy could have reduced the life expectancy of subjects in this study, rather than the Gerson therapy increasing it.

The Gerson therapy is based on a diet of raw fruit and vegetables, with emphasis on consumption of fresh juices. Presently, we will show that a diet of raw vegetables may be a powerful way of controlling cancer. The effects of a raw vegetable diet may be mediated by generation of hydrogen peroxide in cancer cells.

Hydrogen peroxide

"We pretend that modern medicine is a rational science, all facts, no nonsense, and just what it seems. But we have only to tap its glossy veneer for it to split wide open, and reveal to us its roots and foundations, its old dark heart of metaphysics, mysticism, magic and myth." Oliver Sacks

Hydrogen peroxide (H_2O_2) is a heavy, colourless, strongly oxidising liquid. Louis-Jacques Thenard discovered this powerful antiseptic in 1818. When applied to a wound, hydrogen peroxide releases oxygen bubbles. Hydrogen peroxide can kill many kinds of micro-organism, especially those that are anaerobic. However, most aerobic bacteria can destroy hydrogen peroxide, and extracts of plant and animal tissues can break it down to water and oxygen.

Hydrogen peroxide is widespread in the environment; there are even low levels in rainwater. Common drinks contain hydrogen peroxide, including black tea (50 μM/L) and coffee (20-160 μM/L).[613] It is also present in honey, which is sometimes used as a dressing for wounds. The antibacterial activity of some types of honey is similar to those of high concentrations of sugar, but others kinds are 100 times more potent. This high antibacterial activity may be a result of the honey's slow release of hydrogen peroxide.

Almost every disease process involves hydrogen peroxide.[614] It is a core molecule in cell signalling and the control of cancer cell division. Most cells contain enzymes to break hydrogen peroxide down to oxygen and water. These are needed, because low concentrations of hydrogen peroxide are produced during metabolism and are a necessary part of cellular communication.

The use of hydrogen peroxide as a treatment for cancer was proposed soon after the substance was discovered and isolated. Unfortunately, while this substance can kill cancer cells, it is also poisonous to healthy tissues, particularly at high concentrations. Cancer cells generally produce higher levels of hydrogen peroxide than normal cells.[615] Hydrogen peroxide may provide a core mechanism for many conventional anticancer drugs.[616,617]

The use of hydrogen peroxide as a treatment for cancer is highly controversial.[618] However, despite a lack of acceptance by the medical establishment, it continues to be used. There is little doubt that it can be an effective anticancer agent. Claims that it has not been proven or has been shown to be ineffective are inaccurate.[a,619,620] Hydrogen peroxide was rejected as a cancer treatment prematurely, indicating a lack of biological insight. Notably, hydrogen peroxide is not patentable and hence has little potential for profit, factors that weigh against research in this area.

If we wish to discount a potentially powerful treatment for a severe disease, we should demand strong evidence that it offers no benefits. Biologically, it is disturbing that the role of hydrogen peroxide has not been properly researched. Despite this, the supporting evidence for its use is substantial. For example, many conventional treatments depend on increasing levels of oxidants, such as hydrogen peroxide, inside cells.

For decades, specialised physicians have used hydrogen peroxide as a treatment for cancer. Our own interest was partly stimulated by Reginald Holman, who provided us with accounts of his successful long-term treatment of several patients. Physicians like Holman have used it intravenously, even injecting it into the arterial supply of affected organs. Not surprisingly, treatment of cancer with such a toxic substance requires great care.

Hydrogen peroxide should only be considered as a cancer treatment under the supervision of a knowledgeable physician. It is a powerful oxidant; therefore, it should not be taken without appropriate medical advice and supervision.[b] Unless diluted sufficiently, oral doses may cause chemical burns. Intravenous doses can release oxygen bubbles, as the hydrogen peroxide is broken down. Such bubbles could block blood vessels, causing tissue anoxia and death. Blocking a vessel in the heart could lead to cardiac arrest and blocking one in the brain could cause a stroke.

Despite these risks, advocates promote it as a self-cure method for a range of diseases, including cancer. Such support for alternative treatments reflects medicine's lack of progress in fighting these diseases. It may also be a sign that the public disapprove of orthodox medicine's disregard for promising research. Self-treatment usually involves cancer patients drinking dilute hydrogen peroxide, in the hope of increasing

---

[a] Claims based on lack of scientific "proof" are a refuge for scoundrels. There is no such thing as proof in science. It is worth noting that the paper referenced here, claiming to refute hydrogen peroxide, is *anonymous*,[619] although it names the proponents of oxygen therapies.

[b] No treatment of cancer should be undertaken without medical supervision.

their chances of survival. While such an approach is fraught with danger, if the alternative is a death sentence, some may choose this option.

This direct use of hydrogen peroxide is unfortunate. There are alternative methods to increase its concentration within cancer cells, without poisoning normal tissues. These derive from the early ideas of hydrogen peroxide as a cancer treatment and are safer than current cytotoxic chemotherapy. Moreover, they are appropriate for terminal conditions, when conventional treatments offer no hope.

Simple chemistry

Hydrogen peroxide has been used as a bleaching agent since the early 1800s. It is produced in industrial quantities for use in diverse applications, such as bleaching paper or treating water. The core chemistry of hydrogen peroxide is remarkably simple. This small molecule is a combination of water and oxygen. A molecule of water, H_2O, combines with an atom of oxygen, O, to give hydrogen peroxide, H_2O_2. Hydrogen peroxide can split, to release both water and oxygen:

$$2H_2O_2 \rightarrow 2H_2O + O_2$$

The properties of hydrogen peroxide are interesting. The above equation is correct, although it does not include the breakdown of hydrogen peroxide to form highly reactive oxygen species.[c] These free radicals can cause rapid oxidation damage in cells and tissues.

Hydrogen peroxide is used by white blood cells for immune defence. It is also involved in the breakdown of chemicals in the liver and in the metabolism of fats. It is highly poisonous, although it is manufactured by both animal and plant cells. Normal metabolism produces hydrogen peroxide, causing damage to mitochondria and possibly contributing to aging. Cells have a host of antioxidant mechanisms to defend against such damage, but the protection is incomplete and, over time, damage accumulates.

As a controversial cancer treatment, it is taken orally at low concentrations, about 0.5%. The quality of the source is said to be important. For example, hydrogen peroxide obtained from the local pharmacy may include chemical stabilisers, which are alleged to interfere with its action. In higher concentrations, about 3% in water, it is used

---

[c] Hydrogen peroxide damages cells by breaking down into superoxide, the hydroperoxyl radical and the vicious hydroxyl radical. An example breakdown is the Fenton reaction with iron: $H_2O_2 + Fe(II) \rightarrow Fe(III) + OH^- + {\bullet}OH$.

medicinally as an antibacterial agent for cleaning external wounds. At about 5% concentration, it is used as bleach for human hair, leading to the term "peroxide blonde".

At still higher concentrations, industrial applications are more common. Applications in medical laboratories and cleaning of computer chips use concentrations around 30%. A concentration of 35% is used for cleaning industrial waste. At concentrations above 90%, it can be used as a rocket propellant. However, high strength hydrogen peroxide is dangerous to produce and handle. For instance, if a concentrated solution of hydrogen peroxide dries out on a laboratory bench, it could spontaneously ignite and cause a fire.

Energetic reactions

High-test peroxide (HTP) is a rocket fuel, consisting of hydrogen peroxide at high concentration. HTP is used as the oxidant in some rocket motors and has advantages over liquid oxygen and other alternatives. Towards the end of World War Two, it was used by the Germans in the Messerschmitt Me163 rocket-powered interceptor fighter. Hydrogen peroxide was also used for the turbine for the motor pump of the V2 Rocket, used to attack London. Following the war, HTP was used in Britain to fuel both rockets and specialised fighter planes. In order to react quickly in a rocket motor, HTP needs a catalyst to speed the reaction. In the German V2, calcium permanganate was used, resulting in a dirty exhaust. In later engines, a silver mesh was used to catalyse the reaction.

A UK-built rocket interceptor fighter, the Saunders Roe SR53, was powered by HTP. The prototype first flew on 16 May 1957 and went supersonic on 15 May 1958. However, rocket planes have problems with endurance and the SR53 was scrapped, as the English Electric Lightning jet fighter provided an alternative. A plan to increase the Lightning's Mach 2 performance using a HTP-based rocket was proposed, but was cancelled before implementation.

We have described the use of highly concentrated hydrogen peroxide as a rocket fuel to illustrate the potentially high reaction rates of this substance. In the body, the concentration of hydrogen peroxide is kept at low levels by antioxidant enzymes, particularly catalase.

Catalase

In many cells, the breakdown of hydrogen peroxide is facilitated by the action of enzymes, such as catalase. Although hydrogen peroxide is a normal constituent of the body, it is damaging and needs to be detoxified. Catalase is a haem protein, a type of protein that contains iron and is associated with the use of oxygen by the body. Haemoglobin, which transports oxygen in the blood stream, is the best known of these. A related oxygen-binding protein, myoglobin, occurs in muscle. Catalase and peroxidase are haem proteins that have antioxidant actions within the cell.

Raised levels of catalase can lead to prolonged life. The maximum lifespan of mice that have been genetically engineered to express additional catalase increases from 35 months to 40 months.[621] Differences in catalase levels may partly explain the variability in human life expectancy. Shorter-lived humans may have genes for lower levels of catalase, making them vulnerable to higher levels of damaging hydrogen peroxide.

Catalase developed early in evolution, as a way to prevent oxidation damage. This is reflected in its widespread incidence, throughout the plant and animal kingdoms. In 1901, Loew named catalase and showed that it was active, even in low dilutions.[622] He thought, incorrectly, that hydrogen peroxide was too reactive to be found in living cells.[623] He also suggested that the purpose of catalase was to protect the body from the poisonous effects of hydrogen peroxide.

Around 1921, McCloud and Gordon suggested that bacteria could be classified according to their use of catalase and their sensitivity to hydrogen peroxide. Such a classification would place hydrogen peroxide at the centre of microbiology, reflecting its fundamental importance to life on earth. Strictly anaerobic bacteria are killed by oxygen and generally contain no catalase. When oxygen is present, such bacteria can produce toxic levels of hydrogen peroxide and die.

Aerobic bacteria contain high levels of catalase and other antioxidant enzymes, which inactivate hydrogen peroxide. Holman suggests that hydrogen peroxide is never neutral for a cell: its presence is either desirable or undesirable.[623] Catalase is a relatively non-specific enzyme, and will catalyse the oxidation of many chemicals, including hydrogen peroxide.[d,624] This non-specific peroxidase action may be important for cells. Holman suggests that when catalase and hydrogen peroxide are out of balance, all manner of malfunctions may occur,

---

[d] The complete reaction mechanism is not yet known but occurs in two stages: H_2O_2 + Fe(III)-E → H_2O + O=Fe(IV)-E and H_2O_2 + O=Fe(IV)-E → H_2O + Fe(III)-E + O_2. Where Fe-E represents the iron centre of the heme group attached to the enzyme.

driven by internal oxidation. This suggestion relates hydrogen peroxide and catalase to the free radical theory of disease.

Holman was interested in cancer, which he claims is intimately related to the catalase/hydrogen peroxide balance within the cell.[625] In modern terms, we can extend Holman's suggestion to include down-regulation of antioxidant enzymes, such as catalase. When such enzymes are in short supply, oxidants and free radicals can accumulate, leading to production of a generally oxidising environment. Our discussion of catalase in cancer cells can be taken to encompass other antioxidant enzymes, such as superoxide dismutase and peroxidase.

Catalase protects the cell from oxidation

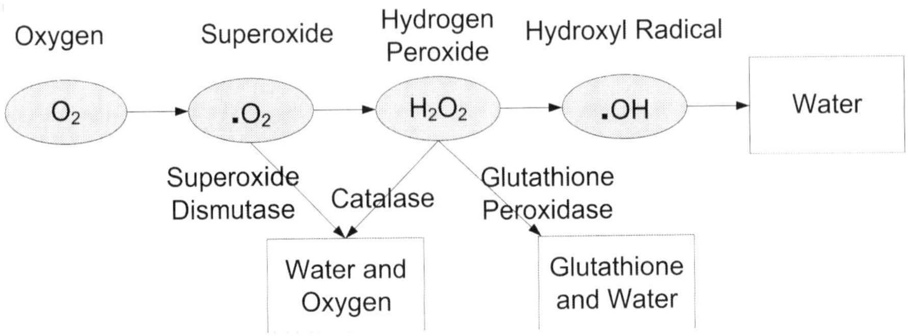

This simplified diagram loosely illustrates the relationship between hydrogen peroxide and catalyse in the cell. Hydrogen peroxide, which can induce oxidation through the hydroxyl radical, is converted safely to hydrogen peroxide and water by catalase.

Catalase and cancer

Holman believed that studying the processes within cells was more likely to be fruitful in understanding cancer than investigating the nature of chemical carcinogens. This follows from the work of Nobel Prize winner, Otto Warburg, who argued that cancer cells derive their energy from anaerobic biochemistry. This is in contrast to healthy cells, which gain most of their energy by using oxygen.

The Krebs cycle is the dominant metabolic pathway for glucose oxidation, although there are several alternative mechanisms. Ultimately, glucose is combined with oxygen and carbon dioxide is released.[626] However, before glucose enters the Krebs cycle, it goes through a series of reactions called glycolysis, which provide energy without the need for oxygen. Muscle cells can use glycolysis to generate energy during vigorous exercise, when oxygen is short. This anaerobic pathway bypasses the Krebs cycle and generates lactic acid.

Warburg argued that the reversion of cancer cells to anaerobic respiration caused them to grow malignantly. If this were true, then all cancers would have a defective respiration and would not use oxygen for energy. In support of Warburg's observations, scientists reported that cancer could be induced in rat fibroblasts by depriving them of oxygen for intermittent periods.[625] However, others have shown that fibroblast cells can survive under these conditions without becoming malignant. Furthermore, immortal cell lines, derived from cancer cells, can survive for long periods in relatively high oxygen concentrations. It seems clear that although anaerobic metabolism may be present to a degree in many cancer cells, this is a result of the microevolution of malignancy, rather than its primary cause.

There are many possible causes of cancer. Any mechanism that increases error-prone cell multiplication and decreases cell death could cause cancer. Cell division and controlled cell death are complicated events, which numerous mechanisms could disrupt. However, an increase in free radicals, including hydrogen peroxide, is intimately involved in both cell division and cell death.

Holman considered that cells low in catalase, which destroys hydrogen peroxide, would be in danger of becoming malignant. By contrast, cells rich in catalase would be resistant to cancer-causing agents. He also suggests that metastases might occur preferentially in tissues with high levels of catalase, such as the liver. Cancer cells, which are deficient in catalase, might be better suited to invade an environment high in catalase, as this would provide protection against local hydrogen peroxide.

Genetically similar rats, kept in the same environment, have different cancer cure rates, because of biological variation. Individual animals also differ in their response to hydrogen peroxide, as do different body tissues. Holman suggested that variations in the concentration or activity of catalase might provide an explanation for these rats' varying resistance to cancer. This is an intriguing idea, which could throw a light on the phenomenon of spontaneous remission.

Occasionally, cancer patients get better for no apparent reason. One possible mechanism is that the patient has consumed an innocuous catalase inhibitor, which has increased the level of hydrogen peroxide sufficiently to kill many of the cancer cells. The immune system could then complete the task. Alternatively, the body's production of catalase might have been decreased, with the same effect. Holman's hypothesis suggests that an interesting study would be to measure hydrogen peroxide and catalase levels when such rare spontaneous remission events occur. This could provide a window for scientific observation of a medical phenomenon that is particularly difficult to examine.

Based on both theoretical considerations and observation, Holman developed the idea that hydrogen peroxide would provide a useful treatment for cancer. He noted that oxygen therapy increases the effectiveness of radiation against cancer. Radiation therapy was claimed to reduce whole body catalase, increasing sensitivity to the toxicity of hydrogen peroxide. If cancer cells contain little or no catalase, then increasing the concentration of hydrogen peroxide should kill these cells preferentially. Healthy cells that contain catalase would be relatively protected from peroxide damage. He tested two subjects, suffering from advanced, inoperable cancer, with 0.2% oral hydrogen peroxide, and found they showed clinical improvements.[627]

Hydrogen peroxide and microevolution

Hydrogen peroxide plays a central role in the body's defence mechanisms: specialised cells produce large amounts of the substance. White blood cells generate oxidants, including hydrogen peroxide, to destroy foreign bodies such as bacteria. Cancer cells can also generate large amounts of hydrogen peroxide.[628]

One mechanism employed by many cancers to increase cell division may be to reduce levels of antioxidant enzymes, thus increasing levels of oxidation.[629,630] In tumours, however, antioxidants either inhibit or promote growth, by modifying the redox environment.[631] Tumour growth depends on cell proliferation being greater than losses from cell death. Cells with the highest oxidant levels are subject to selection by apoptosis, because levels of oxidants, such as hydrogen peroxide, are directly involved in signalling cell suicide.[632] In test tube experiments, persistent oxidative stress has been shown to result in cellular changes, which make the cells highly resistant to hydrogen peroxide.[633,634] As a tumour advances, the level of oxidants may fall, making the cancer more resistant to cell death.[635] A possible explanation is that the cells most sensitive to oxidation-induced suicide have been eliminated by selection pressure, induced by the high levels of oxidants.

Hydrogen peroxide may provide a cancer with a primary control for its development. In healthy cells, hydrogen peroxide levels are low and the small amounts produced are held in check by antioxidants. Within these cells, antioxidants reduce the high oxidant levels that cause cell division. This lessens the risk of microevolution to cancer. A reducing environment may also hinder cell division in early stage cancer cells, by the same inhibitory mechanisms. For healthy cells, or cells undergoing early cancerous changes, antioxidants such as catalase generally inhibit cell division, which involves increased oxidation.

Malignant cancer cells do not react to antioxidants in the same way as healthy cells. High levels of hydrogen peroxide signal the cancer cell to divide. Since redox signalling is a controlling factor in cell division, a raised level of oxidants may be a driving factor for cancer growth.[636,637] Hydrogen peroxide signals also control the energy generation required for growth.[638] The substance also inhibits tumour suppressor and other genes, facilitating abnormal cell growth.[639,640,641] The resulting increase in oxidation enhances cell division. However, it also means the cell is nearer to triggering suicide by apoptosis.

Metaphorically speaking, cancer cells are treading a fine line between increasing their oxidation levels enough to divide rapidly and increasing it too much, so it triggers apoptosis. Surprisingly, antioxidants can tip the balance in favour of the cancer, by reducing the oxidation level just enough to prevent malignant cells from committing suicide. In rapidly dividing cancer cells, hydrogen peroxide levels can be high, close to cytotoxic values. With such cells, antioxidants can lower the high levels of oxidant stress. This lowering of stress could, in this case, inhibit mechanisms that lead to malignant cell death,[642,643,644] and thus promote growth.

According to the microevolutionary model, selection pressure in the development of cancer cells favours high levels of oxidation. Changes in redox signalling promote cell division, along with increased oxidation and free radical damage: important factors for evolving cancer cells. In order to achieve this raised oxidative state, the precancerous cell must inhibit cellular antioxidants. A side effect of this inhibition is that resultant cells are unable to detoxify hydrogen peroxide, as they are deficient in catalase and other antioxidant enzymes.[e,645]

---

[e] A few cancers display an increase in some antioxidant enzymes, particularly manganese superoxide dismutase.

An immediate implication of the microevolutionary model is that dietary and other antioxidants should lower the risk of developing cancer.[646] However, once a cancer has been established and turned malignant, a small intake of antioxidants could be detrimental, as it might lower the rate of apoptosis. Despite this, large intakes of specific antioxidants may have a paradoxical action. If their levels are sufficiently high, antioxidants such as vitamin C can act as oxidants in cancer cells, leading to raised levels of hydrogen peroxide. Cancer cells, as we have seen, are unable to neutralise hydrogen peroxide, so are destroyed.

The use of dietary supplements that can act as antioxidants in healthy cells, and as oxidants when killing malignant cells, provides a fundamentally new approach to the treatment of cancer.[131] We will now examine the science behind these suggestions and the implications for the treatment of this disease.

Vitamin C

"If an elderly but distinguished scientist says that something is possible he is almost certainly right, but if he says that it is impossible he is very probably wrong."
Arthur C. Clark

The aim of cancer research is to find a safe agent that can destroy a wide range of cancers. Recently, we described the action of vitamin C (ascorbate) in such terms.[15] In this chapter, we revise and update our earlier description.

An effective treatment must either prevent cancer cell proliferation or increase cancer cell death.[1] Vitamin C can do both. Evidence suggests that high dose Vitamin C eradicates cancer cells selectively, while leaving healthy cells unharmed.[216,217] In fact, rather than damaging normal cells, the vitamin's antioxidant properties may improve the patient's health.

Vitamin C belongs to a class of chemicals whose importance is only just being realised: effective cancer treatments that have few side effects. The idea that vitamin C could provide a useful treatment for cancer originated over half a century ago.[647] Since then, hundreds of research papers, including laboratory, animal and human studies, have been published on the effects of ascorbate on cancer. The evidence continues to accumulate; for example, Mark Levine's group at the US National Institutes of Health (NIH) recently confirmed reports of the selective cytotoxic action of vitamin C on cancer cells.[648] Levine's report indicates that the cytotoxic actions of vitamin C are no longer considered controversial.

Vitamin C is a widely used nutritional supplement. It has been the subject of large claims, from Linus Pauling and others, as well as numerous attacks by government authorities. The medical establishment's attitude towards vitamin C is, at best, ambivalent. For example, a book on medical blunders includes the establishment's response to Linus Pauling's proposals on vitamin C.[649]

The story of vitamin C is one of the most fascinating in modern medical science.[15] Humans, unlike most other animals, do not synthesise vitamin C in their bodies. As a result, people need to obtain vitamin C from their food to avoid the disease scurvy. For decades, there has been a controversy about the appropriate levels of intake for optimal health.

Our current model, which supports high dose intakes, is called dynamic flow.'[15,470] According to this, frequent doses of vitamin C are required to maintain high blood levels, leading to excretion in the urine. Dynamic flow ensures the body is constantly in a reducing state. Repeated high doses provide a mechanism for humans to restore their physiology to a state close to that of most other animals, which synthesise the vitamin.

Low blood levels of vitamin C lead to sickness and ill health,[650] while higher levels are associated with prevention of cancer and general good health.[651] Dynamic flow level oral intakes or intravenous doses of sodium ascorbate may be beneficial in treating cancer.[652,675,676,693,694,695,697,698,701,703,704,710,711,712,713,721] Intravenous administration is generally more effective as a treatment for disease,[653] although it is possible that oral doses, sufficient to maintain dynamic flow, could be an effective treatment. Frequent, gram level oral doses can lead to sustained plasma concentrations, at up to 250 μM/L. Intravenous doses can produce higher, but transient, plasma levels in the millimolar range (i.e. above 1000 μM/L).[652] However, as we shall see, certain synergistic supplements increase the potency of vitamin C against cancer cells. These supplements may allow an effective oral therapy for many forms of this disease.

Further indications that vitamin C could be an effective treatment arise from the finding that ascorbate prevents cancer cells from growing. Ki Won Lee and colleagues, from the University of Seoul, have described a mechanism by which vitamin C stops cancer cells proliferating.[654,655] Normal cells stop growing when they receive signals from neighbouring cells; these signals are induced by hydrogen peroxide and other oxidants. Cancers can continue spreading, because they do not respond to the signals that inhibit growth in normal cells. Ki Won Lee showed that vitamin C enabled cancer cells to receive and act on such messages to stop growth.

Reduction and oxidation

Vitamin C normally acts as a reducing agent in the body. It is the main water-soluble antioxidant in the diet and the most important in the extracellular fluid. Despite these properties, under certain circumstances, it can also act as an oxidant. It shares this ability to act as either an oxidising or a reducing agent with many other dietary antioxidants, including vitamin E.

People critical of high doses have used the possibility that vitamin C acts as a pro-oxidant as a justification for not taking supplements.

However, both antioxidant and oxidant properties of vitamin C offer health benefits. In particular, the oxidant properties of vitamin C are believed to be a central feature of its anticancer action.[217,656] Ironically, the property that detractors use as a key point against the use of high doses of vitamin C could form the basis of one of its clinical benefits.

Ascorbate kills cancer cells

Ascorbate has powerful effects on the metabolism of cancer cells.[217,657] It has been shown to inhibit the growth of several cancer cell lines,[217,658] and is effective in inhibiting tumour growth in animal experiments.[659] Vitamin C kills cancer cells by oxidation and, more specifically, by generating hydrogen peroxide inside the cell body.[660,661,662,663,664,665,666] Furthermore, the hydrogen peroxide produced in this way breaks down to give additional oxidants, such as the hydroxyl radical, which can severely compromise cancer cells.[718]

In cancer, when vitamin C cycles between ascorbate and dehydroascorbate, hydrogen peroxide is produced.[667,853,668,669,670] These effects are increased by the presence of free iron and copper,[671] and a combination of vitamin C and copper has been suggested as a possible cancer treatment.[217] Cancer cells contain high levels of iron and some other metals.[672,673] As we have explained, the cells are short of antioxidant enzymes, including catalase, which converts hydrogen peroxide to oxygen and water.[674,675] Consequently, they are unable to detoxify the large quantities of hydrogen peroxide produced by high levels of vitamin C. This hydrogen peroxide damages and kills the cancer cells.

Intravenous administration

Intravenous ascorbate can kill cancer cells,[711,676,1028] but the blood levels obtainable with oral doses have generally been shown to be less effective. When vitamin C is given intravenously, high levels (up to about 15 mM/L) can be achieved. This is far greater than the concentration required to kill cancer cells. A thin layer of cancer cells may succumb to even a short exposure at levels of 1 mM/L.

For many years, there has been confusion about the difference between intravenous and oral doses of ascorbate. Intravenous doses of sodium ascorbate are reported to be more effective, consistent with the increased blood levels obtained. Following an intravenous injection, the high blood levels of ascorbate fall rapidly. Typically, therefore, the dose is infused over a period of several hours, to maintain blood levels.

The intravenous doses of sodium ascorbate required to be cytotoxic to cancer cells can exceed 100 grams. This is a massive dose, even when compared with the "megadose" or gram level intakes suggested by Linus Pauling and others. Typically, these massive infusions will be accompanied by an oral intake of more than 10 grams per day.

Oral doses

It is sometimes claimed that it is not possible to reach cytotoxic levels of vitamin C with oral intakes. However, this assertion is based on incomplete and inappropriate data. In a well-nourished individual, the background blood plasma level is about 70 μM/L. Above this concentration, ascorbate is rapidly excreted from the blood, with a half life of about 30 minutes. Large oral doses raise blood levels to a peak in about 2-3 hours, after which the level decays back to the baseline. Frequent oral doses can sustain blood plasma levels of, perhaps, 250 μM/L, in a dynamic flow. Such levels are toxic to some cancer cell lines.

Blood plasma levels of ascorbate

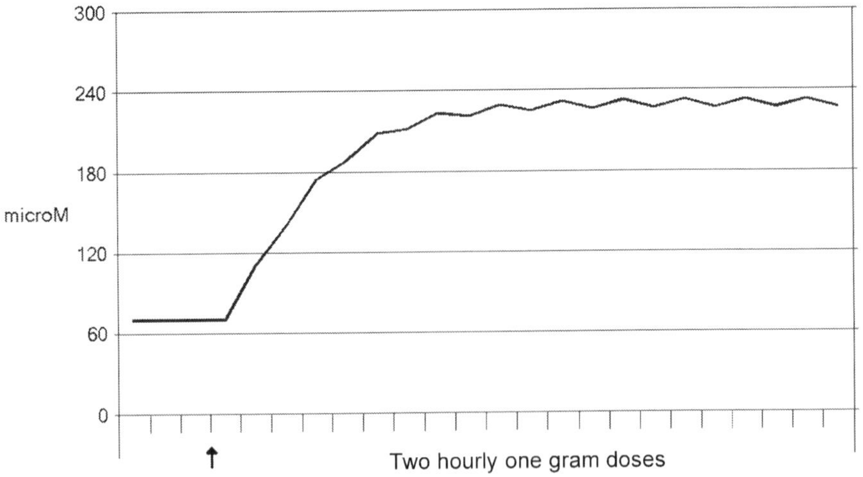

This graph shows computed plasma levels for gram level doses of ascorbate, taken orally, at two-hour intervals. The first dose is given in the third hour.

Repeated large oral doses can establish plasma vitamin C levels equivalent to those that, in test tube studies, kill some types of cancer when administered for even a short period. It is not yet known what the result of sustained high oral doses would be on cancer in humans. Oral doses have been reported to be effective in clinical cases of cancer.[15,581,677,] However, Robert Cathcart, a physician highly experienced in the use of vitamin C, reports that he has never succeeded in curing cancer with oral doses and warns against hubris in this area.[678] It is possible that dynamic flow intakes could be effective in destroying cancer, but this hypothesis has yet to be subjected to clinical trials.

Bowel tolerance levels, as described by Cathcart,[679] produce blood levels in the region of 250 μM/L; these can be sustained by repeated dosing at short intervals. In a healthy adult, bowel tolerance is typically reached by a single oral dose above about five grams. Intakes of approximately 20 grams per day, in frequent divided doses, can sustain these high plasma levels. In sick people, the bowel tolerance increases greatly. A mild cold can increase the tolerance level to above 50 grams per day, while influenza and other severe viral infections can raise the level to approaching 200 grams.

Uptake by cancer cells

Like people with infections, cancer sufferers have an increased bowel tolerance to vitamin C. This probably reflects higher requirements for ascorbate. Cancer patients' tissue vitamin C levels are lower than are those of healthy people. Patients undergoing conventional treatments have even lower values: the body's ascorbate reserves appear to be inversely related to the intensity of conventional treatment.[581]

Hugh Riordan's research group have described an interesting example, which relates to this finding.[680] A 70 year-old man from Kansas, with cancer of the pancreas, was given a 15-gram infusion of sodium ascorbate, over a period of one hour. Immediately following the treatment, his blood vitamin C levels were far lower than expected.[a] Our explanation for this finding is that cancer cells absorb and metabolise high levels of ascorbate.[659]

Cancer cells can absorb ascorbate by a different mechanism than is used by most healthy cells. Healthy cells take up vitamin C from the surrounding plasma, using biochemical pumps. Some cells contain

---

[a] The quoted value was 34 mg/dl compared with 120-200 mg/dl for a healthy subject. These figures seem to relate to measurements following intravenous infusion.

specific ascorbate pumps, while others absorb oxidised vitamin C, or dehydroascorbate, using glucose pumps.[681,682,683] Oxidised vitamin C is structurally similar to glucose, so can be transported by glucose pumps. Some specialised white blood cells take in vitamin C by oxidising their local surroundings, absorbing the resulting dehydroascorbate and then reducing it back to ascorbate.[2]

Tumours have a similar mechanism, which can accumulate high levels of ascorbate within cancer cells.[684] When glucose levels are low, tumours absorb more vitamin C. However, high levels of glucose inhibit uptake of both dehydroascorbate and ascorbate.[685] Cancer can inhibit active transport of the vitamin,[686] hence reducing its antioxidant effects. Despite this, cancer cells absorb higher levels of vitamin C than might be expected. The importance of the relationship between glucose and vitamin C, in cancer and its treatment, has been stressed by John Ely.[831]

A general feature of malignant tumours is that they are in an oxidising state. Cancer cells use the oxidising conditions to assist their growth and cell division. Consequently, when ascorbate molecules enter the tumour's environment, they become oxidised to dehydroascorbate and may produce hydrogen peroxide.[687] Glucose transporters in the cancer cells' outer membranes are then able to transport the dehydroascorbate into the cell bodies.[688] Moreover, cancer cells have a higher than normal dehydroascorbate transport rate.[689] In animal models, tumour cells that accumulate only oxidised ascorbate have been shown to take in the vitamin, rapidly.[690]

Malignant melanoma is one of the most aggressive of human cancers; it is derived from melanocyte cells in the skin. It requires large amounts of glucose to power its growth and activities. Like other cancers, melanoma cells have large numbers of glucose transporters. Some melanoma cells actively facilitate the uptake of the dehydroascorbate form of vitamin C.[691] Melanoma cells transport ascorbate with similar levels of efficiency to those of healthy melanocytes. By contrast, melanoma cells transport the oxidised form, dehydroascorbate, 10 times faster than melanocyte cells.[692] This increased rate of dehydroascorbate transport is achieved using glucose transporters. Melanoma cells can concentrate dehydroascorbate to levels 100 times greater than those in the surrounding medium.

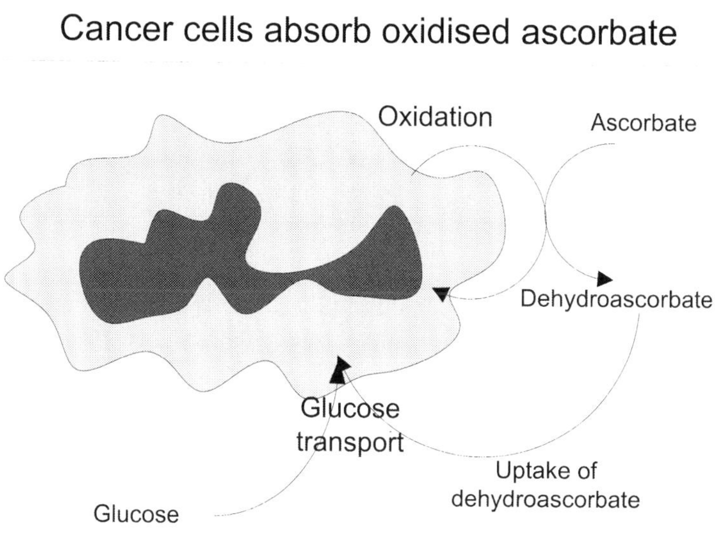

Diagram showing the uptake of dehydroascorbate by glucose pumps in cancer cells.

Clinical studies of cancer

The history of vitamin C as a cancer treatment is a catalogue of errors. Many experiments were poorly performed and lacked controls. Other researchers failed to give the vitamin C by injection, thinking an oral dose would be as effective. Where trials did use oral intakes, the doses were given infrequently and were not adequate to sustain the patients' blood levels.[15] Scientists on both sides of the debate have been subjected to personal abuse and had their motives challenged.[b] However, we are not interested in debate or argument, only in the facts.

The idea that vitamin C could be used to treat cancer was popularised in 1976, by Linus Pauling and Ewan Cameron, a Scottish surgeon. Cameron was a respected and established cancer specialist, who believed that the critical factor in recovery from cancer was the person's biological response. At first, he was sceptical that vitamin C could work against cancer. However, he realised that his patients had little to lose. Some were in the terminal stages and, even if vitamin C were ineffective, it would not do harm. He started giving 10 grams per day of vitamin C to patients with terminal disease, and convinced himself of the benefits.

[b] We considered whether to describe these allegations, but decided they were not pertinent to the discussion.

Initially, Cameron and Pauling published case reports of 50 patients, who were given sodium ascorbate injections, together with oral supplements. They increased the number of patients to 100, and found that cancer patients treated with vitamin C survived three to four times longer than untreated controls. The control group consisted of 1,000 patients, who received no vitamin C. They were matched (10 to 1) with the experimental patients, with respect to age, sex, type of cancer and clinical status of "untreatability". They were treated in the same hospital, by the same staff and were managed identically, except for the vitamin C.

Traditionalists argued that the selection of patients in Cameron's studies could have been biased. They also suggested that the vitamin C and control groups were not properly matched and might have had a different severity of disease. They further claimed that taking control patient details from medical records might introduce additional bias. Such a selection bias, as suggested by the detractors, might explain the positive results, but would need to be extreme.

Cathcart has raised another objection to the Vale of Leven studies, namely, that the dose of vitamin C was too small. Ten grams is at the low end of clinical doses and would not provide relief from a common cold. Animals that manufacture their own vitamin C get cancer. Since these animals often have high tissue levels of ascorbate, this sets a lower limit on the effective dose. Pauling and Cameron's positive findings for such low doses are therefore unexpected, but can be attributed to their choice of the intravenous route, which leads to higher tissue levels.

Pauling and Cameron provided an estimate of the benefits of vitamin C, at these relatively low doses.[581] These figures appear to derive from their experience of clinical case studies and apply to patients with terminal disease considered untreatable by conventional means. As we shall show, higher doses might be expected to give a greater response. The doses used by Pauling and Cameron were only a fraction of the doses normally found to be cytotoxic to tumours. However, we now know that tumours preferentially accumulate ascorbate.[c]

The response of Pauling and Cameron's patients to vitamin C therapy is indicated in the following table:

---

[c] Tumours preferentially take up L-ascorbate by active transport. This form of the vitamin, which is now becoming more available in supplement form, may therefore be a more effective cytotoxic agent.

Response	Percentage of patients
None	20
Minimal	25
Tumour retardation	25
Tumour unchanged	20
Tumour regression	9
Tumour death	1

The patients used to derive these figures would not have reached and sustained plasma levels of vitamin C likely to kill the cancers. Pauling and Cameron could have been conservative in their estimates of the number of patients who might benefit from vitamin C therapies.[15]

A later study was conducted from 1978 to 1982, in the same region of Scotland. This included 294 patients, treated with ascorbate, and 1532 controls. Patients received either vitamin C or palliative care, according to which doctor admitted them. Patients receiving vitamin C had an average survival period of 343 days, almost twice as long as the controls (180 days). Moreover, the supplemented patients appeared to have an improved quality of life. However, the controls were not subject to exactly the same conditions as treated patients. Cameron, being convinced of the efficacy of the treatment, was ethically unable to deny vitamin C treatment to dying patients. Differences in survival figures could reflect unconscious bias by physicians, or errors in diagnosis.

In 1979, Morishige and Murata published a report confirming the results obtained by Cameron.[693] In this Japanese study, the death rate for higher dose ascorbate patients was only one third that of patients receiving lower doses. They studied 99 patients, of whom 44 subjects received four grams of vitamin C or less per day and 55 received five grams or more. Patients receiving the low dose of vitamin C lived an average of 43 days. Those receiving five to nine grams lived 275 days and subjects receiving 10 to 15 grams lived an average of 278 days. Surprisingly, patients receiving the highest doses, 30 to 60 grams, lived an average of only 129 days. This period is three times longer than the lowest dose patients, but only half as long as the 5-15 gram groups. However, the highest doses were given to the patients with the most advanced disease, which could explain these findings.

The Japanese experiment appears to confirm Cameron's study, as the ascorbate treated patients lived far longer. Regrettably, like Cameron's, this study was not a double-blind trial. Detractors used this fact to suggest the study was invalid. Despite this objection, the groups with different dose levels acted as internal controls. Unless there was a large selection bias, this experiment confirmed that vitamin C prolongs life in patients with cancer. Nonetheless, the medical community found the results of the Japanese study questionable and did not accept its validity.[694] It appears that it is necessary for someone suggesting a new approach to cancer to provide comprehensive evidence before the treatment will even be tested.

Pauling's reputation as a world-leading scientist gave Cameron's results publicity and the medical establishment was forced to reply to his findings. The prestigious Mayo Clinic decided to conduct its own controlled trial.[695] However, some writers have suggested that their real aim was to quash the claims for vitamin C as a cancer treatment.[696] Linus Pauling himself thought the Mayo Clinic studies were an example of scientific fraud.[827] The Mayo Clinic study did not answer the problems raised, but fuelled further controversy.[697,698]

In a later study, concerning the anticancer effects of laetrile, Moertel would write,

> *"it would be unconscionable to randomise people between a drug and a standard therapy that would hold a known potential for cure or [life] extension."*[699,700]

This is the same argument used by Ewan Cameron to defend his uncontrolled clinical trials of vitamin C.[d] Such inconsistency supports Pauling's questioning of the motives behind Moertel's investigation.

The Mayo Clinic study was prospective and double-blind, with randomised controls. Ten grams of vitamin C were given *orally* to the subjects, and a placebo to the controls. A single dose of this size would raise blood levels only transiently. The 63 controls matched the 60 supplemented subjects. Both groups survived for about the same length of time, seven weeks. The study concluded that there was no benefit

---

[d] We ignore the fact that there is poor evidence for Moertel's assertion of chemotherapy being curative or extending life expectancy, at least for the majority of solid tumours in adults.

from vitamin C. Following numerous complaints about the adequacy of the initial clinical trial, the Mayo Clinic conducted another study, to confirm the negative result.[701] This study also used oral supplementation and an inappropriately low and infrequent dose. Despite repeated requests, the Mayo Clinic refused to release their raw data for analysis by other scientists. This refusal to release the data suggests the researchers lacked confidence in their published analysis. The limited data available indicate that patients who survived had their treatment changed in such a way as to increase the death rate. A plausible explanation might be that chemotherapy reduced the life expectancy of the surviving patients.

It is unfortunate that Linus Pauling did not explain the crucial difference between intravenous and oral vitamin C. The Mayo Clinic's use of oral doses would clearly have biased the results. In 2002, Gonzalez and colleagues reported that intravenous sodium ascorbate gives results that are more consistent in cancer patients, as higher blood levels are attained.[713] Sebastian Padayatty and Mark Levine, of the US National Institutes of Health, have also reported that intravenous sodium ascorbate is more effective in cancer treatment, since it enables higher blood levels to be reached.[702] They suggest that the change in method of administration could explain the differences between the Pauling and Mayo Clinic studies.

Back in 1969, a study by Dean Burk showed direct killing of cancer cells by ascorbate.[718] Burk suggested that,

"The future of effective cancer chemotherapy will not rest on the use of host-toxic compounds now so widely employed, but upon virtually host-non-toxic compounds, that are lethal to cancer cells, of which ascorbate represents an excellent prototype."

Dean Burk was correct. In the intervening three decades, enough data has accumulated to indicate that both oral and intravenous non-toxic anticancer therapies are a practical possibility.

Further positive reports

Reports on the efficacy of vitamin C as a treatment for cancer have continued. Perhaps most notable are those from Abram Hoffer, the first physician to use a double-blind clinical trial in psychiatry. His results confirm those of Cameron, Morishige and Murata, in showing greatly increased survival times.[703,704] Hoffer's book provides a wealth of information on his studies, including analysis of results and case study presentations. Other physicians who have used vitamin C to treat cancer over recent decades include Robert Cathcart[705] and Selva Kumar. Kumar reports positive results with intravenous sodium ascorbate in renal cell carcinoma and lymphoma.[706]

In an important step, laboratory studies by Riordan and others confirmed that vitamin C is toxic to cancer cells.[707] This finding has been replicated in animal studies,[708] and in humans.[709] Riordan published case studies demonstrating that vitamin C is an effective cancer treatment,[709,710] and has published a detailed mechanism of action.[711,712,713] Derivatives of vitamin C have been shown to have anti-tumour effects,[714,715] and a recent paper proposes encapsulation of vitamin C in micro-particles, for use as an anticancer treatment.[716]

With the accumulation of scientific evidence, the establishment view is changing. Recent papers on vitamin C and cancer from Mark Levine's group at the National Institutes of Health support the anticancer actions of vitamin C.[648,652]

Vitamin C kills cancer

The idea that vitamin C might be effective against cancer originated in the 1940s. By 1969, it was known that vitamin C could kill cancer cells directly.[718] Fred Klenner suggested the use of massive-dose intravenous ascorbate as a treatment for cancer, back in 1971.

It is difficult to extrapolate from laboratory studies to an anti-cancer effect in the body. First, the effect must be demonstrated in test tube studies. Next, researchers must show that the effect can also occur in the body. Finally, they must obtain clinical results to indicate an increased life expectancy. In the case of vitamin C, this has all been done. The only thing that remains is for the experiments to be replicated and suitable double-blind clinical trials to be performed. However, the bulk of clinical trials are by drug companies and, unfortunately for patients, vitamin C offers no profit incentive.[9,11]

Recently, Mark Levine's group has confirmed that vitamin C is toxic to tumour cells in both tissue studies and animal models.[648,652] These experiments involved applying vitamin C to several types of cancer and normal cell lines, for a period of only one hour. As we might expect from the preceding three decades of results, vitamin C killed the cancer cells but did not harm normal cells. Based on these findings, Levine concluded that vitamin C is only effective when used intravenously.[648] This conclusion does not take account of longer periods of exposure, which might allow oral dynamic flow level doses to be effective against cancer.

Levine examined the dose response in human lymphoma cells, which were highly susceptible to the action of vitamin C in a one-hour exposure. Over the following day (18-22 hours later), the number of cells dying was recorded. At high intravenous dose levels (1-5 mM/L) more than 80% of the cancer cells died, mainly by necrosis and apoptosis. This implies a massive poisoning of the cells and the potential for rapid destruction of tumours. However, doses consistent with blood plasma levels in the dynamic flow range of oral doses (200-300 µmol/L) also produced cell death. In this case, exposure for one hour produced death in 10-30% of the cancer cells. Importantly, dynamic flow levels from oral doses can be maintained indefinitely. Levine's results suggest that oral doses of vitamin C, at sustained dynamic flow levels, are likely to be a safe and effective way of treating cancer.

Levine claims that vitamin C acts by generating hydrogen peroxide in the plasma surrounding the cancer cells, rather than inside the cells. The evidence presented for this suggestion is flawed. The cancer cells were apparently preloaded with a low dose of vitamin C. Levine assumes that a low dose of vitamin C in the extracellular fluid saturates cancer cells to a maximum internal concentration, so they cannot absorb more.[e,2] No data is provided to support this idea, which is probably incorrect. Since the initial low dose does not kill the cells, but a subsequent higher dose (2 mM/L) does, Levine assumes the action must be extracellular. Levine's results are generally consistent with earlier descriptions by Holman, Riordan and others, in which healthy cells have a full complement of antioxidant enzymes to prevent the build up of hydrogen peroxide. For example, red cells in blood prevent the formation of hydrogen peroxide. In tumours, where the catalase and other antioxidant systems are deficient, hydrogen peroxide can accumulate and destroy sensitive cells.

A new treatment?

The first step in developing a new treatment is to show that a chemical kills tumour cells at lower concentrations than it harms human cells. In most cases, a new therapy has to go through detailed toxicity testing and must be capable of delivery to the target tumours. With vitamin C, many of these issues simply do not apply, as ascorbate is a normal and essential part of the body's biochemistry. It is important,

---

[e] Levine believes that normal body cells contain milimolar amounts of vitamin C, which overestimates the true value by about an order of magnitude.

however, to show that a suitably high concentration of vitamin C can reach the cancer.

Riordan tested samples of human serum from patients receiving intravenous ascorbate. His measurements confirm that the levels obtained are equivalent to those that are cytotoxic to tumour cells in experimental studies. Riordan also showed that samples of blood plasma, taken following ascorbate injections, killed cancer cells in tissue culture.

Riordan and others have proposed that ascorbate does not kill cells directly, but acts by producing hydrogen peroxide.[717] Riordan confirmed that the concentration of the enzyme catalase, which breaks down hydrogen peroxide, is up to one hundred times greater in normal cells than in tumour cells. This suggestion agrees with the findings of Holman, who discovered the acute sensitivity of cancer to hydrogen peroxide, back in the 1950s.

As we have explained previously, Holman and others reported that preparations of hydrogen peroxide selectively kill cancer cells in test tube and animal experiments. Such preparations also slow growth and destroy tumours in humans. The results with vitamin C are in harmony with this earlier research. Both lines of evidence are consistent with the mechanisms of action of current radiation and chemotherapy treatments. They also agree with the known biochemistry of cell signalling and cell division.

Agents that cause hydrogen peroxide to be generated in cancer should be toxic to cancer cells, while being safe for normal tissues. Both ascorbate and metabolites of ascorbate have anti-tumour activity in isolated tissues. Thus, sufficient doses of vitamin C should kill tumour cells, without toxic effects to healthy cells. The evidence suggests that vitamin C could be the elusive magic bullet.

Dean Burk's research at the US National Cancer Institute indicates that ascorbate is highly toxic to carcinoma cells.[718] Notably, the toxicity increases if a catalase inhibitor is present. Bram reported that vitamin C is preferentially toxic to experimental cell lines of malignant melanoma (skin cancer).[719] He also found that copper increases this toxicity. The presence of copper and some other metals can cause ascorbate to act as a pro-oxidant, rather than an antioxidant, leading to greater production of hydrogen peroxide in cancer.

In 1990, Helgestad reported that a new malignant lymphoma T-cell line was sensitive to ascorbate in culture, at concentrations attainable in human blood.[720] Additional research confirms that several leukaemia cell cultures are sensitive to vitamin C concentrations achievable in the human body, while normal blood-forming cells are not suppressed.[853,721,722,723] Research into the joint actions of vitamin C and selenium on gastric cancer has suggested that that the combination may be a useful treatment.[724]

Occasionally, it has been reported that high dose intravenous ascorbate can destroy a tumour too quickly for the patient to be able to cope with the resulting mass of dead tissue. These clinical observations reinforce the obvious need for medical supervision. However, on the plus side, these effects show that vitamin C certainly can destroy tumours within the human body.

Case study

Hugh Riordan has described a number of case studies on the use of vitamin C in cancer. In 1995, he saw a female patient with metastatic end-stage breast cancer.[680] She presented with cancer in "nearly every bone in her skeleton" and blood clots in both of the large veins that run behind the collarbones, draining the arms and head. One of the bed-ridden woman's arms was badly inflamed, because of the lack of blood flow. The blood clots were treated with an anticlotting drug,[f] and she was given an initial daily infusion of 30 grams of sodium ascorbate. This was increased to a five-hour infusion of 100 grams per day. Within a week, she was reportedly able to walk about the hospital and looked like a new person. She was discharged from the hospital and her treatment continued, with 100-gram infusions of sodium ascorbate, three times a week. Three months after starting the therapy, several tumours in her skull were no longer visible on x-ray. Sadly, six months later, she fell while walking in a shopping mall and died from complications of the resulting fractures.

[f] Activase was used to dissolve the clots.

Conclusions

We have described evidence that vitamin C can kill cancer. Indeed, the properties of ascorbate reflect, perfectly, the requirements for a chemotherapeutic agent.

- The effects of vitamin C on cancer can be explained theoretically.
- The results can be predicted from earlier research on hydrogen peroxide.
- Cell culture studies demonstrate the killing of cancer cells.
- Animal studies show anticancer activity.
- In humans, measurements show that blood levels that would be expected to kill cancer cells can be reached, with intravenous injection or oral dynamic flow intakes.
- The blood plasma of people injected with vitamin C can kill cancer cells.
- A large and increasing number of clinical reports find that ascorbate is an effective treatment.
- Case studies report that people with metastatic cancer have been cured.

The only item missing from this list is a randomised, double-blind, clinical trial, showing that vitamin C is an effective treatment or cure. Such trials can be prohibitively expensive to perform and most are conducted by pharmaceutical companies. The absence of such trials with vitamin C and associated redox agents may simply reflect the lack of potential financial reward.

With powerful evidence of efficacy and a large degree of safety, lives may be saved with little financial or health risk to the patient. In the following chapters, we provide evidence that the beneficial effects of vitamin C in cancer have been underestimated. This suggests that large and frequent oral doses vitamin C could form the basis of a powerful new approach to cancer therapy.

Sugar

"Pure, white and deadly" John Yudkin

In well-designed scientific studies, a "sugar pill" is a placebo or dummy, which is given to control subjects in a double-blind trial. Early scientists presumably assumed that sugar was safe and would have minimal effects on the body, making it ideal for comparison with the actions of a drug or treatment. However, sugar has profound effects: as far as the body is concerned, it contains pure calories. Much of the energy intake of a modern diet is in the form of sugars and starches, which supply nutrient-free energy. High sugar consumption leads to weight gain, particularly when combined with low exercise levels.

Diets high in sugar and carbohydrates lead to poor health.[725] A recent self-reported comparison of health and nutrition in 37,053 British people indicates the relationship of sugar to health.[726] People who consumed large amounts of sugar-based snacks were about five times more likely to report poor health. Other dietary intakes associated with sugar intake, such as wheat, dairy, refined foods, tea and coffee, were also linked with reduced health. Additional foods related to feelings of illness were red meat and salt. Foods associated with good health had a relatively high ratio of effective antioxidants to carbohydrate; these include fresh fruit, vegetables, salad, seeds, nuts and oily fish.

Overeating is a major risk factor for cancer.[725] Obese people are more likely to develop tumours, whereas people who eat less are more resistant. The problem of obesity is rising throughout the western world. The average American is now overweight,[727] and two percent or more are severely obese.[728] Lack of exercise and poor diet are leading causes of premature death.[729] 14% of cancer deaths in men and 20% of those in women are related to obesity.[730] This includes deaths from cancer of the breast, cervix, colon, oesophagus, gallbladder, kidney, liver, ovary, pancreas, prostate, stomach (male), and uterus.

The appeal of sugar is not limited to humans; it has a broad biological basis. Sugars provide easily assimilated energy for many organisms. As a result, the exchange of sugar is involved in many relationships between organisms.[731] For example, the sweet nectar of flowering plants is used to attract insects for pollination. Some plants, such as a species of eucalyptus, excrete large amounts of sugars to attract ants, which provide protection against herbivores. Ants themselves farm caterpillars that excrete sugar. Keeping the caterpillars requires more energy than they excrete in the form of sugars, which indicates the importance of the sugar supply. In some cases, such as with the caterpillar *Maculinea arion*, the farmed animal will even eat ants.[732] The relationship between cancer and its host also depends on the provision of sugars and other essential nutrients, to fuel abnormal growth.

This chapter looks at the relationship between sugar and cancer. We will consider prevention of cancer separately from treatment. This is because a diet to help people avoid cancer may differ from a diet that is optimal for a patient with advanced disease. As we consider the role of glucose, it will become apparent that it is intimately related to the actions of vitamin C in the cell. This interaction provides a means for preventing tumour growth and selectively killing cancer cells.

Restricting calories

Those who wish to avoid cancer or to live longer, if they already have the disease, are often advised to cut down on sugar.[725,733,749] Glucose deprivation generates oxidative stress within tumours, because of a possible defect in the metabolism of cancer cells.[734] One anticancer diet is called Calorie Restriction with Optimal Nutrition (CRON). This consists of eating 70–80% of the calories required to maintain body weight, while getting an optimal level of vitamins and other nutrients. The core restriction is the total energy consumed.

Research has shown that a reduced calorie diet can extend average lifespan in many species, while also reducing the chronic disease load.[735] In mice, energy restriction reduces the number of spontaneous tumours.[736] Such deprivation inhibits the growth of mammary cancer, suppresses the growth of implanted tumours and increases survival rates.[737,738] In Sweden, researchers found that women who had been treated for anorexia nervosa before they were 40 had a 53% lower incidence of breast cancer.[739] Although this result is consistent with a large effect of calorie restriction, it is worth noting that these women's intake of specific vitamins and nutrients was also reduced.

Glucose metabolism

Sugar is the ultimate junk food; it is high in energy but contains no other nutrient value. Sugars and refined flour products make up a large portion of the western diet. Indeed, high carbohydrate diets seem to dominate official recommendations for healthy eating.

One approach to estimating the effect of these foods on the body is the glycaemic index (GI). The glycaemic index is a measure of the metabolic effects of dietary carbohydrates.[740,741] The GI measures how a particular food increases blood sugar levels in the body. The higher the GI value, the greater are the body's glucose and insulin responses to the food in question. Since the body's response varies with the amount of food, this measure can be misleading. For this reason, a modification called the glycaemic load (GL) has been developed,[740] to take into account the amount of food consumed.

For many people, this is a rather unworkable form of dietary advice. To illustrate the problem, we refer the reader to the publication of an international table of the glycaemic index and load of foods. In 2002, the table contained 750 entries.[742] Such a table is huge: even a simple meal could require a substantial amount of arithmetic, not to mention guesswork as to the effects of combining different foods. In the longer term, following such a diet may be impractical for the majority of people. However, the approach may be productive in helping to construct general dietary advice guidelines, or for studying the health impacts of manufactured food.

Several projects have investigated the association between glycaemic load and cancer. Controlled studies have shown that a high glycaemic load carries an increased risk of cancer in the higher gut,[743,744] uterus,[745] ovary,[746] and colon.[747] Studies of glycaemic index and glycaemic load provide valuable research tools, as high levels of glucose in the blood promote cancer.[748,749,820] Indicators of diabetes, such as increased blood glucose, are associated with higher risk of colon cancer.[750] Diabetes is associated with increased risk of cancer of the bowel,[751,752,753,754,755,756] uterus,[757] and pancreas.[754,758] Furthermore, diabetes may be an independent predictor of mortality from cancer of the colon, pancreas, liver, bladder and female breast.[754]

Healthy people can benefit from learning the concept of glycaemic load and its relationship to different types of food. Prospective studies have shown an increased risk of cancer with high glycaemic load.[759,760,761] Other studies suggest this risk is specific to sedentary, overweight subjects.[762,763,764,765] As is often the case with such studies, others found no increase in risk, providing contradictory evidence.[766,767,768,769]

Despite these conflicting findings, people wanting to maintain good health would be wise to limit their intake of foods with a high glycaemic index. To a first approximation, the result of reducing the glycaemic load is similar to a low carbohydrate diet.

Treatments

There are many nutrient-based "treatments" for cancer. Several assume that making sure that the person is well nourished will allow them to fight the cancer. Although this may sound like common sense, there is no logical reason for thinking that such an approach will produce a favourable outcome. If true, the nutrients would need to benefit the person's normal cells more than they would the cancer cells. Conversely, if the cancer's growth was previously limited by the nutrient supply, then ensuring good nutrition could be catastrophic. For a nutrient-based approach to increase survival, differences between the host and cancer cells must be exploited.

We have mentioned that microevolution favours anaerobic cancer cells. As soon as anaerobic metabolism is mentioned, a biologist will think of sugar, the preferred energy source for such cells. Glucose is the main food supply for many cells in the body, including neurons in the brain. Most cells gain some of their energy from glucose, using glycolysis, which is the breakdown of glucose in a series of biochemical steps. Glycolysis is the early stage of glucose metabolism and is not dependent on oxygen. Cells typically have the ability to use glucose as a source of energy. However, the majority of healthy cells are more able than cancer cells to gain energy by breaking down fats and protein.

Since cancer cells use glucose metabolism as their preferred method of generating energy, changing the balance of food consumed could be beneficial. A reduced calorie diet, based on fibre, proteins and fats, rather than glucose and starch, might starve the cancer. Healthy cells, on the other hand, would be better adapted to accommodate the modified fuel source.

A possible basis for an anticancer diet might be one that reduces carbohydrate intake. Indeed, at first glance, the low-carb diets popularised by the late Robert Atkins might have the unexpected benefit of generally lowering cancer incidence or, in some rare cases, providing a cure. Such a suggestion may aggravate many dieticians, nutritionists and vegetarians. Objections to the Atkins diet in this context are that it does not limit total energy intake and may not provide sufficient vitamins and minerals. While not promoting the Atkins diet, we can risk the wrath of dietary experts and investigate the low-carb idea, to see if it might be productive.

Glucose deprivation could slow the growth of cancer cells, or even kill them. Glucose reduction leads to oxidative stress, which is preferentially toxic to cancer cells.[770,771] Such oxidative stress is caused by the generation of hydrogen peroxide and superoxide in the cell body.[772,773] The increased level of oxidants implies that glucose deprivation could be synergistic with the actions of substances such as vitamin C and others, which increase hydrogen peroxide in cancer cells.

Increased rates of cancer, which are associated with high blood glucose levels, are consistent with the microevolutionary model. Higher levels of glucose provide energy for anaerobic cells, facilitating cancer growth. Since glucose appears to be crucial for tumour cell microevolution, selection pressure should favour the development of mechanisms to increase its absorption. Cancer cells with the ability to pump glucose into their bodies are likely to have a selective advantage, leading them to dominate the colony. Supporting evidence includes the finding that glucose transporters are increased in cancer cells.[774,775,776,777,778]

Glucose molecules enter cancer cells preferentially, hence modified glucose molecules are used in taking medical images of cancers.[779,780,781] Interestingly, patients may be asked to fast prior to such investigations, lowering their blood sugar, in order to increase the uptake of these glucose markers.[782] Cancer cells appear to have a scavenging mechanism to meet their glucose needs, which is consistent with the idea that a low-carb diet would slow cancer growth by reducing glucose availability.

The microevolutionary theory predicts that if cancer cells were able to stimulate high levels of glucose in the blood, they would have a selective advantage. Supporting this idea, glucose intolerance has been observed in cancer patients, as long ago as 1919. Such intolerance means the body becomes less efficient at converting carbohydrates into energy, resulting in raised blood glucose levels.

The relationship between cancer and glucose intolerance is unclear. One possibility is that people with pre-existing glucose intolerance may be more subject to cancer. High blood glucose levels are common in cancer patients, and a reduced efficiency of energy conversion contributes to weight loss (*cachexia*).[783,784,785] Even a small tumour may disturb the body's glucose balance.[786] Although the results are not entirely consistent,[787,788,789,790] they imply that glucose metabolism is an important factor in understanding cancer.[791]

These findings suggest that cancer cells are likely to use glucose at a high rate. Research confirms that increased glucose utilisation is a general feature of cancer cells,[792,793,794,795] as has been known since 1930.[212,210] The ability to sustain high rates of glycolysis, even when oxygen is plentiful, is characteristic of cancer cells.[796,797,798] Factors that alter cancer growth rates can act by modifying glucose metabolism and transport, indicating its importance to tumour progression.[799] Glucose has even been suggested as a target for selective drug action.[800,801] Drugs that inhibit the uptake of glucose or its use in cancer cells have been considered as potential treatments.

The microevolutionary model predicts that gene expression and biochemistry in cancer cells might be modified, to facilitate the use of glucose as the main energy source. This gene expression is a direct result of cell selection, acting in an environment where oxygen is scarce. Such adaptations support a greater role for enzymes involved in anaerobic metabolism. Gene expression and biochemical studies of cancer cells confirm this predication.[802,803,804,805,806,807,808,809,810] There is little doubt that changes in glucose biochemistry are found in cancer cells.

If glucose promotes cancer growth, then we might predict that cancer patients with low glucose levels would survive longer than would those with high levels. A recent paper backs up this suggestion,[811] and further results, from patients with cancer of the pancreas, support this research.[812]

Glucose and vitamin C

For many years, biophysicist John Ely has investigated the interactions between glucose and vitamin C. When blood sugar is low, vitamin C enters body cells more easily. The transporters that pump glucose into the cell also transport oxidised ascorbate. When a cancer cell boosts its glucose transport mechanisms, it increases its capacity to absorb vitamin C. Cancer patients are reported to have an enhanced tolerance for large oral supplements of vitamin C.[813] This may result from the cancer cells' capacity to absorb large amounts of the vitamin from the bloodstream.

In 1972, Ely warned Linus Pauling that it was essential to make sure patients had low sugar intakes, if they were benefit fully from megadose supplementation. This is because decreasing the availability of glucose helps increase the absorption of vitamin C into cells. Ely described this competition in his glucose-ascorbate antagonism theory. Later, he showed that the cross-linking of proteins by glucose, which occurs in diabetes, aging and cataracts, could be inhibited by vitamin C.[814,815,816] He also related this cross-linking mechanism to increased risk of heart disease and cancer.[817] Ely proposed that sustained high plasma levels of vitamin C would greatly reduce cross-linking of proteins in the blood of diabetics. According to this idea, two grams of vitamin C every two hours would maintain high levels of vitamin C in the blood,[a] and prevent many of the complications of diabetes.

Ely made his point dramatically with the statement "aging is scurvy." Aging is a process characterised by oxidation and cross-linking.[818] This claim implies that a person who does not have a constantly high level of vitamin C will age rapidly. Ely suggests that high blood glucose levels have other health effects, such as birth defects.[819,820,821] He put forward an essential nutrient theory, which has similarities to Linus Pauling's ideas of orthomolecular medicine.

Ely proposed a role for the interaction of glucose and vitamin C in assisting cellular immunity. Glycolysis is the generation of energy from glucose, in the absence of oxygen. A related biochemical pathway, the pentose pathway or hexose monophosphate shunt, is used to power the synthesis of DNA and provide antioxidant electrons.[725] The speed of this second pathway depends on the concentration of vitamin C in the cell.[822] Ely has pointed out that cell-mediated immunity requires this additional pathway. If vitamin C levels are low, as is common in the elderly, the shunt rate and cell-mediated immunity fall. Then, infectious diseases, cancer and mortality from all causes rise. Further evidence includes the finding that the behaviour of white blood cells is greatly inhibited in diabetics.[823,824,825,826] People with compromised cell-mediated immunity, such as those suppressed following organ transplants, are at greater risk of developing some forms of cancer.

Raw fruit and vegetables

The effect of calorie and sugar intake on cancer is profound. Arthur Robinson, former President and Research Director of the Linus Pauling Institute, became interested in Edie Mae Hunsberger's claim that she had cured herself of cancer by eating a diet of fresh fruit and vegetables.[827] Robinson tested the claim on experimental mice suffering

---

[a] We calculate the blood plasma level in a normal healthy young adult with this level of intake to be about 220 μM/L.

from cancer, which he fed a diet of raw fruit and vegetables.[828] The result was unexpected - a massive reduction in the number of lesions. Pauling described the result as a "bombshell". It appeared that simply reducing calorie intake, by feeding raw fruit and vegetables, had extended the life of the cancer-ridden mice.

Mice given a diet of raw fruit and vegetables for two months showed a marked reduction in tumour growth. This reduction was not as great as that observed with massive doses of vitamin C, but was impressive nonetheless. The number of minor lesions was reduced from 54 in 60 animals (90%), to 11 in 40 animals (28%). Severe lesions were also reduced, from 13 in 60 animals (22%), to just one in 40 animals (2.5%).[b] The effect was largely independent of the type of vegetables provided. However, the benefit was lowered by the addition of seeds and nuts, which contain relatively large amounts of available energy. Adding apples and pears, which contain larger amounts of available sugar, also reduced the effect. Similarly, addition of protein inhibited the effects of raw fruit and vegetables. A combination of fruits and vegetables, taken with a human equivalent intake of 168 grams of vitamin C,[c] gave results similar to those of vitamin C alone (four minor lesions and no severe lesions in 45 animals). The numbers of lesions in the mice are shown in the following table:

Robinson's mice after two months			
Treatment	**Number of Mice**	**Minor Lesions**	**Major Lesions**
Controls	60	54(90%)	13(22%)
Vitamin C 192g	45	7(16%)	0
Raw vegetables	40	11(18%)	1(2.5%)
Raw vegetables + 165g vitamin C	34	4(12%)	0

b Percentages are average (arithmetic mean) lesions per mouse.

c For simplicity, we are using Robinson's suggested conversion to human equivalent intake, ie 1g/Kg approximates to a one gram human dose. No-one is suggesting that this conversion is accurate, but it provides the reader with a basic heuristic.

When the study was repeated, consistent results were obtained. Robinson then carried out further experiments.[828] Mice were irradiated to induce cancer, before being fed the experimental diet for four months. Waiting four months could give a more accurate evaluation of the slower growth of some lesions. Four months is not a long period in human terms, but laboratory mice have a total life expectancy of about three years. The follow-up experiment confirmed the tumour growth reduction and anti-cancer effects of vitamin C and raw vegetables. In 56 control mice, there were 218 minor lesions (3.9 per mouse) and 95 severe lesions (1.7 per mouse). 96 grams per day of vitamin C greatly reduced these numbers, to 56 minor lesions (1.2 per mouse) and 13 severe lesions (0.29 per mouse) in 45 mice. Notably, a combined treatment of raw fruit and vegetables, plus 25 grams of vitamin C, prevented the formation of severe lesions almost completely, resulting in 24 minor (0.7 per mouse) and 1 severe lesion (0.03 per mouse), in 34 mice. The following table presents the results obtained:

Robinson's mice, after four months			
Treatment	**Number of Mice**	**Minor Lesions**	**Major Lesions**
Controls	56	218	95
Vitamin C 96g	45	56	13
Raw vegetables + 25g vit C	34	24	1

These effects can be explained if we assume that the availability of glucose inhibits the action of vitamin C. The results are consistent with anaerobic cancer cells requiring nutrients for growth.[828] An appropriate analogy might be the different responses of infants and adults to reduction in available food. An adult would lose weight but could survive for long periods on limited food intake. However, an infant requires nutrients for growth. Even moderate food deprivation could reduce the growth rate in an infant. Prolonged deprivation might result in a permanent growth restriction, leading to an adult of short stature. Severe food deprivation would kill a child more quickly than an adult. In a similar way, if a malignant cancer is deprived of nutrients, particularly glucose, its growth rate may be reduced. Conversely, good nutrition may benefit a cancer more than it would the host.

The mice that died

The Linus Pauling Institute was set up by Arthur Robinson and Pauling himself, to study the effects of vitamin C and other nutrients. Pauling's popularisation of vitamin C and its effects generated great publicity. However, the medical establishment attacked Pauling, promoting the impression he was a quack.

Robinson carried out important experiments based on Pauling's ideas, during which he fed a meganutrient mix to mice with cancer. His results indicated that when he fed the mice a full supply of vitamins and minerals, their cancers got worse. This result was disturbing and not what might have been expected at the time. If true, the many popular books telling people what to eat when they have cancer might be doing more harm than good. A simplistic interpretation of Robinson's result is that much current nutritional advice for cancer sufferers is wrong.

Then came a further shock. To Robinson's surprise, he found that his mice died sooner if they were treated with nutritional doses of vitamin C. Vitamin C greatly affected the growth of the cancer in mice. As we have seen, later results showed that high intakes, equivalent to 50 grams or more in a human, inhibited cancer growth. However, Robinson's initial results indicated a human equivalent intake of 1.5 - 3 grams per day decreased survival times. The result was a heated and acrimonious dispute between Robinson and Pauling.[827,830] Indeed, it ended their 16-year collaboration. In Robinson's words,

"He [Pauling] was not willing to accept the experimentally proved fact that vitamin C in ordinary doses accelerated the growth rate of squamous cell carcinoma in these mice."

This was a particularly damaging event. Pauling believed that 75% of all cancers could be cured with vitamin C.[15] However, results from his own institute appeared to invalidate his case. Pauling felt publication should be delayed until further results were obtained. Robinson, however, believed the experimental results were clear. We should add that the stakes were higher than simple damage to scientific reputations, as both were aware that the lives of millions could depend on the outcome. It was almost inevitable that the two should fall out over this question.

As we shall now explain, neither side was wrong. The full set of results, later published by Robinson, is both interesting and informative. Robinson completed his series of experiments, using 1846 hairless mice.[830] Disturbingly, he found that diets apparently least suitable for healthy individuals greatly extended the lifespan of mice with cancer. It appeared possible that dietary changes might extend the expected survival time of a cancer patient by a factor of 20. In human terms, this is the difference between having one year to live, compared to twenty years.

The interpretation of these results requires some care. Firstly, mice were an unfortunate choice for the experiment, as they synthesise vitamin C in their bodies and normally have high blood levels. Humans are not in this fortunate position and have to obtain their vitamin C from the diet. If mice are given supplemental vitamin C, their homeostatic control mechanisms might respond by producing less of their own. For this reason, it is difficult to apply the results from mice given low, gram level vitamin C doses to humans. For example, unsupplemented mice might correspond to humans with an intake of several grams a day.

Pauling's objection to Robinson's results may have arisen from his knowledge of the differences between the physiology of humans and mice. He might have suggested that the biggest biochemical difference was that humans cannot manufacture vitamin C so, at these low intakes, the results cannot easily be extrapolated to man. Furthermore, these mice were sick and their requirements are expected to be greater than those of healthy animals. Robinson was also correct in indicating that these results should not be concealed. Despite this, the experiments had not been completed and reporting the results would have been premature. The full results support Pauling's assertion that vitamin C is an effective anticancer agent.

Two months after irradiation, higher intakes of vitamin C were seen to have greatly suppressed the growth of cancers. The suppression effect was marked: no mice given high dose supplements (above 48 human-equivalent grams) had severe lesions. Furthermore, the number of minor lesions was reduced from about 54 lesions in 60 mice to seven lesions in 45 mice. Taken at face value, these results are astounding. They imply that dietary restriction or high dose vitamin C can be powerful methods for the treatment of cancer.

Mice are not people

Robinson's results show that a raw fruit and vegetable diet with supplemental vitamin C will slow cancer growth – in mice. The first objection a physician will state to this experiment is that mice are not people. Some anticancer agents are effective in particular animals but not in humans. Robinson's result could provide false hope for victims and might make them avoid conventional therapy.

Robinson has said,

"if diet restriction were practiced by all cancer patients in the United States, the resulting life-extension might equal or surpass that resulting from the combined efforts of the entire current medical oncology effort"

We might normally dismiss this statement as having little value. However, the lifespan of his experimental mice was greatly extended. When Robinson suggests that some raw vegetable diets may help cancer patients, by reducing the supply of nutrients, he is making a measured statement and understands the biological differences between mice and humans.

In the first place, he embarked on this study because of reported effects in *people.* He was stimulated into studying raw vegetables by reports of their efficacy in human cancers, particularly in the book *How I conquered cancer naturally,* by Edie Mae Hunsberger.[829] Healthy body cells may accommodate a nutrient shortage more effectively than the disturbed metabolism of cancer cells, whose growth depends on an adequate supply of specific nutrients.

Robinson gives the example of a cancer patient with an inoperable throat cancer.[830] The surgeon explained to the patient that he had run out of options and had only a short time to live. However, the patient sought the advice of a nutritionist, Ann Wigmore, and started a strict diet. Notably, he followed Wigmore's early dietary advice, eating nothing but raw vegetables. Later variations on this diet involved a higher calorie intake.

After several months, the patient returned to his surgeon, who noticed three remarkable facts. Firstly, he was still alive and would normally have been dead by this time. Secondly, the patient looked like a concentration camp survivor, emaciated from malnutrition. Finally, the cancer in his throat had completely gone. The surgeon made sure the patient ate food containing essential nourishment and he regained his health, apparently free of cancer. John Ely has reviewed similar results for glucose deprivation and vitamin C supplementation, both in mice and in humans.[831,832]

Restricting carbohydrates in cancer patients may greatly increase the anticancer effects of vitamin C. In a study of 36 patients, carried out in 1957, only five of the patients, those with improved glucose tolerance, were reported to be tumour-free following treatment.[725,833] In 1978 and 1979, two patients with Stage IV breast cancer decided to use ascorbate (10 grams per day or less) with restricted carbohydrate intake. Both patients were reported to be tumour free after six months. Moreover, both patients were still alive 13 years later, in 1992.[831] The patients reportedly suffered no side effects with the high ascorbate and low glucose therapy. Ely claims that this was the first description of such a rapid disappearance of a massive tumour burden, with any form of chemotherapy. His explanation is that the vitamin C caused the cancer cells to redifferentiate or revert to a more normal form. This redifferentiation is an experimentally established attribute of antioxidants.[834,835]

John Ely and co-workers have demonstrated destruction of mouse tumours, by means of vitamin C and glucose restriction.[836] Sugar reduction helps vitamin C to kill cancer cells, both directly and indirectly. Both ascorbate supplementation and nutritional deprivation can increase free radicals within cancer cells. Lack of glucose increases free radical damage within cancer cells. Moreover, nutrient deprivation acts at a basic physical level, preventing the supply of energy required for cancer growth. The fundamental nature of this mechanism would be predicted to hinder adaptation by the cancer cells, as all cells require energy for growth.

A diet to eliminate cancer

Few people are willing to exist indefinitely on a diet with a severely restricted calorie intake. Despite the known risks to health, many people cannot resist the temptation to indulge their cravings for junk food, leading to an obesity epidemic in developed countries. Such people seem unable to overcome what amounts to an addiction to overeating.

Even the most stoical patient would find it difficult to exist in a state of semi-starvation for their remaining lifespan. Many would consider such a cost greater than the benefit of life extension. For this reason, it is impractical to propose that cancer patients should embark on a lifelong strategy of dieting. Such a scheme would involve careful monitoring by qualified physicians and nutritionists, because of the potential for damage to the patient's health. However, the perceived cost-benefit analysis is a question for the patient themselves to answer. Given all the facts, a patient might elect to eat normally and die more quickly, or to have an extended life but feel chronically hungry. A proportion of people will choose the latter option.

Cancer cells, like any biological organism, find it difficult to multiply and grow without an adequate supply of energy or other factors essential to their survival. While different cancers will respond in diverse ways, it will be difficult for any cancer types to continue growing without sufficient food. In some cases, dietary restriction alone could cause the cancer cell population to become extinct.

Restriction of either calories or carbohydrates can slow the growth of established cancer. The extent to which the anticancer effect of restricting calories is explained by lack of available sugar is not clear from the current evidence. It could be that the benefits of calorie restriction can largely be obtained by a more acceptable, low carbohydrate diet. Lowering carbohydrate intake may provide the benefits of dietary restriction, in a more sustainable form.

Hoodia

Currently, the extent of dietary restriction needed to prevent cancer growth is not established. It may be possible to restrict the growth of many cancers with a moderate dietary restriction of carbohydrates and selected dietary nutrients. However, a severe, near-starvation diet, consisting of raw vegetables alone, may be required for many slow growing cancers. Patients who decide it is worth starving themselves to gain the opportunity to live longer will need all the help they can get. Clearly, willpower is not sufficient, or we could all look like supermodels! However, certain substances might make it easier to embark on a highly restricted diet, without feeling hungry all the time. One of these is Hoodia, an appetite suppressant used by the San Bushmen of the Kalahari Desert. The Bushmen eat slices of the *Hoodia Gordonii* cactus on their hunting trips, allowing them to exist for long periods without food.[837]

Hoodia appears to act centrally in the brain to reduce appetite and maintain energy levels.[838] Tom Mangold, a BBC reporter, has described its effects. Mangold visited the San Bushmen to experience the effects of Hoodia for himself.[839] One evening, Mangold and his cameraman each ate a portion the size of half a banana, before driving back to Capetown.

Both felt well on their four-hour journey and did not even think about food. The next day, they did not bother taking breakfast, as they were not hungry. Mangold ate lunch, but without pleasure. Their appetites returned gradually, after 24 hours.

In the 1960s, the South African Council for Scientific and Industrial Research (CSIR) isolated and patented an appetite suppressant molecule (P57) from Hoodia. In 1997, a British company called Phytopharm licensed P57. Following initial tests, Phytopharm sold the licence to Pfizer, one of the pharmaceutical giants. The shares of both companies rose, as investors looked forward to huge profits from this new aid to dieting. However, the companies were accused of exploiting the San Bushmen.

The chief executive of Phytopharm, Richard Dixey, claimed the San Bushmen "had disappeared" but, in fact, about 100,000 of them lived in a region spanning several countries, in southwest Africa. The San sued for a share of the profits, gaining 6% of the royalties from the CSIR license on commercial sales.

In a press release published in 2001, Phytopharm claim to have conducted a double-blind, randomised, placebo-controlled study of P57's effects.[840] 19 overweight males received either P57 or a placebo, twice daily, for 15 days. Nine subjects in each group completed the study. The experimental group showed a significant reduction in their average daily calorie intake and a loss of body fat, compared to the controls.

The enormous commercial possibilities of a safe appetite suppressant resulted in a large number of Hoodia-based products becoming available. However, many appear to contain little or no active ingredient. Use of Hoodia requires standardisation of its active ingredients or its derived appetite suppressant substance, P57. Nevertheless, the existence of Hoodia indicates that it may be possible to undergo calorie restriction with less discomfort than expected.

It appears possible that simple dietary changes could prevent many deaths from cancer. The mechanisms involved are based on oxidation and reduction, and are synergistic with the action of vitamin C. The use of natural supplements of L-ascorbic acid may be more efficient than "synthetic" vitamin C, as this form is more available to the cells. Ascorbate's effectiveness is greatly enhanced if combined with a reduction in sugar, and by using it with other nutrients, such as vitamin K.

Vitamin K

"There are in fact two things, science and opinion; the former begets knowledge, the latter ignorance." Hippocrates.

The actions of Vitamin K are less familiar than are those of vitamins C and E: it is essential for blood clotting and is involved in bone metabolism. More importantly, it can also be an effective anti-cancer agent. Unfortunately, this natural and safe substance has tended to be neglected in the fight against cancer. The anti-cancer activity of vitamin K has been investigated for decades,[841,842] yet its clinical potential remains to be fully realised. Vitamin K provides an introduction into a new world of anticancer agents. Anticancer drugs based on vitamin K may be safe for normal tissues, but deadly for cancer cells.

A Danish researcher, Henrick Dam, won the 1943 Nobel Prize for discovery of vitamin K, which was first isolated in 1939. This substance is needed for blood to clot. Dam named it vitamin K, after *koagulation*, the Danish word for coagulation. Vitamin K exists in several forms, called vitamins K_1, K_2, K_3, and so on. Some of these are natural forms, whereas others, such as K_3, are synthetic.

In this chapter, we describe the properties and pharmacology of this fat-soluble vitamin. Vitamin K is known for its classic action on blood clotting.[843] The anticoagulant drug, warfarin, is a vitamin K antagonist. Originally developed as a rat poison, warfarin interferes with the clotting action of vitamin K and has been used for the last fifty years to "thin the blood" of people with heart disease.[844] As it is both odourless and tasteless, rodents will feed on warfarin-laced food for days, until they accumulate a lethal dose. However, warfarin's effectiveness as a rat poison is declining, as rats have evolved resistance to it.

Warfarin was discovered through observations of a cattle disease, first noticed in the early 1920s. A Canadian veterinary surgeon, Frank Schofield, found that mouldy silage contained an anticoagulant, which made the cattle bleed. About two decades later, chemists from the University of Wisconsin, Karl Link and Harold Campbell, described the toxin as a coumarin derivative. Coumarin is a chemical produced by plants and smelling of freshly mown grass. Link began work on coumarin-based rat poisons, and generated warfarin in 1948.

Vitamin K has other biological functions besides prevention of bleeding.[845,846] It has been used in the treatment of diseases, including osteoporosis,[847,848] vascular calcification and atherosclerosis.[849] It has also been shown to inhibit the growth of cancer.[850,851,852,853,854,855,856,857,858] The anticancer effects of vitamin K are accepted by the scientific community.[859]

Requirements and toxicity

In adults, a shortage of vitamin K may result in nosebleeds and haemorrhaging. However, vitamin K deficiency occurs more often in babies; about one percent of infants suffer the tendency to bleed that comes from deficiency. There are many reasons why babies should be short of this vitamin. These include low transfer across the placenta, lack of vitamin K in breast milk and prematurity of the liver. In addition, the relatively sterile newborn gut lacks the bacteria that manufacture the vitamin in adults. The primary deficiency symptom in infants is bleeding, together with easy bruising. Birth defects, such as underdevelopment of the face, nose, bones and fingers, have also been noted.

The US National Institutes of Health have estimated the daily need of vitamin K for adults at 0.08mg per day. Dietary sources include green vegetables, liver, vegetable oils and milk. It is possible to have too much vitamin K, although the different K vitamins differ in their toxicity. Vitamin K_1 is not toxic at doses up to 500 times the recommended daily intake (0.5 mg/kg/day).[860] However, vitamin K_3 is more toxic: it should not be used to treat deficiency. In infants, an overdose of K_3 can cause haemolytic anaemia, in which the red blood cells are destroyed, or jaundice, when excess bilirubin builds up in the blood and brain, turning the skin yellow.

Mechanism for killing cancer

Although the physiology of vitamin K is interesting, we are mainly concerned with its actions against cancer. Vitamin K is reversibly oxidised and reduced inside cancer cells, which suggests a mechanism by which it might destroy the cells.[844,845,a] In the presence of iron and other metals, vitamin K takes part in a Fenton reaction, generating free radicals or oxidants that are believed to induce DNA breakage and cell death. However, vitamin K may also kill cancer cells in the absence of iron.[861]

---

[a] For chemists: vitamin K (a hydroquinone) cycles from an epoxide to a quinone and back to the hydroquinone for another gamma-carboxylation reaction.

Receptors

Many drugs and hormones act via specific receptors on the surface of cells. They bind with the receptors, stimulating the cell to perform a biochemical action. Vitamin K interacts with such receptors, influencing cell signalling. Vitamin K dependent signalling is involved in cell death, transformation to a state of unlimited growth, and replication.[862,863] These processes are central to the development of cancer cells. Vitamin K influences a gene called Gas6,[b] which prevents cell division. When Gas6 is expressed highly, cell growth is inhibited.[863,864,865,866] One function of this growth-limiting gene is to increase cell survival under conditions where cell division and proliferation are restricted.[867]

Vitamin K may act as a physiological anti-inflammatory agent, disposing of dead or dying cells by helping phagocytic cells to recognize them.[868,869] Its actions may help regulate cell growth and tumour formation, including multiple myeloma and lung cancer.[870,871]

Vitamin K_1

The most abundant member of the family is vitamin K_1, which is involved in the fundamental chemistry of life.[c] It occurs in green plants and algae, and is associated with photosynthesis. Leafy, green vegetables provide vitamin K_1 in the diet.[872] Vitamin K_1's anticancer activity is a good reason to eat your greens.

Several studies have established Vitamin K_1 as an anticancer agent. It can diminish growth in a number of cell lines, including cancer of the breast, liver, colon, lung, nasopharynx, stomach, oral cancer and leukaemia. The concentration of K_1 needed to halve cell growth in these cancers ranges from 2-10 mM/L.[850] However, other cancers may respond to lower physiological concentrations; these include cell lines from brain tumours,[851] other forms of leukaemia,[891] and liver cancer.[858] Clinical trials have supported the anticancer effects of vitamin K_1 in humans.[873,874,875] The fact that vitamin K_1 can slow the growth of tumour cells has been demonstrated, but the concentrations necessary to achieve the effect are high, compared with those of other forms of the vitamin.

---

[b] Gas6 is a Growth Arrest Specific gene.

[c] Vitamin K_1 is also known as phylloquinone, phytonadione or menaphthone.

Vitamin K_2

Vitamin K_2 also has anticancer activity.[d] This second form is chemically similar to K_1 and occurs naturally. It is produced by bacteria, rather than plants, and was originally isolated from putrefied fishmeal. It is possible that it is synthesised in some animals.[876,877,878] In animals, K_2 is the most common form of vitamin K. Intestinal bacteria provide a consistent source for adult humans.

Vitamin K_2 has widespread anticancer actions. Experiments, in both humans and mice, have shown that it is effective against several cancer cell types.[879,930] As with vitamin K_1, a number of cancer cell lines are sensitive to vitamin K_2, including breast, colon, liver, leukaemia, lung, stomach, lymphocyte, nasopharynx, and oral cancers. Vitamin K_2 is effective at lower concentrations than vitamin K_1 (0.8-2 mM/L), but much higher than required with vitamin K_3 (18-45 μM/L).[850]

Vitamin K_2 reduces cancer growth by stopping cells dividing or by killing them; the extent of this action varies with the dose.[880,921,881] Some leukaemic cell lines resist K_2 induced cell death; however, they change to become more like healthy cells.[881] Researchers in Japan claim that culturing cancer cells with 1 μM/L of vitamin K_2, but not K_1, induces cellular differentiation. This suggests K_2 causes cancer cells to revert to a healthier state. Other chemicals that induce differentiation are synergistic with vitamin K_2,[e] and increase its effectiveness.[882,883]

At concentrations as low as 10 μM/L, vitamin K_2 induces cell death in leukaemic cells, within 48 hours.[883] Furthermore, chemicals similar to vitamin K_2, which occur naturally, cause 90% of leukaemic cells to commit suicide, while having little effect on healthy bone marrow cells.[883,884] These results illustrate the selective killing of cancer cells.

Case studies confirm the potential of vitamin K_2 as a therapy. In one study, an eighty-year-old female patient, whose bone marrow did not produce enough blood cells,[f] was given an oral dose of 45 mg per day. After 14 months of treatment, her regular blood transfusions were no longer needed and her blood counts improved.[885]

In a second study, a 72-year-old female, with acute leukaemia, went into remission after one week of conventional treatment and retinoic acid

---

[d] Vitamin K_2 is also known as menaquinone.
[e] Retinoic acid, interferon-gamma and camptothecin.
[f] The women had MDS - myelodysplastic syndrome.

(a derivative of vitamin A).[g] Eight months later, she relapsed. Her initial treatment was repeated, with the addition of vitamin K_2 (20 mg/day), which produced measurable improvements in her white blood cells, after two months. Her bone marrow was examined and was found to suggest complete remission,[886] a result that would normally be unexpected.

Finally, a 65-year-old man with acute myeloid leukaemia was treated with oral vitamin K_2 (90 mg/day). Six weeks later, he showed a significant decrease in his immature white blood cell count, and an increase in platelets. After 10 months of treatment, the dose was reduced to 45 mg/day and the good response was maintained, without side effects.[880] These encouraging results from vitamin K_2 led to cancer therapy trials in Japan.[887,888] Vitamin K_3 is still more effective and its clinical promise is potentially even greater.

Vitamin K_3

Vitamin K_3, or menadione, is an effective killer of cancer. It is considered a synthetic chemical, which acts as a provitamin, and is converted to a vitamin by the body. It has a simpler structure than vitamins K_1 or K_2. It may also be formed from K_1, by bacteria in the gut.[877,889] Although it does not provide the full nutritional benefits of the natural vitamin (K_1 and K_2),[890] K_3 is a more powerful anticancer agent.

Laboratory experiments have demonstrated the anticancer activity of vitamin K_3 in both animal[851,891,892,893,894] and human[852,853,895,896,925,937] cancer cells. This vitamin has been reported as effective against multi-drug resistant leukaemia cells.[925] In rats, vitamin K_3 is active against leukaemia cells that are resistant to a conventional drug.[897, h]

Researchers have demonstrated the anticancer activity of Vitamin K_3 in laboratory, animal and human experiments. Intravenous K_3 has increased the survival time of lung cancer patients, also receiving radiation therapy, from 3.77 months to 5.42 months.[898] Injections of vitamin K_3 in rats with malignant liver cancer increased their survival period to 60 days, compared to 17 days for control animals.[856] In mice with transplanted liver tumours, radiation treatment was more effective if the mice were given oral or intraperitoneal injections of vitamins K_3 and C.[899]

---

[g] The women received all-trans-retinoic acid (60 mg/day), enocitabine (200 mg/day), and daunorubicin (40 mg/day).

[h] Adriamycin.

Vitamin K_3 has been shown to work in combination with conventional chemotherapy. It increases the effectiveness of 5-fluorouracil against liver cancer cells.[894] In a culture of human oral cancer cells, it was shown to be synergistic with conventional cancer agents.[925,i] This beneficial action of vitamin K_3 has been also demonstrated in cells from cancer of the upper throat.[900] Vitamin K_3 allows a 10- to 50-fold reduction in the levels of the anticancer drug, mitomycin C, needed to kill cancer cells.[901] Pre-treatment with vitamin K_3 also increases the toxicity of standard chemotherapy treatments against breast cancer cells.[903]

The promising results of these experiments with vitamin K_3 have resulted in a number of trials, in both animals and humans. Studies in rats indicate that vitamin K_3, at blood levels less than 1 μM/L, increases the action of the anticancer drug, methotrexate, without increasing its toxicity. The combination of methotrexate with vitamin K_3 led to a massive reduction in cancer growth, supporting K_3's potential to increase the effectiveness of this form of chemotherapy.[857] Results from human studies with vitamin K_3 and a different anticancer drug have been less consistent,[j,901,902,903] which may indicate that the drugs have differing mechanisms of action.

Related compounds

We have no information about the anticancer activity of vitamin K_4. A recent study has shown that vitamin K_5 has antitumour activity.[904] We suggest that numerous other natural substances, which are chemically related to vitamin K, especially quinones,[905,906,907,908] could prove to have non-toxic anticancer activity.[909]

The selective actions of vitamin K_3 and other, non-toxic, cancer agents are closely related to biochemical mechanisms occurring in standard chemotherapy. Some synthetic anticancer drugs have redox-active quinone forms, although they are more toxic than vitamin K.[910,911] Such drugs can undergo redox cycling, to generate free radicals.[912] Their anticancer mechanisms have some degree of specificity for diseased cells. Like vitamin K, such drugs can be influenced by ascorbate and other antioxidants.[913]

The idea of using this relatively safe supplement to augment toxic chemotherapy and radiotherapy seems superficially reasonable. If lower

---

[i] 5-FU, bleomycin, cisplatin, dacarbazine, mercaptopurine, cytarabine, hydroxyurea, VP-16, vincristine, doxorubicin, mitoxanthine, mitomycin C, actinomycin D, and thiotepa.

[j] Mitomycin 3.

doses of these conventional therapies produce the same effects when combined with the vitamin, the reductions in side effects are likely to be significant. However, such an approach indicates a failure of imagination. Surely, it would be more appropriate to carry out research to discover the mechanism of action that allows vitamin K to kill cancer cells, without harming normal cells. Physicians could than use this knowledge as the basis for treatments that do not poison their patients.

How does it work?

Vitamin K's cytotoxic effects and cancer growth inhibition have been demonstrated in isolated cell lines, in animals and in humans. Scientists do not yet fully understand how the vitamin works against cancer cells. There are several possible mechanisms, of which the primary model depends on vitamin K's ability to affect cell oxidation and generate free radicals. This production of free radicals could explain how vitamin K_3 can kill cancer cells directly. Some of the effects of vitamins K_1 and K_2, such as redifferentiation and inhibition of the cell cycle, may involve additional mechanisms.

The primary action of vitamin K_3 against cancer is a result of redox cycling.[914] Vitamin K_3 is oxidised and reduced, in a repeated cycle that releases free radicals. This redox cycle is similar to that for vitamin C, described previously. If redox cycling produces more free radicals than a cell can quench by generating antioxidant electrons, the free radicals damage the cell, which eventually dies. At higher doses, vitamin K_3 kills cancer cells more effectively than other forms of vitamin K.[k] Vitamins K_1 and K_2 are less effective at generating free radicals.[l]

Vitamin K_3 can exist in three different forms: a quinone, a semiquinone and a hydroquinone. It switches between these states by oxidation (losing electrons) and reduction (gaining electrons). In its oxidized form, vitamin K_3 is a quinone. Quinones can be reduced, gaining one electron to generate semiquinone radicals, or two electrons to become hydroquinones. In moving from the quinone form to the hydroquinone, K_3 gains two electrons. In the return section of the cycle, the hydroquinone is oxidised by the loss an electron to form a semiquinone. Loss of another electron returns it to its original quinone state.

---

[k] Vitamin K_3 can arylate nucleophiles, such as glutathione, and thus initiate rapid one- or two-electron redox cycling.

[l] K_1 and K_2 have lower rates of redox cycling and a higher proportion of two-electron transfers than vitamin K_3.

Vitamin K_3's redox cycling increases oxidative stress in malignant cells,[915,916,917,918] and can produce the highly reactive hydroxyl radical, •OH. This damaging free radical derives from superoxide by way of a Fenton reaction, involving transition metals such as iron. Antioxidants, such as glutathione, can quench superoxide and hydrogen peroxide, decreasing the oxidative stress caused by vitamin K_3.[919,920] The antioxidant enzymes catalase and superoxide dismutase have a similar effect.

Vitamin K_3's anticancer effect is decreased if oxidative stress is lower, in the presence of some antioxidants, for example.[858] Despite this, vitamin C increases the ability of vitamin K_3 to kill cancer cells. It may be that only antioxidants that are able to reduce quinones directly, as opposed to general antioxidants, have protective effects for the cancer cell.[921] At high doses, vitamin C may act as an additional oxidant, driving the oxidation-reduction cycle and generating even greater amounts of hydrogen peroxide, which kill the cancer cell.

Vitamin K and similar molecules inhibit cancer growth. This inhibition is blocked by sulphur containing antioxidants, but not by other reducing agents, such as vitamin C.[921] Vitamin K_3 may lower the activity of critical cell-growth enzymes, by reacting with sulphur-containing amino acids.[922] Vitamin K_3 can combine directly with sulphur-containing molecules, in a process called arylation.[921,923,924] This refers to the addition of aromatic groups, such as those contained in vitamin K_3, to a sulphur-containing molecule, such as the antioxidant glutathione. Adding glutathione to cancer cells inhibits the anti-cancer actions of vitamin K_3.[925,914,915,926,927]

Vitamin K_3 causes oxidative stress in cancer cells, generating a cascade of events, which leads to cell shrinkage, DNA fragmentation and cell death. It does this mainly by oxidation and through combination with sulphur-containing molecules, lowering the activity of cell growth enzymes. It also stops the cell dividing by a separate mechanism, which may be unrelated to its redox cycling.[928] However, the increased chemotoxicity of K_3 when used in combination with electron donors, such as vitamin C, suggests that the redox cycle is a primary factor in K_3's cancer killing role.[15]

The enzyme catalase, which destroys hydrogen peroxide, inhibits the anticancer effect of K_3, but has less effect on the actions of vitamins K_1 and K_2.[858,929] Both K_1 and K_2 appear to act through additional mechanisms, which do not involve generation of hydrogen peroxide.

Vitamins K_1 and K_2 do not take part in the same redox cycling and may induce cell death using additional, non-oxidative mechanisms.

Low concentrations (20 μM/L) of vitamin K_3 result in limited cell death, whereas higher concentrations (60-150 μM/L) can lead to necrosis. In necrosis, cells die abruptly, in a relatively uncontrolled fashion. Vitamin K_2 has similar effects, but at higher concentrations. Concentrations of K_2 as high as 150 μM/L, for example, may bring about a low frequency of cell death, with no necrosis.[930]

Vitamin K_3 may kill cancer cells by two different mechanisms. At higher levels, oxidation induced by the vitamin leads to necrosis, or another form of cell death, autoschizis.[931,932,933,934] During autoschizis, the cell's internal fluid or cytoplasm is extruded from the cell, leaving an intact nucleus.[935] This involves substantial membrane damage. The cell bleeds cytoplasm, leaving behind the sub-cellular particles and structures. The cell becomes smaller, until its membranes surround a shrunken nucleus and a narrow ring of cytoplasm, which contains the internal particles or organelles. The mitochondria also shrink, but the cell does not appear to die from lack of energy. The process of autoschizis depends on a reduction in DNA synthesis.[936] Moreover, addition of the enzyme catalase can prevent the cell from dying,[853] which indicates the involvement of oxidation.

At lower levels, Vitamin K_3 can prevent cell division and induce cell death by influencing gene transcription factors.[937,938,939,940] This method probably forms the basis for its non-oxidative anticancer action. Transcription factors can induce cell cycle arrest and apoptosis, and alter the expression of oncogenes.[855,941,942,943,944,945,946]

Combining vitamins C and K

Since both vitamins K and C use related mechanisms to kill cancer cells preferentially, it makes sense to ask if they might work in combination. Researchers have shown that vitamin K increases the anticancer properties of ascorbate. Vitamin C, combined with a much smaller amount of vitamin K_3, will kill cancer cells more effectively than either substance alone.[947,948,949,950,951] A combination of vitamins C and K_3 could be used for treatment of cancer,[952,953] even with oral dosing.

In experiments on mice with tumours, researchers found that treatment with oral vitamin C and vitamin K_3 increased the lifespan of the treated mice, whose tumours grew less quickly. Human prostate tumours implanted into mice were found to be selectively sensitive to these vitamins.[954] Oral doses of vitamin C and K_3 (15g/L ascorbate with 0.15g/L K_3, in drinking water) have been shown to be effective in mice with metastatic cancer. The controls had greater numbers of tumours of the lung (19, in 14 of 33 mice) and local lymph nodes (9 in 33 mice). The treated mice had fewer tumours of the lung (10, in just 3 of 19 mice) and less lymph node involvement (3 in 29 mice). Microscopic examination confirmed that, in the treated animals, many tumour cells were undergoing autoschizitic cell death. These results demonstrate clearly that oral vitamin C and K_3 significantly inhibit and kill tumour cells.

Mice are not ideal animals for vitamin C studies, as they synthesise the vitamin internally and do not need it in their diet. The effect may be greater in vitamin C dependent species, such as humans or guinea pigs. Preliminary results in human patients with prostate cancer are consistent with the experimental and animal studies.[955]

Human bladder cancer cells are sensitive to the combined action of vitamins C and K_3.[956] These vitamins also increase the effects of standard anti-cancer chemotherapeutic drugs.[957,958] A Chinese research group carried out laboratory studies, confirming that vitamins C and K_3 kill a range of cancer cells.[959] In these experiments, two types of human cancer were shown to be more sensitive to the vitamins than were normal cells. A Japanese group has demonstrated that the cytotoxic activity of both vitamin C and K_3 can be enhanced by using a plant extract, lignin F.[960]

Vitamins C and K_3 have shown anticancer effects both *in vitro* and *in vivo*. In combination with vitamin C, (dose ratio of 100(C):1(K_3)) vitamin K_3 is cytotoxic in doses 10-50 times less than when administered alone.[853] In laboratory experiments, researchers demonstrated a 50% inhibition of growth of oral carcinoma cells with high concentrations of vitamin K_3 (100 μM/L).[859] However, when combined with vitamin C (5 mM/L), a lower concentration of K_3 (8 μM/L) resulted in an equivalent level of growth inhibition. Thus, the synergistic inhibition of cell growth from combined administration required much less vitamin K_3.

When the concentrations were increased, 100% inhibition of cancer cell growth was achieved. Similar results were found using human endometrial adenocarcinoma cells.[853] High concentrations of vitamins C (10 mM/L) and K (100 μM/L) were even more effective, producing a 93% inhibition. In another short experiment, vitamin K_3 at 8 μM/L, when given with vitamin C (500 μM/L), gave 74% inhibition.

The combination of vitamins C and K_3 (dose ratio of 100:1) has been shown to inhibit a number of urological cancer cell lines.[961,962] Two groups of cell lines were selected, to test their sensitivity. The first group contained cell lines covering several grades of cancer of the kidney and bladder. A combination of vitamin C (89 μM/L) and vitamin K_3 (0.9 μM/L) produced a 10-20 fold increase in toxicity, compared to the individual agents. The second group of cell lines included prostate, bladder and testicular cancer. The combination of vitamin C (212 μM/L) and vitamin K_3 (2.13 μM/L) produced a 7-22 fold increase in potency, compared to either vitamin alone.[961,962] Oral doses can sustain plasma levels of vitamin C indefinitely at up to 250 μM/L.[15] These results indicate that oral treatments may be effective in humans, since tumours absorb ascorbate preferentially.

In a further study on prostate cancer, human tumours implanted into nude mice were treated with vitamins C and K_3, resulting in cell death.[954] Nude mice are a form of hairless mouse, which are unable to generate mature T lymphocytes and therefore cannot mount a full immune response. The mice were implanted with human prostate cancer and then given vitamins K_3 and C, orally or by injection. Treated mice were found to survive 25% longer than controls.[963]

Numerous studies illustrate the combined effects of vitamins K_3 and C in cancer treatment. Mice with liver tumours and abdominal fluid, known as ascites, were resistant to the anticancer drug oncovin. When pretreated with the combination of vitamins K and C, the mice regained sensitivity to the drug, without additional organ toxicity.[958] Furthermore, the combination of vitamins C and K_3 is toxic to oral cancer and human leukaemia, but not to healthy cells, such as fibroblasts or pulp cells.[954]

Conclusions

Vitamin K clearly has anticancer activity. Its most direct mechanism is cell oxidation, caused by the redox cycling of vitamin K_3. This exceeds the cancer cells' capacity for production of antioxidants and leads to cell death.

At this stage of the chapter, you might be wondering why vitamin K, or a derivative of it, is not a central feature of cancer research. There is no scientific answer to this question. The vitamin, which is normally present in the body, has a powerful anticancer action and relatively low toxicity. Any new drug with such properties would surely be vigorously promoted and researched. Such a compound would add greatly to the share price of the company that held the patent.

In practice, the use of vitamin K may have been limited by its inability to be patented and thereby to generate profits. Considering the promise of vitamin K as an anticancer agent, with a high degree of safety, it is disappointing that more research has not been performed on this nutrient.

Alpha-lipoic acid

"The antioxidant miracle." Lester Packer.

Like Vitamins C and K, natural elements that are essential to the normal functioning of the body, alpha-lipoic acid is an effective anticancer agent. Alpha-lipoic acid is a powerful antioxidant, synthesised in plants and animals, and found widely in nature.[964] It has much in common with vitamins, but is not an essential dietary requirement, since humans can create it internally.[a]

Alpha-lipoic acid prevents the development of cancer by acting as an antioxidant. It has the properties of an ideal anticancer drug since, at higher levels, it destroys cancer cells while leaving normal cells unharmed. Despite this, its use as a treatment has been overlooked for many years.

The power of alpha-lipoic acid is illustrated in a case related by Dr Burt Berkson, which also shows the constraints involved in practising medicine. The full story can be found in his book *The Alpha Lipoic Acid Breakthrough.*[965] In 1977, Berkson was a physician in a Cleveland hospital. One weekend, a woman picked wild mushrooms and cooked them for herself and her husband. A short time later, the couple were rushed to hospital where they were treated for a stomach upset, before being sent home. Their son, a paramedic, asked them what they had eaten and, realising it could be mushroom poisoning, they returned to the hospital.

It emerged they had eaten several mushrooms from a group known as destroying angels, in this case *Amanita verna.* The main prognostic indicator for poisoning is the amount of mushroom ingested: if a large amount is eaten, the person usually dies.[966] The husband had consumed several: enough to amount to a death sentence. The wife had eaten less than one, making it was possible that she might live. Berkson was given charge of the two patients and told to give them painkillers and fluids. He was expected to wait and see if they died of liver failure.

As luck would have it, Berkson was a trained scientist, with a PhD in the biology of fungi. He contacted a colleague from the US National Institutes of Health, who suggested that an experimental substance,

---

[a] Alpha-lipoic acid is sometimes called thioctic acid, 2-dithiolane-3-penatanoic acid or 1,2-dithiolane-3-valeric acid.

alpha-lipoic acid, might be helpful for treating the liver failure caused by mushroom poisoning. Berkson gave alpha-lipoic acid to the couple, who recovered in a couple of days. Even the husband recovered most of his liver function. A senior doctor explained to Berkson that, although rare, such recoveries are sometimes reported.

The following week, another couple was admitted to the hospital with mushroom poisoning. Berkson was told they had no chance of survival, as the poisoning was severe. He was instructed *not to use alpha-lipoic acid* on these patients, as it was not a recommended treatment. He was not to order more alpha-lipoic acid until the hospital pharmacy committee had approved its use. By this time, the patients would be dead, killed by bureaucracy or fear of legal action. Berkson could not stand by and watch his patients die. He gave them the remainder of his alpha-lipoic acid supply and they went home, alive and well, 10 days later.

Berkson expected to be in trouble, but he was saved by the National Institutes of Health's interest in the patients' recovery. The story illustrates the extent to which legal and other pressures on the modern-day practice of medicine can override scientific knowledge and the interests of patients. It also suggests that alpha-lipoic acid is a remarkable substance.

Chemical structure

Alpha-lipoic acid contains two sulphur atoms that can be reversibly oxidised and reduced. In addition to its properties as an antioxidant, alpha-lipoic acid is a cofactor for several essential enzymes.

There are two optical isomers of alpha-lipoic acid, called R-alpha-lipoic acid and S-alpha-lipoic acid, which have different properties. The term optical isomer refers to one of the main physical differences between the two forms: they rotate polarised light in opposite directions. Chemically, the two are identical and can be difficult to separate. However, as explained previously, such molecules are mirror images of each other, differing like the right and left hands of a pair of gloves.

When alpha-lipoic acid is synthesised in a laboratory, the result is a 50/50 mixture of the R and S forms.[967] However, in living systems, optical isomers generally occur in only one form. Within the body, the two forms are processed differently. In biological systems, only the R-form appears to occur naturally. In this chapter, unless we wish to be specific about the form, we refer simply to alpha-lipoic acid.

Both plants and animals can synthesise R-alpha-lipoic acid. The method of manufacture is not known, but it may be made in the mitochondria,[991] and some may be derived from bacteria in the gut. Normal synthesis does not lead to large amounts of free alpha-lipoic acid in the bloodstream.[987] The alpha-lipoic acid we obtain from food is usually bound to proteins. Offal such as kidney, heart and liver is rich in alpha-lipoic acid. Spinach, broccoli and tomatoes also contain large amounts, and it is found in rice, peas, Brussel sprouts and bran.[968] The amount absorbed from a typical diet is unknown. Dietary alpha-lipoic acid may be bound to the amino acid lysine, which may explain its apparent absence in the blood of unsupplemented individuals.[987]

Supplements

Alpha-lipoic acid is available as a dietary supplement. Many supplements consist of the synthetic form and contain a mixture of R- and S-alpha-lipoic acid.[b] The naturally occurring form, R-alpha-lipoic acid, is also available as a supplement. Supplements of alpha-lipoic acid are rapidly absorbed, metabolised and excreted. Because of their short retention in the body, supplements should be taken in divided doses throughout the day. The absorption of a 200 mg oral dose is 20-40% that of an intravenous dose.[969,970,971] As with vitamin C, large doses are likely to be more effective if given intravenously. Oral doses can decrease oxidative stress,[972] and alpha-lipoic acid supplements act as an effective antioxidant. A notable action of alpha-lipoic acid is that, like vitamin C, it inhibits replication of the HIV virus and may be beneficial in AIDS.[973]

Synthetic supplements of alpha-lipoic acid are less effective than the natural form. The synthetic form is less well absorbed and the S-alpha-lipoic acid it contains is less biologically active. Outside the mitochondria, but inside the cell, S-alpha-lipoic acid may be reduced to dihydrolipoic acid (DHLA) about twice as fast as R-alpha-lipoic acid. However, this conversion rate may be misleading as, when taken orally, natural R-alpha-lipoic acid is more readily absorbed than S-alpha-lipoic acid,[969] so there is less of the L form to be reduced.

It has been suggested that R-alpha-lipoic acid has antiaging properties. This is because oxidation damage to mitochondria may be an important factor in aging. Within the mitochondria, R-alpha-lipoic acid is reduced to DHLA, which is a more potent antioxidant than alpha-lipoic acid. The conversion is nearly 30 times faster with R-alpha-lipoic acid

---

[b] RS-alpha-lipoic acid is sometimes called racemic or DL-alpha lipoic acid.

than with the S- form. This is consistent with the R- form having a more powerful antioxidant effect within the body.

Alpha-lipoic acid supplementation may be beneficial for diabetics, as it increases glucose uptake and metabolism. This is relevant to cancer treatment, because of the relationship between glucose and cancer, emphasised previously. In insulin-resistant rat skeletal muscle, natural R-alpha-lipoic acid is more effective than the synthetic form at enhancing glucose uptake and metabolism.[1011]

R-alpha-lipoic acid is a critical cofactor for several important enzymes, related to energy metabolism.[974] In the body, R-alpha-lipoic acid binds closely to proteins, forming a lipoamide. One reason for the binding may be to retain alpha-lipoic acid in the body, as the unbound form has a short half-life, and is rapidly excreted. The rate of excretion has implications for the use of alpha-lipoic acid in diabetes and cancer.

Toxicity

There appear to have been no reports of overdose with alpha-lipoic acid. Doses of 600 mg/day are well tolerated. In patients with diabetic nerve damage (neuropathy), higher doses, such as 1,200 mg/day for two years or 1,800 mg/day for three weeks, have been taken without adverse effects.[975] In dogs, the LD50 (lethal dose, leading to death in half the subjects) is 400-500 mg/kg after oral doses.[974,976] In rodents, the LD50 is estimated to be 500-1000 mg/kg. These figures equate roughly to 35-70 g in a 70kg human. However, lower dosages may be fatal in rats that are severely deficient in the B vitamin, thiamine.[977] The safety of use by pregnant or breastfeeding women is not established, as full data are not available for these conditions.

Binding heavy metals

Lipoic acid is able to cross the blood-brain barrier and to bind heavy metals. It is a sulphur-containing molecule, which readily binds to (or chelates) mercury and other metals. For this reason, it has been suggested as a medium for removing mercury or other toxic metals from the brain. In the blood, other sulphur-containing supplements such as MSM (methylsulfonylmethane) also bind mercury and other heavy metals, transferring them safely into the urine for excretion. However, unlike alpha-lipoic acid, MSM does not cross the blood-brain barrier.

Lipoic acid can act to prevent and treat the effects of heavy metal poisoning.[978] Free metal ions, such as copper and iron, induce oxidation by catalysing reactions that produce free radicals. Both alpha-lipoic acid and its reduced form, DHLA, can bind metal ions, preventing the reactions that generate free radicals.[979] Alpha-lipoic acid can act as an indirect antioxidant, by inhibiting enzymes involved in oxidation and by binding to metal atoms.[980]

Alpha-lipoic acid forms stable complexes with copper, manganese and zinc.[981] In animal studies, alpha-lipoic acid has been found to provide protection from the poisonous metallic element, arsenic.[982] Alpha-lipoic acid forms a complex with arsenic, preventing its absorption from the gut, and, if the arsenic has been absorbed, stopping it from interacting with enzymes.[987,983,984] Alpha-lipoic acid may also reduce cadmium damage to the liver,[985] and can bind mercury, to remove it from the kidney.[986]

Alpha-lipoic acid can bind low levels of metal ions, removing them from the body and preventing damaging oxidation. However, in the presence of free copper and iron, which may occur in cancer cells, alpha-lipoic acid can generate free radicals.[c] In cancer, this interaction of alpha-lipoic acid with metal irons may be a primary source of free radical generation, leading to cell death.

An outstanding antioxidant

Alpha-lipoic acid is a powerful antioxidant.[987,988,989] It is particularly valuable, since it is available for use inside the cells, as well as in the surrounding tissue. Free alpha-lipoic acid enters cells rapidly, where it is reduced to DHLA, using electrons derived from metabolism.[990] This reduced form then acts as an antioxidant, preventing oxidative damage.[991]

Since alpha-lipoic acid is soluble in both water and fat, it may have an exceptionally widespread distribution in the tissues of the body. Most other antioxidants are primarily soluble in water (like vitamin C) or lipid (like vitamin E). By contrast, alpha-lipoic acid can act as an antioxidant in water- and lipid-based body compartments.

Alpha-lipoic acid is a potent quencher of free radicals. When the reduced form, dihydrolipoic acid (DHLA), neutralizes a free radical, it is oxidized to form alpha-lipoic acid. Once this has happened, the alpha-lipoic acid is regenerated, being reduced back to DHLA by the cell's metabolism. DHLA is one of the most powerful free radical scavengers

---

[c] Sulphur containing organic molecules (RSH) like lipoic acid can react with copper or iron to form thiol radicals: $RSH + Fe^{3+} \rightarrow RS\bullet + Fe^{2+} + H^+$ and $RSH + Cu^{2+} \rightarrow RS\bullet + Cu^+ + H^+$.

available to the cell and can regenerate a number of other antioxidants, including glutathione, coenzyme Q10, and vitamins C and E.[992,993,994]

In some conditions, such as stroke or coronary thrombosis, body tissues may be deprived of oxygen for a period. When the blood supply is restored, a burst of free radicals can damage the tissues. Alpha-lipoic acid may prevent such reperfusion injuries,[995,996,997,998,999,1000] and increase the rate at which cells can repair oxidative damage.[1001] Although this process has been observed in test tube and animal studies, it remains to be fully confirmed in clinical trials.

The reduced form of alpha-lipoic acid increases levels of an important cellular antioxidant, glutathione, which is synthesized using a sulphur-containing amino acid, called cysteine. DHLA increases the uptake of cysteine into the cell, leading to greater glutathione synthesis. Although increases in intracellular DHLA following supplementation with alpha-lipoic acid may be short-lived, DHLA may improve intracellular antioxidant capacity by inducing glutathione synthesis.[967]

A genetic regulator

Alpha-lipoic acid is an anti-inflammatory agent and acts on mechanisms that control cell growth. Increasingly, inflammation is recognised as a factor in many diseases. A nuclear factor, known as NF-κB,[d] is an important protein in regulating genes that are involved in inflammation and diseases, such as atherosclerosis, cancer and diabetes.[974] In particular, NF-κB regulates cell proliferation, transformation and tumour development.[1002] It does this by altering the way genes are expressed in inflamed cells.[e] When added to cells in culture, alpha-lipoic acid inhibits the activation of NF-κB.[1003] The reduced form, DHLA, inhibits the activity of other transcription factors.[1004]

Diabetes

Because of the involvement of glucose in the development and possible treatment of cancer, we need to describe the action of alpha-lipoic acid on blood glucose. In Germany, high intakes of alpha-lipoic acid have been used in the treatment of diabetes mellitus since the 1960's.[988,1005] Supplements may lower blood glucose levels in diabetics and

---

[d] Pronounced N-F-Kappa-B.

[e] NF-κB is a transcription factor that binds to DNA and alters the rate of expression of certain growth related genes. DHLA inhibits the transcription factor AP-1, by decreasing the expression of the proto-oncogene, c-fos.

subjects with impaired glucose tolerance. Diabetics using alpha-lipoic acid may be able to reduce their dependence on insulin or other treatments; they should therefore monitor their blood glucose levels. People with diabetes also suffer an increased risk of cataracts. In rats, R-alpha-lipoic acid is more effective than the synthetic RS form for preventing cataracts.[1006]

There is evidence that high dose alpha-lipoic acid increases sensitivity to insulin. This has implications for the uptake of vitamin C by tumours, using insulin dependent glucose transporters. Researchers gave intravenous doses of alpha-lipoic acid (600mg and 1,000mg) to patients with type 2 diabetes, and found insulin sensitivity increased (by 27% and 51%, respectively).[1007,1008] In a second study, 20 type 2 diabetics were given twice-daily oral doses of 600mg alpha-lipoic acid for four weeks, whereupon their measures of glucose metabolism improved.[1009] In a further study, 72 type 2 diabetics were given doses of 600 mg/day, 1,200 mg/day or 1,800 mg/day for four weeks. The results showed that oral alpha-lipoic acid increased insulin sensitivity by 25%.[975] Type 2 diabetics given injections of alpha-lipoic acid were reported to show a substantial improvement in glucose uptake.[1008,1010]

These studies were performed using synthetic alpha-lipoic acid; it is possible that R-alpha-lipoic acid is more effective for improving insulin resistance.[1011,1012] In addition, alpha-lipoic acid itself may stimulate glucose uptake by muscle cells, in a similar way to insulin.[1012]

People with either type 1 or 2 diabetes experience high levels of oxidative stress. This may be the principle cause of associated vascular and neurological complications. In one experiment, 33 diabetic patients took alpha-lipoic acid (600 mg/day) for at least three months. The researchers found the supplemented patients had lower levels of oxidised lipid in their blood than 74 controls.[1013] This finding has been confirmed in a trial with 10 diabetic subjects.[1014]

Cancer

Alpha-lipoic acid is a powerful antioxidant and may be an effective anti-cancer agent.[1015] Its antioxidant action indicates its potential in cancer prevention. In this book, however, we are primarily concerned with its use as a cancer treatment. Much research in this area has focussed on its use in combination with conventional treatments.

Preventing side effects of chemotherapy

The platinum-based chemotherapy drug, oxaliplatin, has a side effect of damaging peripheral sensory nerves. This acute but transient neuropathy can appear during the first few treatments, in 85% to 95% of patients receiving the drug.[1016] With continued doses, nerve damage accumulates in about 10-20% of patients, limiting the dose that can be given. Alpha-lipoic acid relieves nerve damage in diabetics,[1017] and can help prevent this restrictive side effect of oxaliplatin.[1018] Intravenous alpha-lipoic acid, 600 mg once a week for three to five weeks, followed by thrice-daily oral doses of 600 mg, prevented the damage in eight out of 15 patients (53%).

Alpha-lipoic acid protects against nerve damage in patients given the anticancer drugs cisplatin and docetaxel.[1019] Cisplatin is a platinum-based therapeutic agent, with a side effect of kidney damage. Alpha-lipoic acid protects rats against cisplatin-induced oxidative damage to their kidneys.[1020] Use of alpha-lipoic acid also produced a synergistic increase in survival in experimental mice, undergoing doxorubicin therapy against leukaemia.[1021] At a low concentrations (1 μM/L), alpha-lipoic acid acted as a cancer growth factor, while at higher concentrations (100 μM/L) it acted as an antiproliferation agent. This concentration-dependent response is analogous to the effects of the related antioxidant, vitamin C.

Cells from a disease[f] that causes increased sensitivity to radiation and oxidative stress were also found to benefit from alpha-lipoic acid.[1022] Since both radiation therapy and most forms of chemotherapy depend on inducing oxidative stress, the role of alpha-lipoic acid as an agent for reducing side effects is easily understood.

Cancer killer

Like vitamins C and K, alpha-lipoic acid kills cancer cells preferentially, leaving healthy cells unharmed. Indeed, its antioxidant properties are beneficial to normal cells. Rather than damaging healthy tissues, as conventional chemotherapy does, alpha-lipoic acid may have general benefits for the health of cancer patients. Human tumour cell lines commit suicide following exposure to alpha-lipoic acid.[g] By contrast, healthy cells are inhibited from cell growth and division[1023] so they may be less affected by chemotherapy. Such properties indicate that alpha-

---

[f] Ataxia-telangiectasia.

[g] Cell lines subject to lipoic acid induced apoptosis include FaDu, Jurkat, and Ki-v-Ras-transformed Balb/c-3T3 murine mesenchymal cells.

lipoic acid could be beneficial in both the prevention and treatment of cancer.

Free radical generation

Alpha-lipoic acid exploits the distinct metabolism of cancer cells, to act as a highly selective anticancer agent. In healthy cells, alpha-lipoic acid acts as a powerful antioxidant.[1024] In cancer cells, it acts as an oxidant, generating free radicals. For example, it has been shown to generate superoxide within human colon cancer cells, causing them to commit suicide. Application of a suitable antioxidant can prevent the cell dying by apoptosis.[1025] A similar induction of cell death in human leukaemic cells, but not in healthy white cells, has been described.[1026] These results indicate the potential for alpha-lipoic acid as an anticancer agent.

Selective killing of cancer cells has been replicated by several laboratories. For example, Lester Packer, a leading alpha-lipoic acid researcher, reports that alpha-lipoic acid (100 μM/L) kills leukaemic cells but not normal white blood cells.[1027]

Synergy with vitamin C

Alpha-lipoic acid boosts the power of vitamin C to destroy cancer cells. Vitamin C might reduce alpha-lipoic acid radicals, generating a redox cycle, resulting in oxidation that damages the cancer cells.[h] Riordan has tested the potency of an ascorbate and alpha-lipoic acid mixture for killing cancer.[1028] The addition of alpha-lipoic acid greatly increased the anti-cancer effect: only about one eighth the amount of ascorbate was required when alpha-lipoic acid was included. The treatment was effective against both rapidly growing and non-proliferating cells. This combination may be more effective than existing drugs.

As with vitamin C, tumours may absorb more alpha-lipoic acid than expected, although the underlying mechanism remains to be established. If this is so, then alpha-lipoic acid may reach cytotoxic levels in cancer tissues at lower blood levels than predicted. Research data supports this suggestion. In cancer patients, alpha-lipoic acid is absorbed more quickly into red blood cells than in healthy patients,[1029] and thus may be more available to tissues. Breast tumour tissues have an increased uptake and binding of alpha-lipoic acid, compared to normal tissue.[1030] Cancer cells accumulate more alpha-lipoic acid in their mitochondria than do healthy cells.[1031] When rats are injected with alpha-lipoic acid, those

[h] The proposed reaction of organic sulphur radicals (RS•)with ascorbate (A) is $AH^- + RS\bullet \rightarrow RSH + \bullet A^-$.

with cancer accumulate more than healthy animals do; repeated injections have been shown to extend the sick rats' lifespan by 25%.[1032,1033]

Conclusions

Like vitamin C and some forms of vitamin K, alpha-lipoic acid can be a powerful, non-toxic anticancer supplement. Its antioxidant action suggests it could prevent cancer development. An equally exciting feature is its potential for use as an anticancer treatment.

The action of alpha-lipoic acid appears to be concentration dependent, as is that of vitamin C. Low levels act as an antioxidant in healthy and cancer cells, benefiting both. At higher concentrations, alpha-lipoic acid causes oxidation and cell death by apoptosis in cancer cells, though not in normal cells. The cancer killing action of alpha-lipoic acid, like that of vitamin C, may be related to the presence of free iron, or similar metals, in cancer cells. Redox cycling of alpha-lipoic acid may be the principle mechanism involved.

As with vitamin K, alpha-lipoic acid's anticancer action is greatly increased by high levels of vitamin C. The combination of R-alpha-lipoic acid and vitamin C can be taken orally for long periods, providing sustained plasma levels. Tumours may accumulate high levels of these antioxidants, promoting redox cycling and destruction of cancer cells. This finding leads us to wonder whether non-toxic anticancer substances may be more frequent in the diet than previously realised.

Anticancer nutrients

"The aim of medicine is to prevent disease and prolong life; the ideal of medicine is to eliminate the need of a physician." William J. Mayo.

We have seen that a number of simple dietary nutrients can kill cancer cells, without damaging healthy cells. This leads us to ask whether there might be further such molecules, awaiting exploitation as potential therapies. Decades of research have largely failed to find effective and safe anticancer treatments, so the orthodox answer to this question might be that these agents are extremely rare. Despite this, the evidence suggests that such substances are surprisingly common.

Vitamin E

Vitamin E is the second best-known dietary antioxidant, after vitamin C. Since vitamin C is an effective anticancer agent, it is worth considering whether vitamin E might have a similar action.

Vitamin E was first recognized in 1922, and was chemically isolated in 1936. Shortly afterwards, it was synthesised in the laboratory. There are two main classes of vitamin E molecule: tocopherols and tocotrienols. The tocopherols are more common and occur frequently in supplements. Both types of vitamin E come in four main categories, called alpha, beta, gamma and delta.[a]

The most common form of vitamin E is alpha-tocopherol. Synthetic vitamin E typically contains a mixture of two forms of alpha-tocopherol, prefixed D- and L-, which have the same molecular formula.[b] The prefix DL-, as in DL-alpha-tocopherol, indicates a mixture of both D and L types. These forms are chemically similar but biologically different. Natural tocopherols come only in the D form, such as D-alpha-tocopherol, and are considered to be nutritionally and biologically superior.

---

[a] The actual chemical description of both forms of vitamin E is more complicated, for example RRR-alpha-tocopherol is alpha tocopherol, but the simplification into alpha, beta, gamma and delta forms is in common use.

[b] Synthetic vitamin E is now often manufactured as all racemic, a mixture of eight different forms, with limited specific biological activity.

Vitamin E succinate

Our first and most important question is whether vitamin E has specific anticancer actions. Research has shown that some forms of vitamin E, such as vitamin E succinate, can kill cancer cells while leaving normal body cells unharmed. The results for this type of vitamin E are unambiguous. Unfortunately, these findings have not been followed up adequately. Despite accumulating evidence of anti-cancer properties, vitamin E succinate has not received significant attention from researchers and clinicians.[1034]

Researchers have shown that the alpha-tocopherol succinate form of vitamin E can induce cancer cells to commit suicide.[1034] The story of how Kedar Prasad and colleagues made this discovery is interesting. In 1979, Prasad showed that a preparation of vitamin E caused cultured cancer cells from mice to revert (differentiate) to a more healthy form.[c] In addition, this vitamin E made rat cancer cells more sensitive to X-rays and anticancer drugs.[1035,1036]

However, Prasad discovered that a solvent used in the vitamin E preparation was itself toxic to melanoma cells.[d] This meant that the anti-cancer effects of this preparation might not have come from the vitamin E at all.[1037] The particular solvent used was a trade secret and could not be revealed. In order to continue his investigation, Prasad dissolved vitamin E, as tocopherol and tocopherol acetate, in alcohol. In the new solvent, he found that vitamin E was not effective against cancer cells.

Many people would have assumed the original solvent was the effective agent, rather than the vitamin E. However, Prasad persevered with his experiments, using two other forms of vitamin E: tocopherol nicotinate and tocopherol succinate. In 1982, he reported that tocopherol succinate induced differentiation, stopped cell division and caused cell death, in mouse melanoma cells.[1037] If the dose required to prevent growth was doubled, the tumour cells died. A concentration of 9.4 μM/L inhibited growth by 50%, without cell death, whereas a concentration of 18.8 μM/L led to almost total cell destruction.

Vitamin E succinate induces human breast cancer cells to commit suicide.[1038,1039,1040,1041] The effect has been demonstrated in a number of experimental cell lines, and also clinically.[1042,1043] It is highly selective for

[c] Aquasol vitamin E was used in this experiment.

[d] This could be another indication of the ubiquity of selective anticancer agents.

cancer cells; for example, it induces human prostate cancer cells to die, while not harming healthy cells.[1044]

By analogy with vitamin C, we might expect this form of vitamin E to increase oxidation levels in cancer cells; there is some evidence to support this suggestion. The cytotoxic action of vitamin E succinate is inhibited by antioxidants, indicating that superoxide and hydrogen peroxide may be involved in its mechanism of action.[1045,1046] Once again, this suggests a redox mechanism for a safe anti-cancer agent.

Although these results are encouraging, oral administration of alpha-tocopherol succinate may not be an effective anti-cancer treatment, because the molecule is not absorbed intact.[1047] Injection, or liposomal preparations, may be needed for vitamin E succinate to be useful as an anticancer agent. Liposomes are small, membrane-bound balls, formed using phospholipids. Drugs that degrade or are poorly absorbed in the stomach can be included in liposomes, to increase their uptake.

Another researcher, Jiri Neuzil, showed that Vitamin E succinate reduces tumour growth by 80%, without side effects, in mice with colon cancer grafts.[1048] Neuzil claimed this epitomized cancer cell killing by a pharmacologically relevant compound, without known side effects. In a later study, he states that vitamin E succinate is a potent and highly specific anticancer agent, of considerable therapeutic potential.[1049]

Vitamin E succinate is potentially a powerful anti-cancer agent, with few, if any, side effects.[1050] It kills many types of cancer cell and prevents cell division.[1051] It also inhibits DNA synthesis in cancer cells.[1052,1053,1054] There is strong evidence that it can signal cancer cells to revert (redifferentiate) to a more healthy state.[1055,1056,1057] Finally, the tumour inhibition observed in isolated cells has been reproduced in the body.[1044,1042,1058,1059,1043]

It is clear that vitamin E succinate provides an alternative, non-toxic approach to the treatment of cancer.[1060] For the sake of cancer sufferers everywhere, clinical trials of its action on cancer should be a matter of urgency. We can only hope such trials will be performed adequately, and soon.

Gamma tocopherol and tocotrienols

Another form of vitamin E, gamma tocopherol, is known to cause apoptosis in cancer cells.[1061,1062,1063] The tocotrienols appear to be even more effective.[1064] Tocotrienols are more readily absorbed, which may explain their increased effectiveness.[1065] Tocotrienols inhibit cancer growth at a dose only 20% of that needed to produce apoptosis.

Vitamins E and C combined

As might be predicted, vitamin C enhances the chemotoxic actions of vitamin E. The proliferation of cancer cells, which is slowed by vitamin E succinate, is decreased further by combining it with vitamin C and forms of vitamin A (beta carotene and retinoic acid).[1066,1067] Once again, these studies indicate that the response of cancer cells to redox supplements can be greatly increased when they are used in combination.

Coenzyme Q10

Since some forms of vitamin E have potent anticancer activity, we now consider another lipid-soluble antioxidant, coenzyme Q10. There is evidence that coenzyme Q10 levels are lower in cancer patients, perhaps reflecting increased oxidative stress.[1068,1069]

Coenzyme Q10 is a member of a class of related molecules, called ubiquinones, because of their chemical structure and the fact that they are found in so many cell types. Coenzyme Q10 is manufactured in the body, although synthesis depends on many other nutrients, including eight vitamins.[1070] Its production therefore depends on good nutrition. Some drugs, such as the statins that are used to lower cholesterol, can prevent synthesis of coenzyme Q10.[e] Ensuring an adequate supply of coenzyme Q10, especially during aging, may be difficult without nutritional supplementation.

The health benefits of coenzyme Q10 are widely reported and it has become a popular nutritional supplement. For years, it has been used to treat heart failure and other cardiovascular disorders.[1071,1072] Coenzyme Q10 has also been used to reduce the tissue damage caused by conventional cancer chemotherapy.[1073]

Karl Folkers, a leading chemist, was a major proponent of coenzyme Q10 and its use in the prevention and treatment of disease.[1074] Working at Merck, he led a research team that was first to isolate vitamin B_{12}, in 1948. In 1951, he gained the Scientific Award of the Board of Directors at Merck.

---

[e] Anyone taking statin drugs needs to supplement with coenzyme Q10 to reduce the risks of severe side effects. However, this advice is generally not provided by the prescribing physician.

This award was given,

"To honor his vision and research acumen, and to honor his many scientific contributions to mankind, especially his research on antibiotics and vitamins, culminating in the isolation and crystallization of vitamin B-12."

Nowadays, it seems surprising that a pharmaceutical company could value basic research into nutrients in this way. In 1956, Folkers was promoted to the position of Executive Director of Fundamental Research at Merck. In 1958, he determined the structure of coenzyme Q10. However, he resigned from Merck in 1963, as he wished to continue his work on coenzyme Q10 and its implications in disease. By this time, the drug company did not agree with the direction of his research. In later years, Folkers published many papers on the benefits of coenzyme Q10 in health and disease.

In one of these, an uncontrolled study, 32 breast cancer patients with poor prognosis were given an antioxidant cocktail, in addition to coenzyme Q10 at 90mg per day. This included vitamin C (2,850 mg), vitamin E (2,500 iu), beta-carotene (32.5 iu), selenium (387 μg), essential fatty acids and other supportive vitamins and minerals. No patient died, although the expected number of deaths was reported as four, and none had additional metastases.[1075] The patients' quality of life was improved, reflected by reduced weight loss and need of painkillers. In addition, six patients appeared to have a partial remission. However, the number of nutrients given makes interpretation of this study difficult.

In a later report, Folkers described how he gave an increased dose of coenzyme Q10 (390 mg per day) to one of these patients, who was showing signs of remission. After one month, the tumour could no longer be felt through the skin. Two months later, it could not be seen using X-ray examination. Encouraged, Folkers treated another case with 300mg per day of coenzyme Q10. This patient had a verified, residual breast tumour, following non-radical surgery. Three months later, the patient was in excellent condition, with no residual tumour tissue.

Three additional remissions from advanced breast cancer were reported.[1076] These patients had undergone conventional therapy, with the addition of coenzyme Q10 at 390mg per day. Over a period of 3-5 years, numerous metastases in the liver of a 44-year-old patient "disappeared", and no new metastases were found in the body. In another case, a 49-year-old patient had a tumour in her pleural cavity. After six months, all signs of this disappeared and her condition was described as excellent. The final patient in this group was a 75-year-old, who had a lumpectomy for cancer of one breast and, after treatment including coenzyme Q10, showed no residual cancer, in either the tumour bed or metastases. Folkers also reported eight patients who survived for a period of five to fifteen years.[1077]

Folkers' evidence is relatively weak. For example, the experiments were poorly controlled and the patients had other treatments that could have been effective. However, even if we assume that these patients were "best case" studies, the findings indicate a need for follow-up research.

In 1999, researchers from Britain and Denmark suggested that coenzyme Q10 might have a role in cancer therapy. As a result of the animal studies and the above clinical observations, they demanded proper clinical trials.[1078] Some cancer patients, especially women, may already be self-supplementing with coenzyme Q10 as an alternative cancer therapy.[1079] John Ely has suggested that if coenzyme Q10, glycaemic reduction and vitamin C supplementation were employed regularly, cancer would cease to be a leading cause of death.[814]

The chemical structure of coenzyme Q10 is similar to that of vitamin K.[f] Like vitamin K, coenzyme Q10 has an oxidation-reduction cycle. There are other similarities. Coenzyme Q10 inhibits cell division and apoptosis.[1080,1081] Moreover, coenzyme Q10 is directly involved in metabolism and affects the oxidation state of mitochondria.[1082,1083] Oxidation of coenzyme Q10 can produce hydrogen peroxide.[1084,1085] Mitochondrial oxidation of coenzyme Q10 has been cited as a primary source of both superoxide and hydrogen peroxide in cells.[1086] This oxidant generation is controlled by the redox state, and is related to the energy available in the mitochondria.[1085] Conditions such as increased oxidative stress, which disturb the binding of coenzyme Q10 to protein, may encourage peroxide production.[1087] Coenzyme Q10 may selectively produce oxidation in cancer cells.

Finally, we have mentioned that dietary restriction can help to slow or kill cancers. It has been proposed that these effects work, at least in part, by increasing levels of coenzyme Q10.[1088] We can conclude that the evidence for coenzyme Q10 is suggestive of a redox-mediated anticancer effect. Indeed, coenzyme Q10 has properties consistent with our requirements for a non-toxic anticancer substance.

---

[f] Both vitamin K and coenzyme Q10 are quinones.

Red wine

French people suffer lower levels of heart disease than many other Europeans do, despite an apparently unhealthy lifestyle that includes smoking and a high fat diet. This inconsistency, known as the French paradox, has been attributed to their consumption of red wine. Red wine contains a number of powerful antioxidants, the best known of which is resveratrol. Interestingly, resveratrol is also an ingredient of Chinese folk remedies. In this section, we investigate whether resveratrol and related molecules might prevent tumour growth or kill cancer cells. If so, we will have a good excuse for drinking red wine!

Resveratrol was first isolated in 1940, from the roots of the white hellebore (*Veratrum grandiflorum O. Loes*).[g] However, since that time it has also been found in numerous plants and plant derivatives, including grapes, berries and peanuts. The main claims for resveratrol concern cardiovascular benefits. However, researchers have also reported that resveratrol can suppress growth or induce cell death in tumour cells, while leaving healthy cells undamaged.[1089,1090] Resveratrol has been shown to be effective against many cancer cell lines, including leukaemia,[1091] myeloma,[1092] breast cancer,[1093,1094,1095,1096] and carcinoma of prostate,[1097] as well as stomach, colon, pancreas, thyroid, squamous cell, ovarian, cervical, lymphoid and myeloid cancers.[1098,1099]

Resveratrol prevents cell division and kills cancer cells in culture.[1100,1101] It has also been shown to be effective against tumours in animals.[1102,1103,1104] Resveratrol's anticancer effect is such that it has been suggested as a potential therapy for prostate cancer,[1105] melanoma,[1106] leukaemia[1107,1108,1109] and lung cancer.[1110] American researchers have proposed it as an agent for purging cancer cells from bone marrow, taken from leukaemia patients, before the marrow is transplanted back into the patient.[1111]

Resveratrol may be an effective therapeutic agent for treating cancer in people. Like vitamin C, it appears to have limited oral absorption and a short half-life in the body.[1112] It should therefore be provided in repeated doses, at short intervals, to provide maximum benefit. We hope that this is realised by doctors, before they start performing clinical trials. A recent animal study, which gave mice only five doses of resveratrol in a week, reported a weak response compared to that expected from test tube results.[1113] Such a result is entirely predictable, considering the short half-life. The resveratrol would only

---

[g] As trans-3,5,4'-trihydroxystilbene

have been in the mice's bloodstreams for a few hours a week: any therapeutic effect would therefore be small.

Anticancer effects have also been demonstrated using naturally occurring molecules, related to resveratrol.[1114] Researchers are currently studying modified forms of resveratrol as anticancer agents,[1098] presumably because, unlike resveratrol itself, the adapted forms can be patented and could potentially make large profits.

Commonplace anticancer agents

Despite the large number of nutrients offering anticancer effects, a recent article suggested that cancer patients should disclose supplement use to their physicians, because nutrients may have unnecessary risks.[1115] The paper pointed out that there were no studies quantifying such risks, but suggested that there were some reports in the literature. Medical practitioners were recommended to ensure that they were able to identify such risks. However, since the report also pointed out that research was needed to clarify what the risks actually were, it is difficult to follow the logic of this advice.

The idea that potential anticancer agents are rare is widespread. Graham Pinn, a physician from Queensland, Australia, has reviewed herbal medicine in oncology. He states that,

> *"Extensive screening of tens of thousands of plants has unfortunately revealed only a handful of potential cancer cures."*[1116]

Presumably, Pinn has not realised that pharmaceutical company screening is performed with the aim of finding patentable substances, and the generation of huge profits. This primary aim of screening would appear to exclude identifying and exploiting the action of nutrients. As we have shown, even a brief look at some commonly used supplements gives a strong indication of non-toxic anticancer effects.

Some people may be amazed to hear that some of these supplements are effective anticancer agents. However, the biology of microevolution suggests that these are just the sort of molecules needed to fight cancer. Control of cell division and cell suicide has been "under development", in plants and animals, for millions of years. Such biological features are essential to the structure of our bodies and to providing resistance from cancer. Potent agents, able to affect cell division and death, are commonplace in nature. Furthermore, given an understanding of microevolution, we should not be surprised to find that such molecules are relatively safe for normal tissues.

Consider government advice to eat more fruit and vegetables. While this advice is excellent, it obscures the underlying biology. Research has revealed that certain fruits and vegetable reduce the risk of getting some cancers. It is often suggested that the vegetables contain chemicals that are not found in supplements. Such assertions lead us to propose clear scientific questions. What are these anticancer substances? How do they work? Can we isolate them and obtain powerful cancer treatments or preventative measures?

It is easy to find non-toxic anticancer agents. Fruit and vegetables contain substances called phytochemicals: many of these have been shown to reduce cancer cell proliferation, hold back growth signalling and inhibit the provision of new blood vessels to tumours. Phytochemicals are anti-inflammatory and can induce apoptosis in cancer cells. They include genistein, diallyl sulfide, S-allyl cysteine, allicin,[1117,1118] lycopene,[1119] capsaicin, 6-gingerol, ellagic acid, ursolic acid, silymarin, anethol, catechins and eugenol.[1120] There are many more such substances,[1121,1122] including fruit extracts,[1123] fermented soy,[1124] pomegranate juice,[1125] curcumin,[1126,1127,1128,1129] which is found in the spice turmeric, and anthrocyanidins, which are found in supplements such as pycnogenol and grape seed extract.[1130,1131,1132] The number, availability and potency of these agents varies. Curcumin, for example, acts as an antioxidant in normal cells but generates free radicals in cancer cells,[1133] leading to cell suicide.[1134]

Orthodox medical reports sometimes suggest that these substances may be used alongside conventional chemotherapy. In some studies, it has been shown that they improve the effectiveness of the conventional treatment. In other cases, they lessen side effects, such as nausea. However, we consider the real value of supplements lies in their direct action on the cancer. Such nutrients could provide hope, when conventional medicine has failed. Most of these anticancer nutrients are remarkably safe and provide potential therapies for cancer, which have been woefully ignored.

A new class of anticancer drug

"The physician is Nature's assistant." Galen (130-200)

Redox agents are a powerful treatment for cancer. Given this, it is inevitable that someone will find a way to make money from them. Sadly, we can forget the idea that the medical industries will consider nutrient-based therapies seriously, since supplements do not offer the potential for vast profits. Furthermore, if widely used, such therapies could devastate the income from current cancer treatments. There is no financial incentive for the development of nutritional therapies.

This cynicism is based on extensive reading of the literature. As described in the previous chapter, numerous substances have been suggested as cures for cancer but have not been fully investigated. Alternatively, follow-up studies have been carried out, but the doses have been too small, too infrequent, or administered via an inappropriate route. The authors then generalise the findings of such flawed studies, claiming to have shown that the nutrient in question is worthless. In the short term, not following up a particular research area properly could be excused as an oversight. In the longer term, repeated failure to follow up promising new lines of research suggests a bias in the philosophy underlying medical research.

Evidence suggests that financial gain is the major driving force in the development of cancer treatments. Nutrient-based therapies are unwelcome competitors to conventional treatments, because they compete strongly on both effectiveness and cost. One way drug companies might try to answer this threat is by finding drugs that work in similar ways to nutrient therapies, by selectively generating a redox cycle in cancer cells, for example. Finding such a drug could allow profits to be maintained. Of course, this countermeasure would fail if purchasers realised they could get the same thing from supplements, for a fraction of the price and with fewer side effects.

A "new" kind of cancer drug

The eventual development of redox-cycling drugs, as part of scientific progress in this area, is likely, if not inevitable. At least one such drug exists and is currently undergoing clinical trials; it is called motexafin gadolinium.[a] The company who developed it deserves credit for championing the redox-cycling approach in the pharmaceutical arena. If the research and development underlying this new drug is successful, motexafin gadolinium could be the forerunner of a new class of selective anticancer drugs. It may also provide a prototype for drugs that can be used in combination with nutrient-based therapies.

The new drug is based on gadolinium, a silvery metal that was isolated relatively recently. Gadolinium, and the associated mineral, gadolinite, were named after Johan Gadolin, a Finnish chemist and geologist. Unlike most rare earth elements, gadolinium is stable in dry air. However, when moisture is present, it tarnishes rapidly and the oxide formed falls off in flakes, exposing further metal for oxidation. Gadolinium is strongly magnetic and is used for making phosphors for colour TV tubes, and in the manufacture of compact discs and computer memory. More importantly, it is used as an intravenous contrast medium, to enhance the images from magnetic resonance scanners.[1135,1136] It does not appear to have a role in normal biological mechanisms, however.

The drug, motexafin gadolinium, was initially used to increase the sensitivity of tumours to radiation therapy.[1137] It was hailed as a minor breakthrough in radiation treatment of tumours, because it could be seen in tumours, using MRI scanning.[1138,1139] When the drug was injected, it would concentrate in the cancer, producing a large change in the magnetic resonance signal, which allowed visualisation of the tumour.[1140]

Researchers used animal models to demonstrate motexafin gadolinium's benefits in enhancing radiation therapy.[1141] Others showed that it changes the redox state of the cell, by oxidising a range of substances and by generating hydrogen peroxide.[b,1142] One group has stated that the drug operates by a "unique mechanism involving futile redox cycling."[1142] Such incorrect assertions, that the redox-cycling mechanism is new, novel or unique, are repeated frequently.[1143]

By now, readers of this book will know that this supposedly "unique" mechanism occurs widely. However, ignoring vitamins, the claim that this drug is the first relatively non-toxic, *synthetic* redox-cycling substance employed in cancer therapy may well be correct.

---

[a] Motexafin gadolinium is also known as gadolinium texaphyrin.

[b] The drug oxidised ascorbate, NADPH (nicotinamide adenine dinucleotide phosphate), glutathione and dihydrolipoate.

From our descriptions of redox cycling in several nutrients, we can predict that motexafin gadolinium should act synergistically with vitamin C to kill cancer cells. This is indeed the case: the drug works with ascorbate, or other antioxidants, to sensitise cancer cells to radiation damage.[1144] As the concentration of vitamin C is increased to a level equivalent to that achieved with low dose supplementation (50-100 μM/L), the cancer cells stop dividing and commit suicide.[1145]

Recently, it has been proposed that redox-active molecules, such as motexafin gadolinium, could have a role in cancer therapy.[1146] Consequently, it was suggested, this mechanism might be fruitful for the development of new drugs.[1147] This "new" approach to cancer treatment has resulted in motexafin gadolinium being the made the subject of clinical trials, as an anticancer agent.[1148] That these trials are preferred to studies of nutrients is not surprising, as about 80% of such clinical trials are privately funded, for drug evaluation and marketing purposes.[9,11] Nonetheless, this "new" approach has been known for over 30 years, in safe nutrient-based supplements. The corresponding nutrients have not had the benefit of appropriate trials, and therefore they are not made available to patients.

Conclusions

Now that drugs employing the mechanism of selective redox cycling in cancer cells have been developed, we can hope that effective treatments for cancer will be forthcoming. However, these redox cycling mechanisms have been established for decades. It is inevitable that, given time, drugs using them will be found and exploited.

Drugs generally have a higher level of toxicity than nutrients. As medicine exploits these new drugs, nutrient therapies might be seen as a threat to profits, leading them to be ignored. Nevertheless, with the medical establishment at last beginning to understand these biochemical mechanisms, there is hope that the use of nutrient therapies will receive consideration. A more scientific treatment of cancer might incorporate nutrients and orthomolecular medicine. Unfortunately, a profit-hungry industry could suppress these alternatives.

Conquering cancer

"One of the first duties of the physician is to educate the masses not to take medicine."
Sir William Osler

Throughout this book, we have referred to different methods of fighting cancer. Although we have discussed these treatment methods separately, some of them work even better in combination. This exciting possibility could allow the development of alternative treatments, with the potential to overcome cancer in the majority of cases.

The microevolutionary model suggests two broad approaches to limiting cancer growth, based on population control and extinction. For population control, the aim would be to establish an environment that inhibits the division and evolution of cancer cells. A sufficiently slow-growing tumour would have a negligible effect on a patient's expected lifespan. If growth can be limited at an early stage, so the tumour remains small, its effect on the patient's health might be inconsequential.

Population control methods could allow patients to live out their normal lifespan, though this outcome is not certain. For example, when treating patients with chemotherapy, physicians consider tumour shrinkage a measure of success. In evolutionary terms, shrinkage reflects a period of negative population growth. Unfortunately, it also corresponds to a period of intense selection pressure on the remaining cancer cells; those that resist the drug are more likely to survive. At the end of the treatment, growth of these more resistant cancer cells can resume. The only way to avoid this re-growth would be to continue therapy until no more cancer cells remained. In this case, the population would be extinct.

Extinction of cancer cells is the most effective therapy and a true cure for cancer. This means keeping the cancer cells in a state of negative population growth, until they all die. This is easier said than done, since the intrinsic biological diversity of a typical cancer means that a proportion of cells are likely to be resistant to any particular therapy. Large tumours contain a variety of cell types, including normal cells, relatively benign cells, invasive cancer cells and infiltrating white blood cells.[1] Furthermore, tumours in secondary locations are even more diverse. This explains why secondary cancers (metastases) can be more difficult to destroy than primary ones.

The toxic nature of many conventional anti-cancer drugs limits their use as agents for the total extinction of cancer cells. Clearly, if the treatment is toxic to the patient, the cancer must become extinct before the patient dies, or is permanently disabled. For this reason, non-toxic therapies have an advantage in the treatment of intractable cancers. Patients can take safe therapies indefinitely, and physicians can combine such treatments for maximal effectiveness. We will consider non-toxic therapies shortly, after a brief reiteration of the development and characteristics of cancer.

Understanding cancer

There are two main approaches to cancer, which differ with respect to the importance accorded to the biological diversity found in cancer cells. Some physicians view cancer as a single disease, with many manifestations. From this viewpoint, the diversity of cancer is secondary to its common features. Others stress the biological diversity, regarding cancer as a collection of diseases, which share common attributes, such as abnormal growth and genetic changes. Cancer cells are highly abnormal, with damaged chromosomes and genetic divergence from healthy cells. A feature of all cancers is the abnormal growth of cells within the body. Malignant cancer is characterised by unrestrained spread to different tissue locations.

Another way of looking at cancer uses the microevolutionary model described in this book. According to this theory, the different manifestations of cancer are the results of a common evolutionary process, in which selection pressures favour the development of 'species' of cancer. In microevolutionary terms, the abnormal spread of cancer cells is the natural response of an independent organism – the cancer.

Reproduction and spread are primary activities of biological organisms, including the single-celled creatures from which animals and plants evolved. The characteristics of living cells have evolved over billions of years, allowing them to survive, initially as single cells and, later, as multicellular organisms.

Cancer is common, because it does not have to evolve from scratch: we all possess the mechanisms for cancer within our cells. Every cell in a multicellular animal retains the biology of its single-celled ancestors. In single-celled creatures, the biological mechanisms for survival stimulate growth, movement and dispersion of cells. The cells of higher animals suppress these actions, in favour of cooperation and structural integrity. They can do this because they have evolved ways to limit the growth of individual cells. However, if something damages these controls, or removes them from a particular cell, then that cell and its descendents will proliferate, reverting to ancient patterns of division and growth. Such cells become what we call cancer.

Causes of cancer

The individual causes of cancer are manifold. Practically any insult, which leads to inflammation or an increase in oxidative stress in the body, can increase the risk of cancer. The minimum requirement for the formation of cancer appears to be an increase in erroneous cell division. Microevolutionary processes then apply selection pressure, which eventually results in the features of malignant cancer. The standard mechanisms of ecology and evolution describe this process well.

How cancer develops

Cancer is an almost inevitable result of microevolution. Whenever cells in the body divide abnormally or erroneously, they accumulate genetic damage and may become polyploid. Researchers have shown that cancer cells are aneuploid, containing abnormal numbers of damaged chromosomes. Polyploidy, the generation of abnormal numbers of sets of chromosomes, is considered a mechanism for the instantaneous generation of new species. Cells that become polyploid, as in cancer, are likely to behave as independent organisms, rather than members of their original species. Such abnormal cells behave in the same way as any other biological organism, reproducing and spreading, in the process known as cancer.

Antioxidant control of cancer

Antioxidants have a central role in the microevolutionary model of cancer development. Redox mechanisms occur throughout nature, being widely used for the control of cell division, differentiation and growth. Their involvement in cancer is not surprising, since they are of fundamental importance to the basic chemistry of all living organisms. Numerous genes, enzymes and small molecules are used in controlling the cell, signalling growth, proliferation and death.

Redox mechanisms play an integral part in all aspects of cancer development. Free radicals are involved from the start of the process. Throughout its progress, increased free radical damage and internal oxidation drives the growth of cancer cells. Raised levels of oxidants initiate cell proliferation and growth. Later, lack of oxygen restricts growth, until the cancer cells evolve mechanisms to stimulate the formation of blood vessels. The resulting blood supply allows the tumour to expand rapidly and spread to distant sites in the body.

Dietary advice to prevent cancer and slow its early development can be stated succinctly. A diet to prevent cancer is low in carbohydrates, with a high proportion of antioxidants. The consumption of a variety of coloured vegetables and fruit in the diet will reduce the incidence of cancer, because of the antioxidants they contain. Supplementation with a range of antioxidant vitamins and minerals will also hinder the initiation and microevolutionary development of cancer.

Anticancer agents are surprisingly common

The toxicity of current chemotherapeutic drugs might lead us to suppose that it is hard to find agents that will kill cancer cells without poisoning the patient. This is a misconception. Numerous common substances exist, which offer the property of selectively killing cancer cells, while leaving normal cells unharmed. Indeed, taking vitamin C as an example, these substances can often be beneficial to healthy cells.

In our discussion of non-toxic anti-cancer therapies, we have concentrated on redox-cycling antioxidants, such as vitamin C. Other antioxidants and nutrients, such as curcumin, may act through cell-signalling, or by modification of gene expression. Cancer occurs because of the way animals have evolved from a single-celled to a multi-celled state. The success of multi-cellular organisms suggests they have evolved ways of dealing with cancer. Hence, the microevolutionary approach predicts that large numbers of agents that prevent or kill cancer cells should occur naturally. This is indeed the case: safe anticancer agents are common and relatively easy to find. However, the medical establishment is apparently unaware of them.

A diet to avoid cancer

Nutrition is one of the most important ways in which a person can alter their behaviour to minimise their risk of cancer.[a] The primary feature of a diet to reduce the incidence of cancer is a high intake of antioxidants. These keep the cells in a reducing environment, whereas the

---

[a] We are assuming that people do not smoke or otherwise actively engage in contact with carcinogens.

development of cancer requires cellular oxidation, to promote error-prone division. Eating a wide range of antioxidants will provide maximal benefit. This agrees with conventional advice, which recommends a high intake of vegetables.

However, the idea that people should increase their intake of vegetables rather than using antioxidant supplements appears to be based on magical thinking, rather than science. Vegetables contain substances that can inhibit cancer growth and development. The identification, isolation and purification of such substances are central features of reductionist science. Once this has been done, the anticancer substances can be made available in concentrated forms, as food supplements.

The advice not to use food supplements might appear holistic or natural, but it is not good science. Historically, the chemical isolation of vitamins and other essential dietary factors represents an increase in knowledge, which has benefited the health of humanity. While there may also be advantages to eating plants that cannot be repackaged in supplement form,[b] this hardly represents a scientific argument for avoiding those factors that can be isolated and used for maximal benefit. Surely, the optimal approach is to eat a wide range of vegetables, while also using safe supplements. In the absence of complete data, we can suggest that the supplementation should cover a range of antioxidants in their natural forms. In particular, a dynamic flow of vitamin C,[c] together with a reduced intake of sugars and starches, is likely to be beneficial.

Supplements that kill cancer

We have described the action of vitamin C in detail. Vitamin C illustrates the basic mechanism for killing and controlling cancer cells. Taken alone, it has been shown to be completely effective in a relatively limited range of cancers; however, it potentiates the action of many other nutrients and drugs.

The cytotoxic action of vitamin C requires high doses. At low doses, cancer cells can use the ascorbate as a nutrient, whereas massive doses kill cancer cells selectively, by generating hydrogen peroxide. The optimal balance of nutrients for elimination or control of cancer has not

---

[b] Herbal remedies are an obvious example of the potential for the use of whole plants and their extracts.

[c] Dynamic flow requires at least 2-3 grams per day, in divided doses, for a health young adult.

yet been established, so we will outline a possible therapeutic regime, based on current scientific knowledge.

Oral therapy

It is possible to produce and sustain high blood plasma levels using frequent doses of oral vitamin C.[15] Such ascorbate levels will kill some susceptible types of cancer cell, but not all. Evidence suggests that tumours accumulate ascorbate to higher levels than healthy tissues, which could explain the increased bowel tolerance seen in cancer patients. Cancer cells take in ascorbate preferentially, especially when glucose is restricted. This suggests that it may be possible to achieve higher ascorbate levels in tumours than previously considered possible. Orally sustained plasma levels may be high enough to allow ascorbate to be accumulated within tumour tissue, and could be cytotoxic. Use of liposomal preparations of vitamin C may increase its effectiveness, blurring the distinction between oral and intravenous therapies.

Oral therapy has practical advantages over intravenous therapy. For example, it could be self-administered by a patient attending an outpatient clinic. However, the principal advantage is that it can be sustained, in a dynamic flow.[15] Patients on oral vitamin C therapy can assault the cancer with a continuous redox pressure, over months, or even years. This means the cancer cells do not have periods of respite, during which the more resistant cells can flourish. Laboratory studies of cytotoxicity generally involve short periods of exposure, often of the order of a day. The effects of continuous exposure on the viability of cancer cells have not been established, but could be far greater than intermittent treatment.

With a limited carbohydrate intake, to deny the cancer the nutrients it requires for easy growth, cancer patients on oral redox-active nutrient therapy might enjoy similar benefits to those described by John Ely. Further restriction of calories and nutrients could provide additional benefits. Arthur Robinson suggests that the expected lifespan of a terminally ill cancer patient could increase by at least an order of magnitude. This means that a patient given one year to live could hope for a further ten years, or more.

We suggest the use of natural vitamin C, as L-ascorbate, for oral administration, as it is more available to cells. Liposomal preparations may enhance the effectiveness of oral vitamin C. In addition, R-alpha-lipoic acid increases the cytotoxic effectiveness of ascorbate, by a factor of more than five. Like ascorbate, it is safe for healthy cells. Oral therapies that combine ascorbic and alpha-lipoic acids move the achievable plasma levels towards the cytotoxic range.[d] Sustained blood levels of L-ascorbate and R-alpha-lipoic acid will lead to accumulation of high concentrations within tumours. Experimental data suggest an intake of R-alpha-lipoic of at least 10% of the ascorbate intake - 100mg or more for each gram of vitamin C. R-alpha-lipoic acid has low toxicity, although safety data on high doses over extended periods is not available. On current evidence, a combination of vitamin C and alpha-lipoic acid is the least toxic approach to oral cancer therapy.

Alternatively, vitamin K_3, or another of the redox-cycling quinones, could be used in place of R-alpha-lipoic acid; these are also synergistic with ascorbate. The combination of vitamins C and K_3 is 10 to 50 times more effective against cancer than is either agent alone.[853] Vitamin K_3 may be more effective than alpha-lipoic acid, but is also more toxic and requires careful administration. Once again, evidence indicates that levels sufficient to kill many types of cancer could be generated using oral doses.

In addition, it is possible to vary the administration of these treatments; for example a period with vitamin C and K_3 could be followed by a period with ascorbate and alpha-lipoic acid. Periods of vitamin C and vitamin E succinate might also be employed.

Redox synergy

It is clear that vitamin C interacts with several different nutrients that create free radicals selectively within cancer cells. Vitamin K_3 and alpha-lipoic acid each increase the cancer killing effectiveness of vitamin C. It is not yet known to what extent use of multiple agents might further improve the effect. The effects of combined use of alpha-lipoic acid with vitamin K_3 are not established. It is reported that alpha-lipoic acid acts as an antioxidant, inhibiting the effects of K_3, but this result may apply only to lower concentrations of alpha-lipoic acid, in the absence of ascorbate.

John Ely has suggested the use of coenzyme Q10 with vitamin C. Since coenzyme Q10 and vitamin K_3 are chemically similar, both being quinones, they may have a similar pharmacology in large doses. The addition of vitamin E succinate to the therapy may destroy cancer cells more effectively and prevent resistance developing. An analogous

---

[d] A one millimolar concentration of ascorbate is at the beginning of the cytotoxic dose response curve for ascorbic acid.

approach is used for treatment of tuberculosis (TB), in which a combination of antibiotics is taken over a period of months, to prevent development of drug resistance.

Levels of oral anticancer nutrients need to be increased slowly. There are reports of cancers that are highly sensitive to the actions of vitamin C and redox cycling. Killing such a cancer quickly may produce a life-threatening shock. Having reached a maximal intake, the supplement regime should be continued indefinitely and the progress of the tumour monitored clinically and with medical imaging.

More than one approach

Viewing cancer as microevolution provides multiple approaches to fighting it. The first is to drive the population of cancer cells to extinction by sustained oxidation, generated by selective pressure from vitamin C and related substances.

If the cancer cells develop resistance to direct redox killing, it is possible to utilise an alternative approach of population growth control, by preventing oxidation within the cancer cells. For example, use of the metal binding supplement, IP6, can remove iron from the tumour, slowing growth. Combined with supplemental antioxidants and gene inhibitors, such as curcumin, cancer growth could be slowed by driving the cancer cells into a reducing state, in which cell division is impeded. This antioxidant approach makes cancer cells less oxidising, to prevent cell growth, rather than generating additional free radicals within the cancer cells. It is likely to be less effective with advanced disease.

Oral therapy for terminal cancer

We can see no reason for physicians not to recommend oral redox therapy for all terminal cancer patients, following conventional treatment. The toxicity is low; indeed, there have been reports of improved quality of life in people on ascorbate, in marked contrast to conventional therapies. Furthermore, the treatment might prolong the patient's life and, in some cases, could even cure the cancer. Such possibilities demand urgent investigation. Failure to carry out such these studies could mean that people go on suffering, or die unnecessarily early deaths.

In the case of patients that have apparently been cured by surgery, oral redox therapy could minimise the risk of recurrence, by destroying small pockets of remaining cancer. There is substantial evidence that this approach would be both non-toxic and effective. What we cannot predict is the number of patients who would benefit, since the required studies have not yet been carried out.

The combined therapy of oral redox-active nutrients and a low carbohydrate diet could be made an option for terminal cancer patients. The risk is small, provided the patient is monitored by competent medical staff. Such patients have little to lose, whereas the potential gain is huge. No one knows what proportion of terminal cancer patients would be cured, although studies are clearly essential. Neither do we know the number of "terminal" patients who might survive for years, eventually dying from an unrelated cause. However, the proportion may well be large.

Terminal patients are being written-off, without the option of a therapeutic approach that is cheap, harmless and stands a good chance of being effective. Who among medical decision makers and politicians will take responsibility for what is potentially a reckless waste of human life?

Intravenous therapy

Intravenous infusions of sodium ascorbate, in combination with vitamin K or alpha-lipoic acid, will kill cancer cells. However, intermittent treatments might cause the cancer cells to become resistant. In microevolutionary terms, if the therapy selects tumour cells for resistance to redox-induced apoptosis, then the gain in life expectancy might be limited.

One strategy to avoid such resistance would be to combine intravenous infusions with frequent large oral doses of vitamin C, to maintain high blood levels. Such a regime is needed to establish a dynamic flow of ascorbate through the body,[15] using intakes close to bowel tolerance levels. This therapy might include a synergistic agent, such as alpha-lipoic acid or, perhaps, motexafin gadolinium. The aim would be to keep the cancer under consistent redox pressure.

A second limitation on intravenous therapy is more practical. Such therapy requires monitoring by a physician, together with appropriate nursing care. Any intravenous treatment has associated dangers. The therapy can take several hours per session, which has an associated cost. Although the substances employed may be inexpensive, the costs of preparation, therapy and supervision may be significant. However, they would still be far lower than would those of the corresponding chemotherapeutic drugs.

Benefits of anticancer nutrition

According to the microevolutionary model, cancer occurs in multi-celled organisms when cells escape the body's controls and start to behave like their single-celled ancestors. Such changes are triggered by oxidation and damage, which result in erroneous cell division. The environment in which cancer cells develop favours anaerobic cells, which use glucose as their main source of energy. This is in contrast to healthy cells, which tend to use aerobic forms of metabolism and can utilise varied food sources.

Over geological timescales, animals and plants have developed ways to stop their cells reverting to primitive forms. Hence, anticancer substances are common throughout nature. Therapies based on these nutrients take advantage of metabolic differences between cancer cells and healthy cells, to destroy the cancer cells while doing no harm to, or even helping, the healthy cells. Clinical trials are urgently needed to test such non-toxic therapies, before more people die unnecessarily.

Fundamental biological research suggests that cancer is a treatable condition. Right now, people who have been given a terminal prognosis may be able to survive in reasonable health, for a prolonged period. Although the current data is not sufficient to indicate the degree of life extension achievable, many "terminal" cancer patients might die of other causes, long before the cancer kills them. Cancer patients deserve to be offered this opportunity.

Glossary

Aerobic: an environment or condition that contains oxygen; organisms able to live only in the presence of air or free oxygen.
Allele: any of the alternative forms of a given gene; one of a number of genes occupying a definite physical position in a specific chromosome.
Allopatric: having separate and mutually exclusive areas of geographical distribution.
Amygdalin; a bitter cyanide-generating extract from the seeds of apricots, plums and bitter almond. Claimed by some to be a vitamin (B_{17}), which can destroy cancer.
Anaerobic: an environment or condition which is free of oxygen; an organism that grows in the absence of oxygen.
Anaplastic: a description of cells that have changed towards a more primitive cell type. Anaplastic cells are often malignant and consistent with an aggressive tumour.
Angiogenesis: formation of new blood vessels.
Aneuploidy: having an abnormal number of chromosomes.
Antioxidant: any substance or action that inhibits or prevents oxidation and the damaging effects of free radicals.
Apoptosis: programmed cell death, the method used by the body to dispose of damaged, unwanted and unneeded cells.
Ascorbate: vitamin C, an essential antioxidant in the diet.
Autoschizis: a form of (cancer) cell death, caused by the action of vitamins C and K_3.
Arylate: introduction of one or more aryl groups into a compound.
Aryl group: any organic group derived from an aromatic hydrocarbon by the removal of a hydrogen atom.
Bacteria: single-celled organisms that multiply by cell division and can cause disease in humans, plants or animals. Other forms of bacteria are essential for human life.
Cachexia: general wasting of the body, due to illness or stress. May be caused by cytokines such as TNF.
Carbohydrate: an organic molecule, or substance, consisting of a chain of carbon atoms to which hydrogen and oxygen are attached in a 2 (H) to 1 (O) ratio. Examples include sugars, starch, glycogen and cellulose.
Carcinoma: a cancer arising in epithelial cells that cover or line the body surface and internal organs.
Catalase: a haem-based enzyme, which catalyses the breakdown of hydrogen peroxide into oxygen and water.
Cell: the fundamental unit of life, which is capable of growth and reproduction.
Chemotherapy: the use of chemical agents as a treatment for cancer.

Cytokine: a small protein molecule used for communication between cells, especially cells of the immune system.
Cytotoxic: poisonous to cells, especially cancer cells.
Diabetes: a disease (diabetes mellitus) in which the body's production and/or use of the hormone insulin is impaired, causing high blood sugar levels.
Differentiation: the process in embryonic development in which unspecialised cells or tissues become specialized for particular functions. In oncology, it is the degree of similarity of tumour cells to the cells from which they arose.
Diploid: having two different sets of chromosomes in the nucleus of a cell; most animals (metazoans) and plants are diploid.
Dysplasia: alteration in the size, shape and organization of cells or tissues.
Endothelium: a layer of cells lining the inside surfaces of body cavities, such as blood and lymph vessels
Enzyme: a protein that can speed up (catalyse) a specific chemical reaction without being changed or consumed in the process.
Eukaryote: a cell or organism with a membrane-bound, structurally discrete nucleus and other subcellular compartments. Most organisms, except viruses, bacteria and blue-green algae, are formed from eukaryotic cells.
Extinction: the complete elimination of a population or species.
False positive: a test result predicting that a person has a condition, when in fact he or she does not.
Fibroblast: a cell that produces the structural (collagen) fibres and ground substance of cartilage and connective tissue.
Free Radical: a molecule or atom with an unpaired electron. Most free radicals are highly reactive; oxidising free radicals are generally damaging to body tissues.
Gene: the fundamental unit of heredity. Consists of one or more lengths of DNA, which function as a unit, to encode RNA and protein.
Genome: all the genetic information possessed by an organism, the entire genetic complement.
Glycolysis: the splitting of glucose to release energy. The most widespread metabolic pathway in living cells, providing the starting point for anaerobic fermentation or aerobic respiration.
Growth: an increase in tissue size or cell number, especially when cell division exceeds cell death.
Haem: a metal-containing molecule or component, consisting of an iron (or similar metal) atom contained in the centre of a large organic ring called a porphyrin. Haemoglobin is an example of a protein that contains a haem component.
Hayflick limit: the number of divisions of which a normal cell is capable.
Haploid: a single set of chromosomes. Half the full set of genetic material of a diploid organism like *Homo sapiens*.
HIV: human immunodeficiency virus, thought to be the cause of AIDS.
Hyaluronidase: an enzyme that splits hyaluronic acid (a connective tissue molecule) and increases the permeability of connective tissue.

Hydrogen peroxide: a chemical substance (H_2O_2). Normally, a clear viscous liquid with strong oxidising properties.
Hydroquinone: a white crystalline substance, $C_6H_4(OH)_2$, obtained by the reduction of quinone.
Hydroxyl radical: fast-reacting free radical that causes tissue damage, consists of one oxygen and one hydrogen atom, $^\bullet OH$.
Hyperplasia: an increase in the number of cells in an organ or tissue.
Interferon: a natural glycoprotein, produced by the cells of the immune system in response to a challenge by a microorganism or tumour cell. Interferons are local signalling molecules (cytokines) and have been produced using genetic engineering and marketed as drugs.
Interleukin: a signalling molecule (cytokine) used by white blood cells for communication.
Krebs cycle: (citric acid cycle), a circular series of chemical reactions involved in aerobic respiration, which occurs naturally in cells.
Laetrile: a substance derived from amygdalin; claimed to be an anticancer agent.
Leukoplakia: a white patch, typically on the cheek, gums, or tongue, which may become cancerous.
Lipid: fat or oil.
Macroevolution: evolution above the species level. The processes that result from differences in species survival or rates of speciation.
Metastases: secondary tumours; the spread of cancer from one part of the body to another, typically by way of the lymphatic system or bloodstream.
Microevolution: subspecies level evolutionary changes, which include adaptation to local environments.
Mitochondria: subcellular bodies, the principal energy source of the cell.
Mitosis: the process of chromosome segregation and nuclear separation that follows replication of the genes in cell division. It ensures that each daughter cell receives an accurate copy of the organism's genome.
Mutualism: a symbiotic interaction between two species, in which both organisms derive benefit.
Mutation: a change (in the number, arrangement or nucleotide sequence) in DNA coding; a permanent, heritable change in a gene or chromosome structure.
Mutagen: a factor, such as a chemical or radiation, which produces mutations.
Necrosis: death of a tissue; a type of cell death in which cells swell, break open, release their internal contents and thus provoke inflammation.
Neoplasia: new cell growth, which may form benign or malignant tumours.
Neoplasm: a growth of new cells, a tumour.
Neuropathy: damage, disease or abnormality of the nerves.
Nucleus: an organelle (or subcellular body) that contains most of the cell's genetic material. The nucleus controls chemical reactions within the cell and stores genetic information (DNA) needed for cellular division

Oncogene: a gene (such as ras) that facilitates the development of cancer, typically by regulating cell growth.
Orthomolecular medicine: the use of natural substances in the diet for good health and the treatment of disease, with emphasis on individual variability.
Oxidation: the addition of oxygen, removal of hydrogen, or the removal of electrons from a molecule. The opposite of reduction.
P53: a common tumour suppressor gene.
Parathyroid: small glands located in the neck, which produce parathyroid hormone to regulate the level of calcium in the blood.
Peroxidase: an enzyme which catalyses the breakdown of peroxides in the body.
Polyp: a mass of tissue that bulges or projects outward from the normal surface level.
Polyploid: having more than the normal two sets of chromosomes.
Population control: limiting the growth of a population, eg by changing the proportion of dividing to dying cells.
Prokaryote: cell or organism, such as a bacterium, that lacks an internal, structurally-discrete nucleus and other subcellular compartments.
Proto-oncogene: a normal gene that has the potential to become an oncogene.
Provitamin: a vitamin precursor; a substance that is converted into a vitamin in the body.
Ras: an oncogene, originally isolated from <u>ra</u>t <u>s</u>arcoma virus.
Risk: the combination of the frequency, or probability, of occurrence and the consequence of a specific dangerous event.
Sarcoma: a cancer of connective or supportive tissue such as bone, cartilage, fat or muscle.
Quinone: a member of a class of aromatic yellow compounds, including several that are biologically important as coenzymes or vitamins.
Reactive Oxygen Species: (ROS) a highly reactive free radical or oxidant involving the oxygen atom.
Reactive Nitrogen Species: (RNS) a highly reactive free radical or oxidant involving the nitrogen atom.
Redox: an abbreviation of REDuction-OXidation, which occur together in biochemical reactions.
Redox cycle: repeated oxidation and reduction of a molecule or group of molecules.
Reduction: addition of hydrogen, removal of oxygen, or addition of electrons to a molecule.
Repressor: a gene that inhibits (represses) cancer growth, a tumour suppressor gene.
Retinoblastoma: a malignant, cancerous tumour of the retina (back of the eye).
Retrovirus: a virus that uses RNA as its genetic material. When introduced into a host cell, it uses RNA as a template to produce viral DNA.
RNA: ribonucleic acid; a (usually) single-stranded relative of DNA. In humans, the primary function of RNA is protein synthesis within cells.

Speciation: the process by which one or more populations of a species become genetically different enough to form a new species.
Species: The basic unit of biological classification; a group of organisms with characteristics that distinguish them from other organisms. If they reproduce sexually, individuals within the same species can produce fertile, viable offspring.
Squamous: covered with or formed of scales; scaly
Src: (pronounced sark) an oncogene associated with the Rous sarcoma virus.
Superoxide: a reactive molecule, •$O2^{-}$. It has an unpaired electron, is unstable and spontaneously decomposes into peroxide and oxygen. White blood cells use superoxide for oxygen-dependent killing mechanisms.
Symbiosis: a close, prolonged association between two or more species.
Sympatric: relating to two or more populations which occupy overlapping geographical areas
Prokaryote: a cell or organism which lacks a membrane-bound, structurally discrete nucleus and other subcellular compartments, e.g. bacteria.
Telomere: the terminal section of chromosomal DNA, which permits consistent replication of the ends of the molecule.
Thioredoxin: a small sulphur-containing protein involved in redox processes. Thioredoxins act as electron donors to peroxidase enzymes.
Translocation: rearrangement of a chromosome; movement of a segment from one location to another, either within the same chromosome or to another chromosome.
Trophoblast: a membrane that forms the wall of the blastocyst in early development and aids implantation in the uterine wall; forms the placenta and chorion (outermost membranous sac, which encloses the embryo in higher vertebrates).
TNF: see tumour necrosis factor.
Tumour: an abnormal tissue growth, or cancerous lump; may be benign or malignant.
Tumour necrosis factor: (also called catchectin) an important cytokine (signalling molecule), involved in inflammation, which can kill cancer cells *in vitro* and *in vivo.*
Tumour suppressor gene: a repressor gene that inhibits cancer growth.
Vascular Endothelial Growth Factor (VEGF): a protein, released by cells, that stimulates new blood vessel formation.
Virus: a tiny organism, consisting of protein and DNA or RNA, which multiples within cells and can cause disease.
Vitamin: an essential, low molecular weight organic compound, required for normal growth and metabolic processes. Lack of one or more vitamins in the nutrient supply results in deficiency diseases.
X-ray: high energy electromagnetic radiation of very short wavelength; used in medical diagnostics and treatment.

Appendix

Cancer terminology

Benign or non-invasive tumours are named by attaching the suffix -*oma* to the name of the tissue or cell from which the cancer arose. An osteoma, for example, is a tumour generated by multiplication of cells related to bone (or osteoid) cells. A lipoma arises from a fat cell: lipid is another name for fat or oil. There are exceptions to this naming rule, notably, a melanoma is a cancer of the pigment cells (or melanocytes) in skin. A melanoma can be a rapidly growing, often fatal disease. In this case, the cancer is malignant, but the name melanoma persists for historical reasons. Sometimes, it is difficult to tell whether a tumour is really benign; in such cases, it is often described as borderline, or as having undetermined malignant potential.

In many cases, the name comes from the cell of origin, but in others it is based on the microscopic or gross appearance of the growth. A tumour that is a mass projecting into a cavity or channel, such as the gut, is called a polyp.

The word "cancer" is Greek for crab, as in the zodiac sign. Malignant tumours invade surrounding tissues and produce roots that resemble the legs of a crab. Surgical removal of a malignant tumour often fails to cure the disease. It may be difficult to cut away all the spreading processes at the site, especially if they have invaded essential tissues. Some malignant cells may move to other parts of the body, to form new tumours.

Malignant or invasive cancers are named according to a convention of *cell type* plus *category* plus *site in body*. For example, an adenocarcinoma of the lung is a cancer that starts in the secretory cells of the lung. The category of carcinomas refers to growths derived from epithelial cells. These line the inner and outer surfaces of the body, including the skin, the blood vessels, the inside of the chest and the abdominal cavity. Tumours originating in these epithelial cells make up over three-quarters of all cancers. There are different types of carcinoma, depending on the particular type of epithelial cell from which the tumour was derived. These include squamous cell carcinoma, adenocarcinoma and transitional cell carcinoma. Since these different types of cells occur in many parts of the body, these cancers are also named for the organ of the body in which they occur. An example is a squamous cell lung cancer, derived from the surface-covering squamous cells of the lung.

The less frequent cancers from supportive or connective tissue, such as muscle or bone, are called sarcomas. The first part of a sarcoma's name often describes the predominant cell type. A liposarcoma arises from a precursor to a lipid-containing fat cell.

Names for malignant tumours of the blood-forming tissue often use the suffix *-aemia* (from the Greek meaning blood). Leukaemia refers to a cancer of white blood cells, which are called leukocytes. In some notable instances, cancers are called after the physician who discovered them. Hodgkin's disease was identified in 1832, by the English physician Thomas Hodgkin, and is a cancer of the lymph glands. The British surgeon, Denis Parsons Burkitt, also described a lymphoma that bears his name.

Index

References

[1] King R.J.B (2000) Cancer Biology, Pearson Education, Prentice Hall, Essex, England.
[2] Hickey S. Roberts H. (2004) Ridiculous Dietary Allowance, Lulu press.
[3] Kuukasjarvi T. Karhu R. Tanner M. Kahkonen M. Schaffer A. Nupponen N. Pennanen S. Kallioniemi A. Kallioniemi O.P. Isola J. (1997) Genetic heterogeneity and clonal evolution underlying development of asynchronous metastasis in human breast cancer, Cancer Res, 57(8), 1597-1604.
[4] Fujii H. Marsh C. Cairns P. Sidransky D. Gabrielson E. (1996) Genetic divergence in the clonal evolution of breast cancer, Cancer Res, 56(7), 1493-1497.
[5] Barrett M.T. Sanchez C.A. Prevo L.J. Wong D.J. Galipeau P.C. Paulson T.G. Rabinovitch P.S. Reid B. (1999) Evolution of neoplastic cell lineages in oesophagus, J.Nat Genet, 22(1), 106-109.
[6] Buerger H. Otterbach F. Simon R. Schafer K.L. Poremba C. Diallo R. Brinkschmidt C. Dockhorn-Dworniczak B. Boecker W. (1999) Different genetic pathways in the evolution of invasive breast cancer are associated with distinct morphological subtypes, J Pathol, 189(4), 521-526.
[7] Kallioniemi O.P. Visakorpi T. (1996) Genetic basis and clonal evolution of human prostate cancer, Adv Cancer Res, 68, 225-255.
[8] Moss R. (1996) The Cancer Industry, New Updated Edition, Equinox Press, New York.
[9] Goonzer M. (2004) The $800 million dollar pill, University of California Press, Berkley.
[10] Moynihan R. Cassels A. (2005) Selling Sickness: How the World's Biggest Pharmaceutical Companies Are Turning Us All into Patients, Nation Books.
[11] Angell M. (2004) The truth about the drug companies, Random House, New York.
[12] Duesberg P. H. (1987) Retroviruses as Carcinogens and Pathogens: Expectations and Reality, Cancer Res, 47, 1199-1220.
[13] Published on Dr Duesberg's website.
[14] News report, (2001) Profile: Peter H. Duesberg, dissident or Don Quixote? Scientific American, 265(2), 30-32.
[15] Hickey S. Roberts H. (2004) Ascorbate: The Science of Vitamin C, Lulu Press.
[16] Wilson E.O. The Diversity of Life, W. W. Norton, New York.
[17] Cairns-Smith A.G. (1990) Seven Clues to the Origin of Life, Cambridge University Press, England.
[18] Bahler D.W. Levy R. (1992) Clonal evolution of a follicular lymphoma: evidence for antigen selection, Proc Natl Acad Sci U S A, 89(15), 6770-6774.
[19] Ishikawa F. (1997) Telomere crisis, the driving force in cancer cell evolution, Biochem Biophys Res Commun, 230(1), 1-6.
[20] Pasqualini J.R. Chetrite G. Nguyen B.L. Maloche C. Delalonde L. Talbi M. Feinstein M.C. Blacker C. Botella J. Paris J. (1995) Estrone sulfate-sulfatase and 17 beta-hydroxysteroid dehydrogenase activities: a hypothesis for their role in the evolution of human breast cancer from hormone-dependence to hormone-independence, J Steroid Biochem, Mol Biol, 53(1-6), 407-412.
[21] Monteiro H.P. Stern A. (1996) Redox modulation of tyrosine phosphorylation-dependent signal transduction pathways, Free Radic Biol Med, 21(3), 323-333.
[22] Droge W. (2002) Free radicals in the physiological control of cell function, 82(1), 47-95.
[23] Yamamoto Y. Yamashita S. (1997) Plasma ratio of ubiquinol and ubiquinone as a marker of oxidative stress, Mol Aspects Med, 18 Suppl, S79-84.
[24] Gray M.W. Burger G. Lang B.F. (2001) The origin and early evolution of mitochondria, Genome Biol, 2(6), WS1018, Epub Jun 5.
[25] Beckman K.B. Ames B.N. (1998) Mitochondrial aging: open questions, Ann N Y Acad Sci, 854, 118-127.
[26] Daban J.R. (2000) Physical constraints in the condensation of eukaryotic chromosomes. Local concentration of DNA versus linear packing ratio in higher order chromatin structures, Biochemistry, 39(14), 3861-3866.
[27] Woodcock C.L. Dimitrov S. (2001) Higher-order structure of chromatin and chromosomes, Curr Opin Genet Dev, 11(2), 130-135.
[28] Mitchel M. (1998) An Introduction to Genetic Algorithms (Complex Adaptive Systems), The MIT Press; Reprint edition.
[29] Goldberg D.E. (1989) Genetic Algorithms in Search, Optimization, and Machine Learning, Addison-Wesley, Boston, USA.
[30] Boland C.R. Ricciardiello L. (1999) How many mutations does it take to make a tumor? Proc Natl Acad Sci U S A, 96(26), 14675–14677.
[31] Malthus T. (1798) An Essay on the Principle of Population, An Essay on the Principle of Population, as it Affects the Future Improvement of Society with Remarks on the Speculations of Mr. Godwin, M. Condorcet, and Other Writers, Printed for J. Johnson in St Pauls Chrch-yard, London , England.

[32] Darwin C. (1876) The Autobiography of Charles Darwin 1809-1882 (Paperback), Nora Barlow (Editor), W. W. Norton & Company, Reissue edition (September 1, 1993).
[33] Darwin C. (1859) Origin of Species, Reprinted 1995, Gramercy, New York.
[34] Noble W.C. Virani Z. Cree R.G. (1992) Co-transfer of vancomycin and other resistance genes from Enterococcus faecalis NCTC 12201 to Staphylococcus aureus, FEMS Microbiol Lett, 72(2), 195-198.
[35] Davies J. (1994) Inactivation of antibiotics and the dissemination of resistance genes, Science, 264(5157), 375-382.
[36] Kruse H. Sørum H. (1994) Transfer of multiple drug resistance plasmids between bacteria of diverse origins in natural microenvironments, Appl Environ Microbiol, 60(11), 4015–4021.
[37] Ainsworth C. (2005) Life's greatest inventions: multicellularity, New Scientist, April 9, 27.
[38] Hamilton W. (1971). Geometry for the selfish herd, Journal of Theoretical Biology, 31(2), 295-311.
[39] Marzluff J. M. Heinrich B. Marzluff C.S. (1996) Communal roosts of common ravens are mobile information centers, Animal Behaviour, 51, 89-103.
[40] Heinrich B. (2004) Winter world: the ingenuity of animal survival, Ecco, New York.
[41] Soltis D.E. Soltis P.S. Tate J.A. (2004) Advances in the study of polyploidy since plant speciation, New Phytologist, 161(1), 173-191.
[42] Yosida T.H. (1983) Karyotype evolution and tumor development, Cancer Genet Cytogenet, 8(2), 153-179.
[43] O'Brien S.J. Menotti-Raymond M. Murphy W.J. Nash W.G. Wienberg J. Stanyon R. Copeland N.G. Jenkins N.A. Womack J.E. Marshall Graves J.A. (1999) The promise of comparative genomics in mammals, Science, 286(5439), 458-462, 479-481.
[44] Digby L. (1912) The citology of Primula kewensis and of other related Primula hybrids, Ann Bot 26, 357-388.
[45] Grant, V. (1971) Plant Speciation, 2nd edition (1981), Columbia Press, New York.
[46] Ramsey J. Schemske D.W. (1998) Pathways, mechanisms, and rates of polyploid formation in flowering plants, Annual Review of Ecology and Systematics, 29, 467-501.
[47] Mayr E. (2002) What Evolution Is, Basic Books, Philadelphia, USA.
[48] Oshimura M. Barrett J.C. (1986) Chemically induced aneuploidy in mammalian cells: mechanisms and biological significance in cancer, Environ Mutagen, 8(1), 129-159.
[49] Carmody R.J. Cotter T.G. (2001) Signalling apoptosis: a radical approach, Redox Report, 6(2), 77-90.
[50] Frade J.M. Michaelidis T.M. (1997) Origin of eukaryotic programmed cell death: a consequence of aerobic metabolism? Bioessays, 19(9), 827-832.
[51] Blackstone N.W. Green D.R. (1999) The evolution of a mechanism of cell suicide, Bioessays, 21, 84–88.
[52] Green D.R. Amarante-Mendes G.P. (1998) The point of no return: mitochondria, caspases, and the commitment to cell death, Results Probl Cell Differ, 24, 45–61.
[53] Ghibelli L. Coppola S. Fanelli C. Rotilio G. Civitareale P. Scovassi A.I. Ciriolo M.R. (1999) Glutathione depletion causes cytochrome c release even in the absence of cell commitment to apoptosis, Faseb J, 13, 2031–2036.
[54] Coppola S. Ghibelli L. (2000) GSH extrusion and the mitochondrial pathway of apoptotic signalling, Biochem Soc Trans, 28, 56–61.
[55] Li P. Nijhawan D. Budihardjo I. Srinivasula S.M. Ahmad M. Alnemri E.S. Wang X. (1997) Cytochrome c and dATPdependent formation of Apaf-1/caspase-9 complex initiates an apoptotic protease cascade, Cell, 91, 479–489.
[56] Krajewski S. Krajewska M. Ellerby L.M. Welsh K. Xie Z. Deveraux Q.L. Salvesen G.S. Bredesen D.E. Rosenthal R.E. Fiskum G. Reed J.C. (1999) Release of caspase-9 from mitochondria during neuronal apoptosis and cerebral ischemia, Proc Natl Acad Sci USA, 96, 5752–5757.
[57] Shimizu S. Eguchi Y. Kamiike W. Waguri S. Uchiyama Y. Matsuda H. Tsujimoto Y. (1996) Bcl-2 blocks loss of mitochondrial membrane potential while ICE inhibitors act at a different step during inhibition of death induced by respiratory chain inhibitors, Oncogene 1996; 13, 21–29.
[58] Proskuryakov S.Y. Gabai V.L. Konoplyannikov A.G. Necrosis is an active and controlled form of programmed cell death, Biochemistry (Moscow), Maik Nauka Interperiodica, 67(4), 387 – 408.
[59] Madeo F. Frohlich E. Frohlich K.U. (1997) A yeast mutant showing diagnostic markers of early and late apoptosis, J Cell Biol, 139, 729–734.
[60] Ameisen J.C. (1996) The origin of programmed cell death, Science, 272, 1278–1279.
[61] Lorenzo H.K. Susin S.A. Penninger J. (1999) Kroemer G. (1999) Apoptosis inducing factor (AIF): a phylogenetically old, caspase-independent effector of cell death, Cell Death and Differentiation, 6, 516-524.
[62] Lennon S.V. Martin S.J. Cotter T.G. (1991) Dose-dependent induction of apoptosis in human tumour cell lines by widely diverging stimuli, Cell Prolif, 24, 203–214.
[63] Albina J.E. Cui S. Mateo R.B. Reichner J.S. (19930 Nitric oxide-mediated apoptosis in murine peritoneal macrophages, J Immunol, 150, 5080–5085.

[64] Dypbukt J.M. Ankarcrona M. Burkitt M. Sjoholm A. Strom K. Orrenius S. Nicotera P. (1994) Different prooxidant levels stimulate growth, trigger apoptosis, or produce necrosis of insulin-secreting RINm5F cells. The role of intracellular polyamines. J Biol Chem, 269, 30553–30560.
[65] Escargueil-Blanc I. Meilhac O. Pieraggi M.T. Arnal J.F. Salvayre R. Negre-Salvayre A. (1997) Oxidized LDLs induce massive apoptosis of cultured human endothelial cells through a calcium-dependent pathway. Prevention by aurintricarboxylic acid, Arterioscler Thromb Vasc Biol, 17, 331–339.
[66] Sandstrom P.A. Mannie M.D. Buttke T.M. (1994) Inhibition of activation induced death in T cell hybridomas by thiol antioxidants: oxidative stress as a mediator of apoptosis, J Leukoc Biol, 55, 221–226.
[67] Sato N. Iwata S. Nakamura K. Hori T. Mori K. Yodoi J. (1995) Thiolmediated redox regulation of apoptosis. Possible roles of cellular thiols other than glutathione in T cell apoptosis, J Immunol, 154, 3194–3203.
[68] McGowan A.J. Bowie A.G. O'Neill L.A. Cotter T.G. (1998) The production of a reactive oxygen intermediate during the induction of apoptosis by cytotoxic insult, Exp Cell Res, 238, 248–256.
[69] McGowan A.J. Ruiz-Ruiz M.C. Gorman A.M. Lopez-Rivas A. Cotter T.G. (1996) Reactive oxygen intermediate(s) (ROI): common mediator(s) of poly(ADP-ribose)polymerase (PARP) cleavage and apoptosis, FEBS Lett, 392, 299–303.
[70] Greenlund L.J. Deckwerth T.L. Johnson Jr E.M. (1995) Superoxide dismutase delays neuronal apoptosis: a role for reactive oxygen species in programmed neuronal death, Neuron, 14, 303–315.
[71] Hockenbery D.M. Oltvai Z.N. Yin X.M. Milliman C.L. Korsmeyer S.J. (1993) Bcl-2 functions in an antioxidant pathway to prevent apoptosis, Cell, 75, 241–251.
[72] Slater A.F. Nobel C.S. Orrenius S. (1995) The role of intracellular oxidants in apoptosis, Biochim Biophys Acta, 1271, 59–62.
[73] Muschel R.J. Bernhard E.J. Garza L. McKenna W.G. Koch C.J. (1995) Induction of apoptosis at different oxygen tensions: evidence that oxygen radicals do not mediate apoptotic signalling, Cancer Res, 55, 995–998.
[74] Jacobson M.D. Raff M.C. (1995) Programmed cell death and Bcl-2 protection in very low oxygen, Nature, 374, 814–816.
[75] Esposti M.D. Hatzinisiriou I. McLennan H. Ralph S. (1999) Bcl-2 and mitochondrial oxygen radicals. New approaches with reactive oxygen species-sensitive probes, J Biol Chem, 274, 29831–29837.
[76] Slater A.F. Stefan C. Nobel I. van den Dobbelsteen D.J. Orrenius S. (1995) Signalling mechanisms and oxidative stress in apoptosis, Toxicol Lett, 82/83, 149–153.
[77] Greenhalf W. Stephan C. Chaudhuri B. (1996) Role of mitochondria and C-terminal membrane anchor of Bcl-2 in Bax induced growth arrest and mortality in Saccharomyces cerevisiae, FEBS Lett, 380, 169–175.
[78] Madeo F. Frohlich E. Ligr M. Grey M. Sigrist S.J. Wolf D.H. Frohlich K.U. (1999) Oxygen stress: a regulator of apoptosis in yeast, J Cell Biol, 145, 757–767.
[79] Xiang J. Chao D.T. Korsmeyer S.J. (1996) BAX-induced cell death may not require interleukin 1 beta-converting enzyme-like proteases, Proc Natl Acad Sci USA, 93, 14559–14563.
[80] Carmody R.J. Cotter T.G. (2000) Oxidative stress induces caspase independent retinal apoptosis in vitro, Cell Death Differ, 7, 282–291.
[81] Creagh E.M. Carmody R.J. Cotter T.G. (2000) Heat shock protein 70 inhibits caspase-dependent and -independent apoptosis in Jurkat T cells, Exp Cell Res, 257, 58–66.
[82] Lesage S. Steff A.M. Philippoussis F. Page M. Trop S. Mateo V. Hugo P. (1997) CD4+ CD8+ thymocytes are preferentially induced to die following CD45 cross-linking, through a novel apoptotic pathway, J Immunol, 159, 4762–4771.
[83] Okuno S. Shimizu S. Ito T. Nomura M. Hamada E. Tsujimoto Y. Matsuda H. (1998) Bcl-2 prevents caspase independent cell death, J Biol Chem, 273, 34272–34277.
[84] Aw T.Y. Ookhtens M. Kaplowitz N. (1984) Inhibition of glutathione efflux from isolated rat hepatocytes by methionine, J Biol Chem, 259, 9355–9358.
[85] Ghibelli L. Coppola S. Rotilio G. Lafavia E. Maresca V. Ciriolo M.R. (1995) Non-oxidative loss of glutathione in apoptosis via GSH extrusion, Biochem Biophys Res Commun, 216, 313–320.
[86] van den Dobbelsteen D.J. Nobel C.S.I. Schlegel J. Cotgreave I.A. Orrenius S. Slater A.F. (1996) Rapid and specific efflux of reduced glutathione during apoptosis induced by anti-Fas/APO-1 antibody, J Biol Chem, 271, 15420–15427.
[87] Beaver J.P. Waring P. (1995) A decrease in intracellular glutathione concentration precedes the onset of apoptosis in murine thymocytes, Eur J Cell Biol, 68, 47–54.
[88] Ghibelli L. Fanelli C. Rotilio G. Lafavia E. Coppola S. Colussi C. Civitareale P. Ciriolo M.R. (1998) Rescue of cells from apoptosis by inhibition of active GSH extrusion, Faseb J, 12, 479–486.
[89] Fernandes R.S. Cotter T.G. (1994) Apoptosis or necrosis: intracellular levels of glutathione influence mode of cell death, Biochem Pharmacol, 48, 675–681.
[90] Ameisen J.C. Estaquier J. Idziorek T. De Bels F. (1995) Programmed cell death and AIDS pathogenesis: significance and potential mechanisms, Curr Top Microbiol Immunol, 200, 195–211.

[91] Sian J. Dexter D.T. Lees A.J. Daniel S. Agid Y. Javoy-Agid F. Jenner P. Marsden C.D. (1994) Alterations in glutathione levels in Parkinson's disease and other neurodegenerative disorders affecting basal ganglia, Ann Neurol, 36, 348–355.
[92] Jenner P. Olanow C.W. (1996) Oxidative stress and the pathogenesis of Parkinson's disease, Neurology, 47, S161–S170.
[93] Wullner U. Loschmann P.A. Schulz J.B. Schmid A. Dringen R. Eblen F. Turski L. Klockgether T. (1996) Glutathione depletion potentiates MPTP and MPP+ toxicity in nigral dopaminergic neurones, Neuroreport, 7, 921–923.
[94] Gatenby R.A. (1995) Models of tumor-host interaction as competing populations: implications for tumor biology and treatment, J Theor Biol, 176(4), 447-455.
[95] Heim S. Mandahl N. Mitelman F. (1988) Genetic convergence and divergence in tumor progression, Cancer Res, 48(21), 5911-5916.
[96] Nowell P.C. (1976) The clonal evolution of tumor cell populations, Science. 194(4260), 23-28.
[97] Shapiro J.R. Yung W.K. Shapiro W.R. (1981) Isolation, karyotype, and clonal growth of heterogeneous subpopulations of human malignant gliomas, Cancer Res, 41(6), 2349-2359.
[98] Heppner G.H. Miller F.R. (1998) The cellular basis of tumor progression, Int Rev Cytol, 177, 1-56.
[99] Beisson J. and Sonneborn T. M. (1965) Cytoplasmic inheritance of the organization of the cell cortex in Paramecium Aurelia, Proc Natl Acad Sci USA, 53, 275-282.
[100] Jablonka E. Lamb M.J. (1995) Epigenetic Inheritance and Evolution, Oxford University Press.
[101] Mates J.M. Sanchez-Jimenez F.M. (2000) Role of reactive oxygen species in apoptosis: implications for cancer therapy, Int J Biochem Cell Biol, 32(2), 157-170.
[102] Cerutti P.A. (1985) Prooxidant states and tumor promotion, Science, 227(4685), 375-381.
[103] Conour J.E. Graham W.V. Gaskins H.R. (2004) A combined in vitro/bioinformatic investigation of redox regulatory mechanisms governing cell cycle progression, Physiol Genomics, 18(2), 196-205.
[104] McEligot A.J. Yang S. Meyskens Jr F.L. (2005) Redox regulation by intrinsic species and extrinsic nutrients in normal and cancer cells, Annu Rev Nutr, 25, 261-295.
[105] Smith J. Ladi E. Mayer-Proschel M. Noble M. (2000) Redox state is a central modulator of the balance between self-renewal and differentiation in a dividing glial precursor cell, Proc Natl Acad Sci U S A, 97(18), 10032-1007.
[106] Bae Y.S. Kang S.W. Seo M.S. Baines I.C. Tekle E. Chock P.B. Rhee S.G.(1997) Epidermal growth factor (EGF)-induced generation of hydrogen peroxide. Role in EGF receptor-mediated tyrosine phosphorylation, J Biol Chem, 272(1), 217-221.
[107] Sundaresan M. Yu Z. Ferrans V.J. Irani K. Finkel T. (1995) Requirement for generation of HO for platelet-derived growth factor signal transduction, Science, 270(5234) 296-299.
[108] Lo Y.Y. Cruz T.F. (1995) Involvement of reactive oxygen species in cytokine and growth factor induction of c-fos expression in chondrocytes, J Biol Chem, 270(20), 11727-11730.
[109] Pani G. Colavitti R. Borrello S. Galeotti T. (2000) Endogenous oxygen radicals modulate protein tyrosine phosphorylation and JNK-1 activation in lectin-stimulated thymocytes, Biochem J, 347(1), 173-181.
[110] Hecht D. Zick Y. (1992) Selective inhibition of protein tyrosine phosphatase activities by H2O2 and vanadate in vitro, Biochem Biophys Res Commun. 188(2), 773-779.
[111] Fialkow L. Chan C.K. Downey G.P (1997) Inhibition of CD45 during neutrophil activation The Journal of Immunology, 158(11), 5409-5417.
[112] Irani K. Xia Y. Zweier J.L Sollott S.J. Der C.J. Fearon E.R. Sundaresan M. Finkel T. (1997) Mitogenic signaling mediated by oxidants in ras-transformed fibroblasts, Science, 275(5306), 1649-1652.
[113] Maeda H. Akaike T. (1998) Nitric oxide and oxygen radicals in infection, inflammation, and cancer, Biochemistry (Mosc), 63(7), 854-865.
[114] Burdon R.H. (1995) Superoxide and hydrogen peroxide in relation to mammalian cell proliferation, Free Radic Biol Med, 18(4), 775-794.
[115] Pani G.Colavitti R. Bedogni B. Anzevino R. Borrello S. Galeotti T. (2000) A redox signaling mechanism for density-dependent inhibition of cell growth, J Biol Chem, 275(49), 38891-38899
[116] Oberley T.D. (2002) Oxidative damage and cancer, American Journal of Pathology, 160, 403-408.
[117] Oberley T.D. Oberley L.W. (1997) Antioxidant enzyme levels in cancer, Histol Histopathol, 12(2), 525-535.
[118] Coursin D.B. Cihla H.P. Sempf J. Oberley T.D. Oberley L.W. (1996) An immunohistochemical analysis of antioxidant and glutathione S-transferase enzyme levels in normal and neoplastic human lung, Histol Histopathol, 11(4), 851-860.
[119] Oberley T.D. Sempf J.M. Oberley M.J. McCormick M.L. Muse K.E. Oberley L.W. (1994) Immunogold analysis of antioxidant enzymes in human renal cell carcinoma, Virchows Arch, 424(2), 155-164.
[120] Oberley T.D. Sempf J.M. Oberley L.W. (1996) Immunogold analysis of antioxidant enzymes in common renal cancers, Histol Histopathol, 11(1), 153-160.

[121] Xu Y. Krishnan A. Wan X.S. Majima H. Yeh C.C. Ludewig G. Kasarskis E.J. St Clair D.K. (1999) Mutations in the promoter reveal a cause for the reduced expression of the human manganese superoxide dismutase gene in cancer cells, Oncogene, 18(1), 93-102.
[122] Oberley T.D. Zhong W. Szweda L.I. Oberley L.W. (2000) Localization of antioxidant enzymes and oxidative damage products in normal and malignant prostate epithelium, Prostate, 44(2), 144-155.
[123] Mantovani G. Maccio A. Madeddu C. Mura L. Gramignano G. Lusso M.R. Mulas C. Mudu M.C. Murgia V. Camboni P. Massa E. Ferreli L. Contu P. Rinaldi A. Sanjust E. Atzei D. Elsener B. (2002) Quantitative evaluation of oxidative stress, chronic inflammatory indices and leptin in cancer patients: correlation with stage and performance status, Int J Cancer, 98(1), 84-91.
[124] Menon S.G. Sarsour E.H. Spitz D.R. Higashikubo R. Sturm M. Zhang H. Goswami P.C.(2003) Redox regulation of the G1 to S phase transition in the mouse embryo fibroblast cell cycle, Cancer Res, 63(9), 2109-2117.
[125] Powis G. Mustacich D. Coon A. (2000) The role of the redox protein thioredoxin in cell growth and cancer, Free Radic Biol Med, 29(3-4), 312-322.
[126] Menon S.G. Sarsour E.H. Spitz D.R. Higashikubo R. Sturm M. Zhang H. Goswami P.C. (2003) Redox Regulation of the G1 to S Phase Transition in the Mouse Embryo Fibroblast Cell Cycle, Cancer Res, 63, 2109–2117.
[127] Rhee S.G. (1999) Redox signaling: hydrogen peroxide as intracellular messenger, Experimental and Molecular Medicine, 31(2), 53-59.
[128] Okamoto K. Toyokuni S. Uchida K. Ogawa O. Takenewa J. Kakehi Y. Kinoshita H. Hattori-Nakakuki Y. Hiai H. Yoshida O. (1994) Formation of 8-hydroxy-2'-deoxyguanosine and 4-hydroxy-2-nonenal-modified proteins in human renal-cell carcinoma, Int J Cancer, 58(6), 825-829.
[129] Portakal O. Ozkaya O. Erden Inal M. Bozan B. Kosan M. Sayek I. (2000) Coenzyme Q10 concentrations and antioxidant status in tissues of breast cancer patients, Clin Biochem, 33(4), 279-284.
[130] Toyokuni S. Okamoto K. Yodoi J. Hiai H. (1995) Persistent oxidative stress in cancer, FEBS Lett, 358(1), 1-3.
[131] Loo G. (2003) Redox-sensitive mechanisms of phytochemical-mediated inhibition of cancer cell proliferation (review), J Nutr Biochem,14(2), 64-73.
[132] He T. Weintraub N.L. Goswami P.C. Chatterjee P. Flaherty D.M. Domann F.E. Oberley L.W. (2003) Redox factor-1 contributes to the regulation of progression from G0/G1 to S by PDGF in vascular smooth muscle cells, Am J Physiol Heart Circ Physiol, 285(2), H804-812.
[133] Burdon R.H. Rice-Evans C. (1998) Free radicals and the regulation of mammalian cell proliferation, Free Radic Res Commun, 6(6), 345-358.
[134] Bostwick D.G. Alexander E.E. Singh R. Shan A. Qian J. Santella R.M. Oberley L.W. Yan T. Zhong W. Jiang X. Oberley T.D. (2000) Antioxidant enzyme expression and reactive oxygen species damage in prostatic intraepithelial neoplasia and cancer, Cancer, 89(1), 123-134.
[135] Tanaka T. Iwasa Y. Kondo S. Hiai H. Toyokuni S. (1999) High incidence of allelic loss on chromosome 5 and inactivation of p15INK4B and p16INK4A tumor suppressor genes in oxystress-induced renal cell carcinoma of rats, Oncogene, 18(25), 3793-3797.
[136] Hiroyasu M. Ozeki M. Kohda H. Echizenya M. Tanaka T. Hiai H. Toyokuni S. (2002) Specific allelic loss of p16 INK4A tumor suppressor gene after weeks of iron-mediated oxidative damage during rat renal carcinogenesis, American Journal of Pathology, 160, 419-424.
[137] Shackelford R.E. Kaufmann W.K. Paules R.S.(2000) Oxidative stress and cell cycle checkpoint function, Free Radic Biol Med. 28(9), 1387-1404.
[138] Shackelford R.E. Kaufmann W.K. Paules R.S. (1999) Cell cycle control, checkpoint mechanisms, and genotoxic stress, Environ Health Perspect, 107(Suppl 1), 5-24.
[139] Laurent A. Nicco C. Chereau C. Goulvestre C. Alexandre J. Alves A. Levy E. Goldwasser F. Panis Y. Soubrane O. Weill B. Batteux F. (2005) Controlling tumor growth by modulating endogenous production of reactive oxygen species, Cancer Res, 65(3), 948-956.
[140] Adler V. Yin Z. Tew K.D. Ronai Z. (1999) Role of redox potential and reactive oxygen species in stress signaling, 18(45), 6104-6111.
[141] Burdon R.H. (1995) Superoxide and hydrogen peroxide in relation to mammalian cell proliferation, Free Radic Biol Med, 18(4), 775-794.
[142] Burkitt M.J. Gilbert B.C. (1990) Model studies of the iron-catalysed Haber-Weiss cycle and the ascorbate-driven Fenton reaction, Free Radic Res Commun, 10(4-5), 265-280.
[143] Valko M. Morris H. Cronin M.T. (2005) Metals, toxicity and oxidative stress, Curr Med Chem, 12(10), 1161-1208.
[144] Samuni A. Aronovitch J. Godinger D. Chevion M. Czapski G. (1983) On the cytotoxicity of vitamin C and metal ions. A site-specific Fenton mechanism, Eur J Biochem, 137(1-2), 119-124.

145 Jones W. Li X. Qu Z.C. Perriott L. Whitesell R.R. May J.M. (2002) Uptake, recycling, and antioxidant actions of alpha-lipoic acid in endothelial cells, Free Radic Biol Med, 33(1), 83-93.
146 Hayflick, L. The limited in vitro lifetime of human diploid cell strains, Exp. Cell Res, 37, 614–636 (1965).
147 Shay J.W. Wright W.E. (2000) Hayflick, his limit, and cellular ageing, Nature Reviews, Molecular Cell biology, 1, 72-76.
148 Olovnikov A.M. (1996) Telomeres, telomerase and aging: origin of the theory, Exp. Gerontol, 31, 443–448.
149 Harley C.B. Futcher A.B. Greider C.W. (1990) Telomeres shorten during ageing of human fibroblasts, Nature, 345, 458–460.
150 Kim N.W. Piatyszek M.A. Prowse K.R. Harley C.B. West M.D. Ho P.L. Coviello G.M. Wright W.E. Weinrich S.L. Shay J.W. (1994) Specific association of human telomerase activity with immortal cells and cancer, Science, 266(5193), 2011-2015.
151 Hochachka P.W. Rupert J.L. Goldenberg L. Gleave M. Kozlowski P. (2002) Going malignant: the hypoxia-cancer connection in the prostate, Bioessays, 24(8), 749-757.
152 Gatenby R.A. Gawlinski E.T. (2003) The glycolytic phenotype in carcinogenesis and tumor invasion: insights through mathematical models, Cancer Res, 63(14), 3847-3854.
153 Gatenby R.A. (1991) Population ecology issues in tumor growth, Cancer Res, 51(10), 2542-2547.
154 Leather S.R. Helden A.J. (2005) Roundabouts: our neglected nature reserves? Biologist, 52(2), 102-106.
155 Wayt Gibbs W. (2003) Roots of cancer, Scientific American, 289, 56-65.
156 Hahn W.C. Counter C.M. Lundberg A.S. Beijersbergen R.L. Brooks M.W. Weinberg R.A. (1999) Creation of human tumour cells with defined genetic elements, Nature, 400(6743), 464-468. Comment in Nature. 400(6743), 401-402.
157 Kacser H. Burns J.A. (1981) The molecular basis of dominance, Genetics, 97(3-4), 639-666.
158 Chung D.C. Rustgi A.K. (2003) The hereditary nonpolyposis colorectal cancer syndrome: genetics and clinical implications, Annals of Internal Medicine, 138, 560-570.
159 Gottlieb T.M. Oren M. (1998) p53 and apoptosis, Semin Cancer Biol, 8, 359–368.
160 Buzek J. Latonen L. Kurki S. Peltonen K. Laiho M. (2002) Redox state of tumor suppressor p53 regulates its sequence-specific DNA binding in DNA-damaged cells by cysteine 277, Nucleic Acids Res, 30(11), 2340-2348.
161 Meplan C. Richard M.J. Hainaut P. (2000) Redox signalling and transition metals in the control of the p53 pathway, Biochem Pharmacol, 59(1), 25-33.
162 Parks D. Bolinger R. Mann K. (1997) Redox state regulates binding of p53 to sequence-specific DNA, but not to non-specific or mismatched DNA, Nucleic Acids Res, 25(6), 1289-1295.
163 Polyak K. Xia Y. Zweier J.L. Kinzler K.W. Vogelstein B. (1997) A model for p53-induced apoptosis, Nature, 389(6648), 300-305. Comment, Nature 1997 389(6648), 237-238.
164 Johnson T.M. Yu Z.X. Ferrans V.J. Lowenstein R.A. Finkel T. (1996) Reactive oxygen species are downstream mediators of p53-dependent apoptosis, Proc Natl Acad Sci U S A, 93(21), 11848-11852.
165 Li P.F. Dietz R. von Harsdorf R. (1999) p53 regulates mitochondrial membrane potential through reactive oxygen species and induces cytochrome c-independent apoptosis blocked by Bcl-2, EMBO J, 18, 6027–6036.
166 Johnson T.M. Yu Z.X. Ferrans V.J. Lowenstein R.A. Finkel T. (1996) Reactive oxygen species are downstream mediators of p53-dependent apoptosis, Proc Natl Acad Sci USA, 93,11848–11852.
167 Brash D.E. Havre P.A. (2002) New careers for antioxidants, PNAS, 99(22), 13969-13971.
168 Jayaraman L. Murthy K.G. Zhu C. Curran T. Xanthoudakis S. Prives C. (1997) Identification of redox/repair protein Ref-1 as a potent activator of p53, Genes Dev, 11, 558–570.
169 Rainwater R. Parks D. Anderson M.E. Tegtmeyer P. Mann K. (1995) Role of cysteine residues in regulation of p53 function, Mol. Cell. Biol, 15(7), 3892-3903.
170 Hainaut P. Milner J. (1993) Redox modulation of p53 conformation and sequence-specific DNA binding in vitro, Cancer Res, 53, 4469–4473.
171 Merwin J.R. Mustacich D.J. Muller E.G.D. Pearson G.D. Merrill G.F. (2002) Reporter gene transactivation by human p53 is inhibited in thioredoxin reductase null yeast by a mechanism associated with thioredoxin oxidation and independent of changes in the redox state of glutathione, Carcinogenesis, 23(10), 1609-1616.
172 Moss R. W. (2000) Antioxidants and Cancer, Equinox Press, PA. USA.
173 Hockenbery D. Nunez G. Milliman C. Schreiber R.D. Korsmeyer S.J. (1990) Bcl-2 is an inner mitochondrial membrane protein that blocks programmed cell death, Nature, 348, 334–336.
174 Kelekar A. Thompson C.B. (1998) Bcl-2-family proteins: the role of the BH3 domain in apoptosis, Trends Cell Biol, 8, 324–330.
175 Oltvai Z.N. Milliman C.L. Korsmeyer S.J. (1993) Bcl-2 heterodimerizes in vivo with a conserved homolog, Bax, that accelerates programmed cell death, Cell, 74, 609–619.
176 Mirkovic N. Voehringer D.W. Story M.D. McConkey D.J. McDonnell T.J. Meyn R.E. (1997) Resistance to radiation-induced apoptosis in Bcl-2-expressing cells is reversed by depleting cellular thiols, Oncogene, 15, 1461–1470.

[177] Voehringer D.W. McConkey D.J. McDonnell T.J. Brisbay S. Meyn R.E. (1998) Bcl-2 expression causes redistribution of glutathione to the nucleus, Proc Natl Acad Sci USA, 95, 2956–2960.
[178] Preston-Martin S. Pike M.C. Ross R.K. Jones P.A. Henderson B.E. (1990) Increased cell division as a cause of human cancer, Cancer Res, 50(23), 7415-7421.
[179] Meraldi P. Honda R. Nigg E.A. (2004) Aurora kinases link chromosome segregation and cell division to cancer susceptibility, Curr Opin Genet Dev, 14(1), 29-36.
[180] Pihan G.A. Purohit A. Wallace J. Knecht H. Woda B. Quesenberry P. Doxsey S.J. (1998) Centrosome defects and genetic instability in malignant tumors, Cancer Res, 58(17), 3974-3985.
[181] Ghadimi B.M. Sackett D.L. Difilippantonio M.J. Schrock E. Neumann T. Jauho A. Auer G. Ried T. (2000) Centrosome amplification and instability occurs exclusively in aneuploid, but not in diploid colorectal cancer cell lines, and correlates with numerical chromosomal aberrations, Genes Chromosomes Cancer, 27(2), 183-190.
[182] Saunders W.S. Shuster M. Huang X. Gharaibeh B. Enyenihi A.H. Petersen I. Gollin S.M. (2000) Chromosomal instability and cytoskeletal defects in oral cancer cells, Proc Natl Acad Sci U S A, 97(1), 303-308.
[183] Bartek J. Lukas J. Bartkova J. (1999) Perspective: defects in cell cycle control and cancer, J Pathol, 187(1), 95-99.
[184] Nigg E.A. (2002) Centrosome aberrations: cause or consequence of cancer progression? Nat Rev Cancer, 2(11), 815-825.
[185] Krämer A Ho A.D. (2001) Centrosome Aberrations and Cancer, Onkologie, 24(6) 538-544.
[186] Duesberg P. Rausch C. Rasnick D. Hehlmann R. (1998) Genetic instability of cancer cells is proportional to their degree of aneuploidy, Proc Natl Acad Sci U S A, 95(23), 13692–13697.
[187] Sen S. (2000) Aneuploidy and cancer, Curr Opin Oncol, 12(1), 82-88.
[188] Mitelman F. Mertens F. Johansson B. (1997) A breakpoint map of recurrent chromosomal rearrangements in human neoplasia, Nat Genet, Apr;15 Spec No:417-474.
[189] Gebhart E. Liehr T. (2000) Patterns of genomic imbalances in human solid tumors (Review), Int J Oncol, 16(2), 383-399.
[190] Atkin N.B. Baker M.C. (1990) Are human cancers ever diploid--or often trisomic? Conflicting evidence from direct preparations and cultures, Cytogenet Cell Genet, 53(1), 58-60.
[191] Loeb K.R. Loeb L.A. (2000) Significance of multiple mutations in cancer, Carcinogenesis, 21(3), 379-385.
[192] Andreassen P.R. Martineau S.N. Margolis R.L. (1996) Chemical induction of mitotic checkpoint override in mammalian cells results in aneuploidy following a transient tetraploid state, Mutat Res, 372(2), 181-194.
[193] Tucker J.D. Preston R.J. (1996) Chromosome aberrations, micronuclei, aneuploidy, sister chromatid exchanges, and cancer risk assessment, Mutat Res, 365(1-3), 147-159.
[194] Oshimura M. Barrett J.C. (1986) Chemically induced aneuploidy in mammalian cells: mechanisms and biological significance in cancer, Environ Mutagen, 8(1), 129-159.
[195] Galloway S.M. Buckton K.E. (1978) Aneuploidy and ageing: chromosome studies on a random sample of the population using G-banding, Cytogenet Cell Genet, 20(1-6), 78-95.
[196] Li R. Sonik A. Stindl R. Rasnick D. Duesberg P. (2000) Aneuploidy vs. gene mutation hypothesis of cancer: Recent study claims mutation but is found to support aneuploidy, Proc Natl Acad Sci U S A, 97(7), 3236–3241.
[197] Duesberg P. Stindl R. Hehlmann R. (2000) Explaining the high mutation rates of cancer cells to drug and multidrug resistance by chromosome reassortments that are catalyzed by aneuploidy, Proc Natl Acad Sci U S A, 97(26), 14295–14300.
[198] Rasnick D. Duesberg P.H. (1999) How aneuploidy affects metabolic control and causes cancer, Biochem J, 340 (3), 621-630.
[199] Duesberg P. Rasnick D. (2000) Aneuploidy, the somatic mutation that makes cancer a species of its own, Cell Motil Cytoskeleton, 47(2), 81-107.
[200] Li R. Yerganian G. Duesberg P. Kraemer A. Willer A. Rausch C. Hehlmann R. (1997) Aneuploidy correlated 100% with chemical transformation of Chinese hamster cells, Proc Natl Acad Sci U S A, 94(26), 14506–14511.
[201] Duesberg P. Li R. Rasnick D. Rausch C. Willer A. Kraemer A. Yerganian G. Hehlmann R. (2000) Aneuploidy precedes and segregates with chemical carcinogenesis, Cancer Genet Cytogenet, 119(2), 83-93.
[202] Beard J. (1902) Embryological aspects and etiology of carcinoma, The Lancet, June 21, 1758-1761.
[203] Westerman B.A. Poutsma A. Maruyama K. Schrijnemakers H.F. van Wijk I.J. Oudejans C.B. (2002) The proneural genes NEUROD1 and NEUROD2 are expressed during human trophoblast invasion, Mech Dev, 113(1), 85-90.
[204] Southall P.J. Boxer G.M. Bagshawe K.D. Hole N. Bromley M. Stern P.L. (1990) Immunohistological distribution of 5T4 antigen in normal and malignant tissues, Br J Cancer, 61(1), 89-95.

[205] Yamashita K. Nakamura T. Shimizu T. Ohno H. (1986) Monoclonal antibodies to human choriocarcinoma, Am J Reprod Immunol Microbiol, 11(4), 130-134.
[206] Gurchot C. (1975) The trophoblast theory of cancer (John Beard, 1857-1924) revisited, Oncology, 31(5-6), 310-333.
[207] Makino Y. Sakagami H. Takeda M. (1999) Induction of cell death by ascorbic acid derivatives in human renal carcinoma and glioblastoma celllines, Anticancer Res, 19, 3125-3313.
[208] Smith J. Ladi E. Mayer-Proschel M. Noble M. (2000) Redox state is a central modulator of the balance between self-renewal and differentiation in a dividing glial precursor cell, Proc Natl Acad Sci U S A, 97(18), 10032-10037.
[209] Esposito F. Agosti V. Morrone G. Morra F. Cuomo C. Russo T. Venuta S. Cimino F. (1994) Inhibition of the differentiation of human myeloid cell lines by redox changes induced through glutathione depletion, Biochem J, 301(3), 649-53.
[210] Warburg O. (1938) The Catalytic Activity of Living Tissues, Springer, Berlin.
[211] Weber G. (1977) Enzymology of cancer cells, N Engl J Med, 296, 541-558.
[212] Warburg O. (1930) Metabolism of tumors (translation by Dickens F.), Arnold Constable, London.
[213] Con C.F. and Con G.T. (1925) The carbohydrate metabolism of tumours, J Bioi Chem, 65, 397-405.
[214] Warburg O. (1956) A review, Science, 123, 309-315.
[215] Szent-Gyorgyi A. (1960) The living state and cancer. In: Submolecular Biology and Cancer, Ciba Foundation Symposium 67, Excerpta Medica, New York.
[216] Gonzalez M.J. Miranda-Massari J.R. Mora E.M. Guzman A. Riordan N.H. Riordan H.D. Casciari J.J. Jackson J.A. Roman-Franco A. (2005) Orthomolecular oncology review: ascorbic acid and cancer 25 years later, Integr Cancer Ther, 4(1), 32-44.
[217] Gonzalez M.J. Miranda-Massari J.R. Mora E.M. Jimenez I.Z. Matos M.I. Riordan, Casciari J.J. Riordan N.H. Rodriguez M.Y.S. Guzman A. (2002) Orthomolecular oncology: a mechanistic view of ascorbate's chemotherapeutic activity, P R Health Sci J, 21(1), 39-41.
[218] Green S. (1998) Oxygenation Therapy: Unproven Treatments for Cancer and AIDS, Scientific Review of Alternative Medicine, Spring/Summer.
[219] Aisenberg A. (1961) The glycolysis and respiration of tumors. A review, Academic Press , New York.
[220] Vaupel P. Kallinowski F. Okunieff P. (1989) Blood flow, oxygen and nutrient supply, and metabolic microenvironment of human tumors: a review, Cancer Res, 49(23), 6449-6465.
[221] Kung A.L. Wang S. Klco J.M. Kaelin W.G. Livingston D.M. (2000) Suppression of tumor growth through disruption of hypoxia-inducible transcription, Nature Medicine, 6(12), 1335-1340.
[222] Koong A.C. Denko N.C. Hudson K.M. Schindler C. Swiersz L. Koch C. Evans S. Ibrahim H. Le Q.T. Terris D.J. Giaccia A.J. (2000) Candidate genes for hypoxic tumor phenotype, Cancer Res, 60, 883–887.
[223] Allen K.G. Venkatraj V.S. (1992) Oxidants and antioxidants in development and differentiation, J Nutr, 122, 631-635.
[224] Gonzalez M.J. Schemmel R.A. Dugan L. Gray J.I. Welsch C.W. (1993) Dietary fish oil inhibition of human breast carcinoma growth: a function of increased lipid peroxidation, Lipids, 28, 827-832.
[225] Helmlinger G. Yuan F. Dellian M. Jain R.K. (1997) Interstitial pH and pO2 gradients in solid tumors in vivo: high-resolution measurements reveal a lack of correlation, Nat Med, 3(2), 177-182.
[226] Brown J.M. Giaccia A.J. (1998) The unique physiology of solid tumors: opportunities (and problems) for cancer therapy, Cancer Res, 58(7), 1408-1416.
[227] Jain R.K. (1988) Determinants of tumor blood flow: a review, Cancer Res, 48(10), 2641-2658.
[228] Freiberg R.A. Green S.L. Giaccia A.J. (2002) Hypoxia and cell cycle, Cell Cycle Checkpoints and Cancer, Ed Blagosklonny, M.V. Landes Bioscience, Georgetown, Texas.
[229] Liotta L.A. Kohn E.C. (2001) The microenvironment of the tumour–host interface, Nature, 411, 375 – 379.
[230] Carmeliet P. Jain R.K. (2000) Angiogenesis in cancer and other diseases, Nature, 14, 407(6801), 249-257.
[231] Huso D.L. Dillehay L. Cheong I. Agrawal N. Borzillary S. McCaffery J.M. Watson E.L. Lin K. Bunz F. Baidoo K. Pomper M.G. Kinzler K.W. Vogelstein B. Zhou S. Bettegowda C. Dang L.H. Abrams R. (2003) Overcoming the hypoxic barrier to radiation therapy with anaerobic bacteria, PNAS, 100(25), 15083-15088.
[232] Jain R.K. Forbes N.S. (2001) Can engineered bacteria help control cancer? PNAS, 98(26), 4748-14750.
[233] Rofstad E.K. (2000) Microenvironment-induced cancer metastasis, Int J Rad Biol, 76(5), 589 – 605.
[234] Tannock I.F. Rotin D. (1989) Acid pH in tumors and its potential for therapeutic exploitation, Cancer Res, 49(16) 4373-4384.
[235] Schwickert G. Walenta S. Sundfor K. Rofstad E.K. Mueller-Klieser W. (1995) Correlation of high lactate levels in human cervical cancer with incidence of metastasis, Cancer Res, 55(21), 4757-4759.
[236] Kallinowski F. Vaupel P. Runkel S. Berg G. Fortmeyer H.P. Baessler K.H. Wagner K. Mueller-Klieser W. Walenta S. (1988) Glucose uptake, lactate release, ketone body turnover, metabolic micromilieu, and pH distributions in human breast cancer xenografts in nude rats, Cancer Res, 48(24) 7264-7272.

[237] Barefoot R.R. Reich C.M. (2002) The Calcium Factor: The Scientific Secret of Health and Youth, Bokar Consultants, Wickenburg, Arizona.
[238] Barefoot R.R. (2002) Death by Diet, Deonna Enterprises, Wickenburg, Arizona.
[239] D'Amato R.J. Loughnan M.S. Flynn E. Folkman J. (1994) Thalidomide is an inhibitor of angiogenesis, Proc Natl Acad Sci U S A, 91(9), 4082-4085.
[240] Bauer K.S. Dixon S.C. Figg W.D. (1998) Inhibition of angiogenesis by thalidomide requires metabolic activation, which is species-dependent, Biochem Pharmacol, 55(11), 1827-1834.
[241] Tseng S. Pak G. Washenik K. Pomeranz M.K. Shupack J.L. (1996) Rediscovering thalidomide: a review of its mechanism of action, side effects, and potential uses, J Am Acad Dermatol, 35(6), 969-979.
[242] Ehrenpreis E.D. Kane S.V. Cohen L.B. Cohen R.D. Hanauer S.B. (1999) Thalidomide therapy for patients with refractory Crohn's disease: an open-label trial, Gastroenterology, 117(6),1271-1277.
[243] Calabrese L. Fleischer A.B. (2000) Thalidomide: current and potential clinical applications, Am J Med, 108(6), 487-495.
[244] Reyes-Teran G. Sierra-Madero J.G. Martinez del Cerro V. Arroyo-Figueroa H. Pasquetti A. Calva J.J. Ruiz-Palacios G.M. (1996) Effects of thalidomide on HIV-associated wasting syndrome: a randomized, double-blind, placebo-controlled clinical trial, AIDS, 10(13), 1501-1507.
[245] Singhal S. Mehta J. Desikan R. Ayers D. Roberson P. Eddlemon P. Munshi N. Anaissie E. Wilson C. Dhodapkar M. Zeddis J. Barlogie B. (1999) Antitumor activity of thalidomide in refractory multiple myeloma, N Engl J Med, 341(21), 1565-1571.
[246] Fine H.A. Figg W.D. Jaeckle K. Wen P.Y. Kyritsis A.P. Loeffler J.S. Levin V.A. Black P.M. Kaplan R. Pluda J.M. Yung W.K. (2000) Phase II trial of the antiangiogenic agent thalidomide in patients with recurrent high-grade gliomas, J Clin Oncol, 18(4), 708-715.
[247] Parman T. Wiley M.J. Wells P.G. (1999) Free radical-mediated oxidative DNA damage in the mechanism of thalidomide teratogenicity, Nat Med, 5(5), 582-585.
[248] Sauer H. Gunther J. Hescheler J. Wartenberg M. (2000) Thalidomide inhibits angiogenesis in embryoid bodies by the generation of hydroxyl radicals, Am J Pathol, 156(1), 151-158.
[249] Wells P.G. Kim P.M. Laposa R.R. Nicol C.J. Parman T. Winn L.M. (1997) Oxidative damage in chemical teratogenesis, Mutat Res, 396(1-2), 65-78.
[250] Hansen J.M. Harris K.K. Philbert M.A. Harris C. (2002) Thalidomide modulates nuclear redox status and preferentially depletes glutathione in rabbit limb versus rat limb, J Pharmacol Exp Ther, 300(3), 768-776.
[251] Brown N.S. Jones A. Fujiyama C. Harris A.L. Bicknell R. (2000) Thymidine phosphorylase induces carcinoma cell oxidative stress and promotes secretion of angiogenic factors, Cancer Res, 60(22), 6298-6302.
[252] Chandel N.S. McClintock D.S. Feliciano C.E. Wood T.M. Melendez J.A. Rodriguez A.M. Schumacker P.T. (2000) Reactive oxygen species generated at mitochondrial complex III stabilize hypoxia-inducible factor-1alpha during hypoxia: a mechanism of O2 sensing, J Biol Chem, 275(33), 25130-25138.
[253] Richard D.E. Berra E. Pouyssegur J. (2000) Nonhypoxic pathway mediates the induction of hypoxia-inducible factor 1alpha in vascular smooth muscle cells, J Biol Chem, 275(35), 26765-26771.
[254] Polytarchou C. Papadimitriou E. (2004) Antioxidants inhibit angiogenesis in vivo through down-regulation of nitric oxide synthase expression and activity, Free Radical Research, 38(5), 501-508.
[255] Ashino H. Shimamura M. Nakajima H. Dombou M. Kawanaka S. Oikawa T. Iwaguchi T. Kawashima S. (2003) Novel function of ascorbic acid as an angiostatic factor, Angiogenesis, 6(4), 259-269.
[256] Pugh C.W. Ratcliffe P.J. (2003) Regulation of angiogenesis by hypoxia: role of the HIF system, Nat Med, 9(6), 677-684.
[257] Knowles H.J. Raval R.R. Harris A.L. Ratcliffe P.J. (2003) Effect of ascorbate on the activity of hypoxia-inducible factor in cancer cells, Cancer Res, 63, 1764-1768.
[258] Koike T. Kimura N. Miyazaki K. Yabuta T. Kumamoto K. Takenoshita S. Chen J. Kobayashi M. Hosokawa M. Taniguchi A. Kojima T. Ishida N. Kawakita M. Yamamoto H. Takematsu H. Suzuki A. Kozutsumi Y. Kannagi R. (2004) Hypoxia induces adhesion molecules on cancer cells: a missing link between Warburg effect and induction of selectin-ligand carbohydrates, PNAS, 101(21), 8132-8137.
[259] Kannagi R. (2004) Molecular mechanism for cancer-associated induction of sialyl Lewis X and sialyl Lewis A expression-The Warburg effect revisited, Glycoconj J, 20(5), 353-364.
[260] Kimura N. Mitsuoka C. Kanamori A. Hiraiwa N. Uchimura K. Muramatsu T. Tamatani T. Kansas G.S. Kannagi R. (1999) Reconstitution of functional L-selectin ligands on a cultured human endothelial cell line by cotransfection of alpha-1-3-fucosyltransferase VII and newly cloned GlcNAcbeta:6-sulfotransferase cDNA, Proc Natl Acad Sci U S A, 96(8), 4530-4535.
[261] Jelinek F. (2003) Spontaneous tumours in guinea pigs, Acta Vet Brno, 72, 221–228.
[262] Nakanishi M. Kuwamura M. Ueno M. Yasuda K. Yamate J. Shimada T. (2003) Pulmonary adenocarcinoma with osteoblastic bone metastases in a cat, J. Small Animal Practice, 44(10), 464-466.
[263] Labelle P. De Cock H.E.V. (2005) Metastatic tumors to the adrenal glands in domestic animals, Vet Pathol, 42, 52-58.

[264] Egner P.A. Kensler T.W. (1985) Effects of a biomimetic superoxide dismutase on complete and multistage carcinogenesis in mouse skin, Carcinogenesis, 6, 1167-1172.
[265] Dermime S. Armstrong A. Hawkins R.E. Stern P.L. (2002) Cancer vaccines and immunotherapy, British Medical Bulletin, 62, 149-162.
[266] Carswell E.A. Old L.J. Kassel R.L. Green S. Fiore N. Williamson B. (1975) An endotoxin-induced serum factor that causes necrosis of tumors, Proc Natl Acad Sci, 72, 3666-3670
[267] Malik S.T.A. Griffin D.B. Fiers W. Balkwill F.R. (1989) Paradoxical, effects of tumour necrosids factor in experimental ovarian cancer, Int J Cancer, 44, 918–925.
[268] Malik S.T.A. Naylor S. East N. Oliff A. Balkwill F.R. (1990) Cells secreting tumour necrosis factor show enhanced metastasis in nude mice, Eur J Cancer, 26, 1031–1034.
[269] Roberts R.A. Kimber I. (1999) Cytokines in non-genotoxic hepatocarcinogenesis, Carcinogenesis, 20, 1397–1401.
[270] Erickson S.L. de Sauvage F.J. Kikly K. Carver-Moore K. Pitts-Meek S. Gillett N. Sheehan K.C. Schreiber R.D. Goeddel D.V. Moore M.W. (1994) Decreased sensitivity to tumour-necrosis factor but normal T-cell development in TNF receptor-2-deficient mice, Nature, 372(6506), 560-563.
[271] Brigati C. Noonan D.M. Albini A. Benelli R. (2002) Tumors and inflammatory infiltrates: friends or foes? Clinical and Experimental Metastasis, 19(3), 247 – 258.
[272] Dvorak H.F. (1986) Tumors: wounds that do not heal, N Engl J Med, 315, 1650–1659.
[273] Balkwill F. Mantovani A. (2001) Inflammation and cancer: back to Virchow? Lancet, 357, 539–545.
[274] Tricot G. (2000) New insights into role of microenvironment in multiple myeloma, Lancet, 355, 248–250.
[275] Thun M.J. Namboodiri M.M. Calle E.E. Flanders W.D. Heath C.W. Jr. (1993) Aspirin use and risk of fatal cancer, Cancer Res, 53, 1322–1327.
[276] Langman M.J. Cheng K.K. Gilman E.A. Lancashire R.J. (2000) Effect of anti-inflammatory drugs on overall risk of common cancer: casecontrol study in general practice research database, BMJ, 320, 1642–1646.
[277] Reddy B.S. Rao C.V. Seibert K. (1996) Evaluation of cyclooxygenase-2 inhibitor for potential chemopreventive properties in colon carcinogenesis, Cancer Res, 56, 4566–4569.
[278] Haghnegahdar H. Du J. Wang D. Strieter R.M. Burdick M.D. Nanney L.B. Cardwell N. Luan J. Shattuck-Brandt R. Richmond A. (2000) The tumorigenic and angiogenic effects of MGSSA/GRO proteins in melanoma, J Leukoc Biol, 67, 53–62.
[279] Wang J.M. Chertov O. Proost P. Li J.J. Menton P. Xu L. Sozzani S. Mantovani A. Gong W. Schirrmacher V. Van Damme J. Oppenheim J.J. (1998) Purification and identification of chemokines potentially involved in kidney-specific metastasis by a murine lymphoma variant: induction of migration and NFkappaB activation, Int J Cancer, 75(6), 900-907.
[280] Fung T. Hu F.B. Fuchs C. Giovannucci E. Hunter D.J. Stampfer M.J. Colditz G.A. Willett W.C. (2003) Major dietary patterns and the risk of colorectal cancer in women, Arch Intern Med, 163, 309-314.
[281] Ames B.N. Wakimoto P. (2002) Are vitamin and mineral deficiencies a major cancer risk? Nature Reviews Cancer, 2, 694 -704.
[282] Platz E.A. Willett W.C. Colditz G.A. Rimm E.B. Spiegelman D. Giovannucci E. (2000) Proportion of colon cancer risk that might be preventable in a cohort of middle-aged US men, Cancer Causes Control, 11, 579-588.
[283] WCRF/AICR (1997) Food, nutrition and the prevention of cancer: a global perspective: World Cancer Research Fund / American Institute for Cancer Research.
[284] Riboli E. Norat T. (2003) Epidemiologic evidence of the protective effect of fruit and vegetables on cancer risk, Am J Clin Nutr, 78, 559S-569S.
[285] Bingham S.A. Luben R. Welch A. Wareham N. Khaw K.T. Day N. (2003) Are imprecise methods obscuring a relation between fat and breast cancer? Lancet, 362, 212-214.
[286] Donaldson M.S. (2004) Nutrition and cancer: a review of the evidence for an anti-cancer diet, Nutrition Journal, 3(1), 19.
[287] Key T.J. Schatzkin A. Willett W.C. Allen N.E. Spencer E.A. Travis R.C. (2004) Diet, nutrition and the prevention of cancer, Public Health Nutr, 7(1A), 187-200.
[288] Holmes M.D. Liu S. Hankinson S.E. Colditz G.A. Hunter D.J. Willett W.C. (2004) Dietary carbohydrates, fiber, and breast cancer risk, Am J Epidemiol, 159, 732-739.
[289] Michels K.B. Giovannucci E. Joshipura K.J. Rosner B.A. Stampfer M.J. Fuchs C.S. Colditz G.A. Speizer F.E. Willett W.C. (2000) Prospective study of fruit and vegetable consumption and incidence of colon and rectal cancers, J Natl Cancer Inst, 92(21), 1740-1752.
[290] Slattery M.L. Curtin K.P. Edwards S.L. Schaffer D.M. (2004) Plant foods, fiber, and rectal cancer, Am J Clin Nutr, 79, 274-281.
[291] Shamsuddin AM, Vucenik I. (1999) Mammary tumor inhibition by IP6: a review, Anticancer Res, 19(5A), 3671-3674.

[292] Jariwalla R.J. (1999) Inositol hexaphosphate (IP6) as an anti-neoplastic and lipid-lowering agent, Anticancer Res, 19(5A), 3699-3702.
[293] Shamsuddin A.M. (1999) Metabolism and cellular functions of IP6: a review, Anticancer Res, 19(5A), 3733-3736.
[294] Shamsuddin A.M. Vucenik I. Cole K.E. (1997) IP6: a novel anti-cancer agent, Life Sci, 61(4), 343-354.
[295] Vucenik I, Shamsuddin A.M. (2003) Cancer inhibition by inositol hexaphosphate (IP6) and inositol: from laboratory to clinic, J Nutr, 133(11 Suppl 1), 3778S-3784S.
[296] Sakamoto K. Vucenik I. Shamsuddin A.M. (1993) [3H]phytic acid (inositol hexaphosphate) is absorbed and distributed to various tissues in rats, J Nutr, 123(4), 713-720.
[297] Vucenik I. Shamsuddin A.M. (1994) [3H]inositol hexaphosphate (phytic acid) is rapidly absorbed and metabolized by murine and human malignant cells in vitro, J Nutr, 124(6), 861-868.
[298] Shamsuddin A.M. Baten A. Lalwani ND. (1992) Effects of inositol hexaphosphate on growth and differentiation in K-562 erythroleukemia cell line, Cancer Lett, 64(3), 195-202.
[299] Deliliers G.L. Servida F. Fracchiolla N.S. Ricci C. Borsotti C. Colombo G. Soligo D. (2002) Effect of inositol hexaphosphate (IP(6)) on human normal and leukaemic haematopoietic cells, Br J Haematol, 117(3), 577-587.
[300] Sakamoto K. Venkatraman G. Shamsuddin A.M. (1993) Growth inhibition and differentiation of HT-29 cells in vitro by inositol hexaphosphate (phytic acid), Carcinogenesis, 14(9), 1815-1819.
[301] Yang G.Y. Shamsuddin A.M. (1995) IP6-induced growth inhibition and differentiation of HT-29 human colon cancer cells: involvement of intracellular inositol phosphates, Anticancer Res, 15(6B), 2479-2487.
[302] Vucenik I. Tantivejkul K. Zhang Z.S. Cole K.E. Saied I. Shamsuddin A.M. (1998) IP6 in treatment of liver cancer. I. IP6 inhibits growth and reverses transformed phenotype in HepG2 human liver cancer cell line, Anticancer Res, 18(6A), 4083-4090.
[303] Shamsuddin A.M. Yang G.Y. Vucenik I. (1996) Novel anti-cancer functions of IP6: growth inhibition and differentiation of human mammary cancer cell lines in vitro, Anticancer Res, 16(6A), 3287-3292.
[304] Shamsuddin A.M. Yang G.Y. (1995) Inositol hexaphosphate inhibits growth and induces differentiation of PC-3 human prostate cancer cells, Carcinogenesis, 16(8), 1975-1979.
[305] Fox C.H. Eberl M. (2002) Phytic acid (IP6), novel broad spectrum anti-neoplastic agent: a systematic review, Complement Ther Med, 10(4), 229-234.
[306] Graf E. Eaton J.W. (1990) Antioxidant functions of phytic acid, Free Radic Biol Med, 8(1), 61-69.
[307] Vucenik I. Yang G.Y. Shamsuddin A.M. (1995) Inositol hexaphosphate and inositol inhibit DMBA-induced rat mammary cancer, Carcinogenesis, 16(5), 1055-1058.
[308] Glade M.J. (1999) Food, nutrition, and the prevention of cancer: a global perspective. American Institute for Cancer Research/World Cancer Research Fund, American Institute for Cancer Research, 1997, Nutrition, 15(6), 523-526.
[309] Sieri S. Krogh V. Pala V. Muti P. Micheli A. Evangelista A. Tagliabue G. Berrino F. (2004) Dietary patterns and risk of breast cancer in the ORDET cohort, Cancer Epidemiol Biomarkers Prev, 13, 567-572.
[310] Hung H.C. Joshipura K.J. Jiang R. Hu F.B. Hunter D. Smith-Warner S.A. Colditz G.A. Rosner B. Spiegelman D. Willett W.C. (2004) Fruit and vegetable intake and risk of major chronic disease, J Natl Cancer Inst, 96(21), 1577-1584.
[311] Slattery M.L. Boucher K.M. Caan B.J. Potter J.D. Ma K.N. (19980 Eating patterns and risk of colon cancer, Am J Epidemiol, 148, 4-16.
[312] Bingham S.A. Hughes R. Cross A.J. (2002) Effect of white versus red meat on endogenous N-nitrosation in the human colon and further evidence of a dose response, J Nutrition, 132, 3522S-3525S.
[313] Chen J. Stampfer M.J. Hough H.L. Garcia-Closas M. Willett W.C. Hennekens C.H. Kelsey K.T. Hunter D.J. (1998) A prospective study of N-acetyltransferase genotype, red meat intake, and risk of colorectal cancer, Cancer Res, 58, 3307-3311.
[314] Fung T. Hu F.B. Fuchs C. Giovannucci E. Hunter D.J. Stampfer M.J. Colditz G.A. Willett W.C. (2003) Major dietary patterns and the risk of colorectal cancer in women, Arch Intern Med, 163, 309-314.
[315] Giovannucci E. Rimm E.B. Stampfer M.J. Colditz G.A. Ascherio A. Willett W.C. (1994) Intake of fat, meat, and fiber in relation to risk of colon cancer in men, Cancer Res, 54, 2390-2397.
[316] Norat T. Lukanova A. Ferrari P. Riboli E. (2002) Meat consumption and colorectal cancer risk: dose-response meta-analysis of epidemiological studies, Int J Cancer, 98, 241-256.
[317] Dwyer J.T. (1988) Health aspects of vegetarian diets, Am J Clin Nutr, 48(3 Suppl), 712-738.
[318] Fraser G.E. (1999) Associations between diet and cancer, ischemic heart disease, and all-cause mortality in non-Hispanic white California Seventh-day Adventists, Am J Clin Nutr, 70(3 Suppl), 532S-538S.
[319] Frentzel-Beyme R. Chang-Claude J (1994) Vegetarian diets and colon cancer: the German experience, Am J Clin Nutr, 59(5 Suppl), 1143S-1152S.
[320] Slattery M.L. Boucher K.M. Caan B.J. Potter J.D. Ma K.N. (1998) Eating patterns and risk of colon cancer, Am J Epidemiol, 148, 4-16.

[321] Fung T. Hu F.B. Fuchs C. Giovannucci E. Hunter D.J. Stampfer M.J. Colditz G.A. Willett W.C. (2003) Major dietary patterns and the risk of colorectal cancer in women, Arch Intern Med, 163, 309-314.
[322] Block G. Patterson B. Subar A. (1992) Fruit, vegetables, and cancer prevention: review of the epidemiological evidence, Nutr Cancer, 18, -29.
[323] Steinmetz K.A. Potter J.D. (1996) Vegetables, fruit, and cancer prevention: a review, J Am Diet Assoc, 96, 1027-1039.
[324] Fahey J.W. Zhang Y. Talalay P. (1997) Broccoli sprouts: an exceptionally rich source of inducers of enzymes that protect against chemical carcinogens, Proc Natl Acad Sci U S A, 94, 10367-10372.
[325] Shapiro T.A. Fahey J.W. Wade K.L. Stephenson K.K. Talalay P. (2001) Chemoprotective glucosinolates and isothiocyanates of broccoli sprouts: metabolism and excretion in humans, Cancer Epidemiol Biomarkers Prev, 10, 501-508.
[326] Fowke J.H. Chung F.L. Jin F. Qi D. Cai Q. Conaway C. Cheng J.R. Shu X.O. Gao Y.T. Zheng W. (2003) Urinary isothiocyanate levels, brassica, and human breast cancer, Cancer Res, 63, 3980-3986.
[327] Michaud D.S. Spiegelman D. Clinton S.K. Rimm E.B. Willett W.C. Giovannucci E.L. (1999) Fruit and vegetable intake and incidence of bladder cancer in a male prospective cohort, J Natl Cancer Inst, 91, 605-613.
[328] Zhang S.M. Hunter D.J. Rosner B.A. Giovannucci E.L. Colditz G.A. Speizer F.E. Willett W.C. (2000) Intakes of fruits, vegetables, and related nutrients and the risk of non-Hodgkin's lymphoma among women, Cancer Epidemiol Biomarkers Prev, 9, 477-485.
[329] Cohen J.H. Kristal A.R. Stanford J.L. (2000) Fruit and vegetable intakes and prostate cancer risk, J Natl Cancer Inst, 92, 61-68.
[330] Kolonel L.N. Hankin J.H. Whittemore A.S. Wu A.H. Gallagher R.P. Wilkens L.R. John E.M. Howe G.R. Dreon D.M. West D.W. Paffenbarger R.S. Jr (2000) Vegetables, fruits, legumes and prostate cancer: a multiethnic case-control study, Cancer Epidemiol Biomarkers Prev, 9, 795-804.
[331] London S.J. Yuan J.M. Chung F.L. Gao Y.T. Coetzee G.A. Ross R.K. Yu M.C. (2000) Isothiocyanates, glutathione S-transferase M1 and T1 polymorphisms, and lung-cancer risk: a prospective study of men in Shanghai, China, Lancet, 356, 724-729.
[332] Fleischauer A.T. Poole C. Arab L. (2000) Garlic consumption and cancer prevention: meta-analyses of colorectal and stomach cancers, Am J Clin Nutr, 72, 1047-1052.
[333] Fleischauer A.T. Arab L. (20010 Garlic and cancer: a critical review of the epidemiologic literature, J Nutr, 131, 1032S-1040S.
[334] Hsing A.W. Chokkalingam A.P. Gao Y.T. Madigan M.P. Deng J. Gridley G. Fraumeni J.F. Jr (2002) Allium vegetables and risk of prostate cancer: a population-based study, J Natl Cancer Inst, 94, 1648-1651.
[335] Reizenstein P. (1991) Iron, free radicals and cancer, Med Oncol Tumor Pharmacother, 8(4), 229-233.
[336] Mainous A.G. 3rd, Wells B.J. Koopman R.J. Everett C.J. Gill J.M. (2005) Iron, lipids, and risk of cancer in the Framingham Offspring cohort, Am J Epidemiol, 161(12), 1115-1122.
[337] Crawford R.D. (1998) The case for iron repletion as a promoter in testicular cancer, Med Hypotheses, 51(2), 129-132.
[338] Weinberg E.D. (1992) Roles of iron in neoplasia. Promotion, prevention, and therapy, Biol Trace Elem Res, 34(2), 123-140.
[339] Syrjakoski K. Fredriksson H. Fredriksson H. Ikonen T. Kuukasjarvi T. Autio V. Matikainen M.P. Tammela T.L. Koivisto P.A. Schleutker J. (2005) Hemochromatosis gene mutations among Finnish male breast and prostate cancer patients, Int J Cancer, Jul 7; [Epub ahead of print]
[340] Kwok J.C. Richardson D.R. (2002) The iron metabolism of neoplastic cells: alterations that facilitate proliferation? Crit Rev Oncol Hematol, 42(1), 65-78.
[341] Weinberg E.D. (1983) Iron in neoplastic disease, Nutr Cancer, 4(3), 223-33.
[342] Muller C.I. Miller C.W. Kawabata H. McKenna R.J.Jr, Marchevsky A.M. Koeffler H.P. (2005) Do cancer cells selectively mutate HFE to increase their intracellular iron? Oncol Rep, 14(2), 299-303.
[343] Alberghini A. Recalcati S. Tacchini L. Santambrogio P. Campanella A. Cairo G. (2005) Iron metabolism in VHL-defective renal carcinoma cells, Loss of the von Hippel Landau tumour suppressor disrupts iron homeostasis in renal cancinoma cells, JBC Papers in Press. June 28, manuscript M500971200.
[344] Deugnier Y. (2003) Iron and liver cancer, Alcohol, 30(2), 145-150.
[345] Desoize B. (2004) Metals and metal compounds in cancer treatment, Anticancer Res. 24(3a), 1529-1544.
[346] Elliott R.L. Elliott M.C. Wang F. Head J.F. (1993) Breast carcinoma and the role of iron metabolism. A cytochemical, tissue culture, and ultrastructural study, Ann N Y Acad Sci, 698, 159-166.
[347] Weinberg E.D. (1999) Iron therapy and cancer, Kidney Int Suppl, 69, S131-4.
[348] Weinberg E.D. (1996) The role of iron in cancer, Eur J Cancer Prev, 5(1), 19-36.
[349] Weinberg E.D. (1992) Iron depletion: a defense against intracellular infection and neoplasia, Life Sci, 50(18), 1289-1297.
[350] Weinberg E.D. (1984) Iron withholding: a defense against infection and neoplasia, Physiol Rev, 64(1), 65-102.

[351] Shen L. Zhao H.Y. Du J. Wang F. (2005) Anti-tumor activities of four chelating agents against human neuroblastoma cells, In Vivo, 19(1), 233-236.
[352] Buss J.L. Greene B.T. Turner J. Torti F.M. Torti S.V. (2004) Iron chelators in cancer chemotherapy, Curr Top Med Chem, 4(15), 1623-1635.
[353] Lovejoy D.B. Richardson D.R. (2003) Iron chelators as anti-neoplastic agents: current developments and promise of the PIH class of chelators, Curr Med Chem, 10(12), 1035-1049.
[354] Mukhtar H. Ahmad N. (2000) Tea polyphenols: prevention of cancer and optimizing health, Am J Clin Nutr, 71, 1703S-1694S.
[355] Cos P. De Bruyne T. Hermans N. Apers S. Berghe D.V. Vlietinck A.J. (2004) Proanthocyanidins in health care: current and new trends, Curr Med Chem, 11, 1345-1359.
[356] Roomi M.W. Ivanov V. Kalinovsky T. Niedzwiecki A. Rath M. (2005) Antitumor Effect of a Combination of Lysine, Proline, Arginine, Ascorbic Acid, and Green Tea Extract on Pancreatic Cancer Cell Line MIA PaCa-2, Int J Gastrointest Cancer, 35(2), 97-102.
[357] Muse K.E. Oberley T.D. Sempf J.M. Oberley L.W. (1994) Immunolocalization of antioxidant enzymes in adult hamster kidney, Histochem J, 26(9), 734-753.
[358] Oberley T.D. Verwiebe E. Zhong W. Kang S.W. Rhee S.G. (2001) Localization of the thioredoxin system in normal rat kidney, Free Radic Biol Med, 30(4), 412-424.
[359] Goodstine S.L. Zheng T. Holford T.R. Ward B.A. Carter D. Owens P.H. Mayne S.T. (20030 Dietary (n-3)/(n-6) fatty acid ratio: possible relationship to premenopausal but not postmenopausal breast cancer risk in U.S. women, J Nutrition, 133, 1409-1414.
[360] Simonsen N. van't Veer P. Strain J.J. Martin-Moreno J.M. Huttunen J.K. Navajas J.F. Martin B.C. Thamm M. Kardinaal A.F. Kok F.J. Kohlmeier L. (1998) Adipose tissue omega-3 and omega-6 fatty acid content and breast cancer in the EURAMIC study, European Community Multicenter Study on Antioxidants, Myocardial Infarction, and Breast Cancer, Am J Epidemiol, 147, 342-352.
[361] Maillard V. Bougnoux P. Ferrari P. Jourdan M.L. Pinault M. Lavillonniere F. Body G. Le Floch O. Chajes V. (2002) N-3 and N-6 fatty acids in breast adipose tissue and relative risk of breast cancer in a case-control study in Tours, France, Int J Cancer, 98, 78-83
[362] Bagga D. Anders K.H. Wang H.J. Glaspy J.A. (2002) Long-chain n-3-to-n-6 polyunsaturated fatty acid ratios in breast adipose tissue from women with and without breast cancer, Nutr Cancer, 42, 180-185.
[363] Pala V. Krogh V. Muti P. Chajes V. Riboli E. Micheli A. Saadatian M. Sieri S. Berrino F. (2001) Erythrocyte membrane fatty acids and subsequent breast cancer: a prospective Italian study, J Natl Cancer Inst, 93, 1088-1095.
[364] Gago-Dominguez M. Yuan J.M. Sun C.L. Lee H.P. Yu M.C. (2003) Opposing effects of dietary n-3 and n-6 fatty acids on mammary carcinogenesis: The Singapore Chinese Health Study, Br J Cancer, 89, 1686-1692.
[365] Brouwer I.A. Katan M.B. Zock P.L. (2004) Dietary alpha-linolenic acid is associated with reduced risk of fatal coronary heart disease, but increased prostate cancer risk: a meta-analysis, J Nutr, 134, 919-922.
[366] Brooks J.D. Ward W.E. Lewis J.E. Hilditch J. Nickell L. Wong E. Thompson L.U. (2004) Supplementation with flaxseed alters estrogen metabolism in postmenopausal women to a greater extent than does supplementation with an equal amount of soy, Am J Clin Nutr, 79, 318-325.
[367] Thompson L.U. Rickard S.E. Orcheson L.J. Seidl M.M. (1996) Flaxseed and its lignan and oil components reduce mammary tumor growth at a late stage of carcinogenesis, Carcinogenesis, 17, 1373-1376.
[368] Thompson L.U. Seidl M.M. Rickard S.E. Orcheson L.J. Fong H.H. (1996) Antitumorigenic effect of a mammalian lignan precursor from flaxseed, Nutr Cancer, 26, 159-165.
[369] Yan L. Yee J.A. Li D. McGuire M.H. Thompson L.U. (1998) Dietary flaxseed supplementation and experimental metastasis of melanoma cells in mice, Cancer Lett, 124, 181-186.
[370] Li D. Yee J.A. Thompson L.U. Yan L. (1999) Dietary supplementation with secoisolariciresinol diglycoside (SDG) reduces experimental metastasis of melanoma cells in mice, Cancer Lett, 142, 91-96.
[371] Chen J. Tan K.P. Ward W.E. Thompson L.U. (2003) Exposure to flaxseed or its purified lignan during suckling inhibits chemically induced rat mammary tumorigenesis, Exp Biol Med, 228, 951-958.
[372] Tan K.P. Chen J. Ward W.E. Thompson L.U. (2004) Mammary gland morphogenesis is enhanced by exposure to flaxseed or its major lignan during suckling in rats, Exp Biol Med, 229, 147-157.
[373] Lin X. Gingrich J.R. Bao W. Li J. Haroon Z.A. Demark-Wahnefried W. (2002) Effect of flaxseed supplementation on prostatic carcinoma in transgenic mice, Urology, 60, 919-924.
[374] Demark-Wahnefried W. Price D.T. Polascik T.J. Robertson C.N. Anderson E.E. Paulson D.F. Walther P.J. Gannon M. Vollmer R.T. (2001) Pilot study of dietary fat restriction and flaxseed supplementation in men with prostate cancer before surgery: exploring the effects on hormonal levels, prostate-specific antigen, and histopathologic features, Urology, 58, 47-52.
[375] Chernomorsky S. Segelman A. Poretz R.D. (1999) Effect of dietary chlorophyll derivatives on mutagenesis and tumor cell growth, Teratog Carcinog Mutagen, 19, 313-322.

376 Sarkar D. Sharma A. Talukder G. (1994) Chlorophyll and chlorophyllin as modifiers of genotoxic effects, Mutat Res, 318, 239-247.
377 Egner P.A. Wang J.B. Zhu Y.R. Zhang B.C. Wu Y. Zhang Q.N. Qian G.S. Kuang S.Y. Gange S.J. Jacobson L.P. Helzlsouer K.J. Bailey G.S. Groopman J.D. Kensler T.W. (2001) Chlorophyllin intervention reduces aflatoxin-DNA adducts in individuals at high risk for liver cancer, Proc Natl Acad Sci U S A, 98, 14601-14606.
378 Egner P.A. Stansbury K.H. Snyder E.P. Rogers M.E. Hintz P.A. Kensler T.W. (2000) Identification and characterization of chlorine(4)ethyl ester in sera of individuals participating in the chlorophyllin chemoprevention trial, Chem Res Toxicol, 13, 900-906.
379 Duffield-Lillico A.J. Reid M.E. Turnbull B.W. Combs G.F. Jr Slate E.H. Fischbach L.A. Marshall J.R. Clark L.C. (2002) Baseline characteristics and the effect of selenium supplementation on cancer incidence in a randomized clinical trial: a summary report of the Nutritional Prevention of Cancer Trial, Cancer Epidemiol Biomarkers Prev, 11, 630-639.
380 Clark L.C. Combs G.F. Jr, Turnbull B.W. Slate E.H. Chalker D.K. Chow J. Davis L.S. Glover R.A. Graham G.F. Gross E.G. Krongrad A. Lesher J.L. Jr, Park H.K. Sanders B.B. Jr, Smith C.L. Taylor J.R. (1996) Effects of selenium supplementation for cancer prevention in patients with carcinoma of the skin. A randomized controlled trial. Nutritional Prevention of Cancer Study Group, JAMA, 276, 1957-1963.
381 Li H. Stampfer M.J. Giovannucci E.L. Morris J.S. Willett W.C. Gaziano J.M. Ma J. (2004) A prospective study of plasma selenium levels and prostate cancer risk, J Natl Cancer Inst, 96, 696-703.
382 van den Brandt P.A. Zeegers M.P. Bode P. Goldbohm R.A. (2003) Toenail selenium levels and the subsequent risk of prostate cancer: a prospective cohort study, Cancer Epidemiol Biomarkers Prev, 12, 866-871.
383 Brooks J.D. Metter E.J. Chan D.W. Sokoll L.J. Landis P. Nelson W.G. Muller D. Andres R. Carter H.B. (2001) Plasma selenium level before diagnosis and the risk of prostate cancer development, J Urol, 166, 2034-2038.
384 Helzlsouer K.J. Huang H.Y. Alberg A.J. Hoffman S. Burke A. Norkus E.P. Morris J.S. Comstock G.W. (2000) Association between alpha-tocopherol, gamma-tocopherol, selenium, and subsequent prostate cancer, J Natl Cancer Inst, 92, 2018-2023.
385 Yoshizawa K. Willett W.C. Morris S.J. Stampfer M.J. Spiegelman D. Rimm E.B. Giovannucci E. (1998) Study of prediagnostic selenium level in toenails and the risk of advanced prostate cancer, J Natl Cancer Inst, 90, 1219-1224.
386 Criqui M.H. Bangdiwala S. Goodman D.S. Blaner W.S. Morris J.S. Kritchevsky S. Lippel K. Mebane I. Tyroler H.A. (1991) Selenium, retinol, retinol-binding protein, and uric acid. Associations with cancer mortality in a population-based prospective case-control study, Ann Epidemiol, 1, 385-393.
387 Ghadirian P. Maisonneuve P. Perret C. Kennedy G. Boyle P. Krewski D. Lacroix A. (2000) A case-control study of toenail selenium and cancer of the breast, colon, and prostate, Cancer Detect Prev, 24, 305-313.
388 van den Brandt P.A. Goldbohm R.A. van't Veer P. Bode P. Dorant E. Hermus R.J. Sturmans F. (1994) Toenail selenium levels and the risk of breast cancer, Am J Epidemiol, 140, 20-26.
389 Hunter D.J. Morris J.S. Stampfer M.J. Colditz G.A. Speizer F.E. Willett W.C. (1990) A prospective study of selenium status and breast cancer risk, JAMA, 264, 1128-1131.
390 van 't Veer P. van der Wielen R.P. Kok F.J. Hermus R.J. Sturmans F. (1990) Selenium in diet, blood, and toenails in relation to breast cancer: a case-control study, Am J Epidemiol, 131, 987-994.
391 van Noord P.A. Collette H.J. Maas M.J. de Waard F. (1987) Selenium levels in nails of premenopausal breast cancer patients assessed prediagnostically in a cohort-nested case-referent study among women screened in the DOM project, Int J Epidemiol, 16, 318-322.
392 Reid M.E. Duffield-Lillico A.J. Garland L. Turnbull B.W. Clark L.C. Marshall J.R. (2002) Selenium supplementation and lung cancer incidence: an update of the nutritional prevention of cancer trial, Cancer Epidemiol Biomarkers Prev, 11, 1285-1291.
393 van den Brandt P.A. Goldbohm R.A. van 't Veer P. Bode P. Dorant E. Hermus R.J. Sturmans F. (1993) A prospective cohort study on selenium status and the risk of lung cancer, Cancer Res, 53, 4860-4865.
394 Nishizawa Y. Yamamoto T. Terada N. Fushiki S. Matsumoto K. (1997) Effects of methylcobalamin on the proliferation of androgen-sensitive or estrogen-sensitive malignant cells in culture and in vivo, Int J Vitam Nutr Res, 67, 164-170.
395 Nishizawa Y. Goto H.G. Tanigaki Y. Fushiki S. (2001) Induction of apoptosis in an androgen-dependent mouse mammary carcinoma cell line by methylcobalamin, Anticancer Res, 21, 1107-1110.
396 Tsao C.S. Myashita K. (1993) Influence of cobalamin on the survival of mice bearing ascites tumor, Pathobiology, 61, 104-108.
397 Shimizu N. Hamazoe R. Kanayama H. Maeta M. Koga S. (1987) Experimental study of antitumor effect of methyl-B12, Oncology, 44, 169-173.

[398] Choi S.W. Friso S. Ghandour H. Bagley P.J. Selhub J. Mason J.B. (2004) Vitamin B-12 deficiency induces anomalies of base substitution and methylation in the DNA of rat colonic epithelium, J Nutrition, 134, 750-755.
[399] Wu K. Helzlsouer K.J. Comstock G.W. Hoffman S.C. Nadeau M.R. Selhub J. (1999) A prospective study on folate, B12, and pyridoxal 5'-phosphate (B6) and breast cancer, Cancer Epidemiol Biomarkers Prev, 8, 209-217.
[400] Zhang S.M. Willett W.C. Selhub J. Hunter D.J. Giovannucci E.L. Holmes M.D. Colditz G.A. Hankinson S.E. (2003) Plasma folate, vitamin B6, vitamin B12, homocysteine, and risk of breast cancer, J Natl Cancer Inst, 95, 373-380.
[401] Harnack L. Jacobs D.R.Jr, Nicodemus K. Lazovich D. Anderson K. Folsom A.R. (2002) Relationship of folate, vitamin B-6, vitamin B-12, and methionine intake to incidence of colorectal cancers, Nutr Cancer, 43, 152-158.
[402] Blount B.C. Mack M.M. Wehr C.M. MacGregor J.T. Hiatt R.A. Wang G. Wickramasinghe S.N. Everson R.B. Ames B.N. (1997) Folate deficiency causes uracil misincorporation into human DNA and chromosome breakage: implications for cancer and neuronal damage, Proc Natl Acad Sci U S A , 94, 3290-3295.
[403] Lashner B.A. Heidenreich P.A. Su G.L. Kane S.V. Hanauer S.B. (1989) Effect of folate supplementation on the incidence of dysplasia and cancer in chronic ulcerative colitis. A case-control study, Gastroenterology, 97, 255-259.
[404] Freudenheim J.L. Graham S. Marshall J.R. Haughey B.P. Cholewinski S. Wilkinson G. (1991) Folate intake and carcinogenesis of the colon and rectum, Int J Epidemiol, 20, 368-374.
[405] Benito E. Stiggelbout A. Bosch F.X. Obrador A. Kaldor J. Mulet M. Munoz N. (1991) Nutritional factors in colorectal cancer risk: a case-control study in Majorca, Int J Cancer, 49, 161-167.
[406] Ferraroni M. La Vecchia C. D'Avanzo B. Negri E. Franceschi S. Decarli A. (1994) Selected micronutrient intake and the risk of colorectal cancer, Br J Cancer, 70, 1150-1155.
[407] Slattery M.L. Schaffer D. Edwards S.L. Ma K.N. Potter J.D. (1997) Are dietary factors involved in DNA methylation associated with colon cancer? Nutr Cancer, 28, 52-62.
[408] Su L.J. Arab L. (2001) Nutritional status of folate and colon cancer risk: evidence from NHANES I epidemiologic follow-up study, Ann Epidemiol, 11, 65-72.
[409] Konings E.J. Goldbohm R.A. Brants H.A. Saris W.H. van den Brandt P.A. (2002) Intake of dietary folate vitamers and risk of colorectal carcinoma: results from The Netherlands Cohort Study, Cancer, 95, 1421-1433.
[410] La Vecchia C. Negri E. Pelucchi C. Franceschi S. (2002) Dietary folate and colorectal cancer, Int J Cancer, 102, 545-547.
[411] Satia-Abouta J. Galanko J.A. Martin C.F. Potter J.D. Ammerman A. Sandler R.S. (2003) Associations of micronutrients with colon cancer risk in African Americans and whites: results from the North Carolina Colon Cancer Study, Cancer Epidemiol Biomarkers Prev, 12, 747-754.
[412] Martinez M.E. Henning S.M. Alberts D.S. (2004) Folate and colorectal neoplasia: relation between plasma and dietary markers of folate and adenoma recurrence, Am J Clin Nutr, 79, 691-697.
[413] Giovannucci E. Rimm E.B. Ascherio A. Stampfer M.J. Colditz G.A. Willett W.C. (1995) Alcohol, low-methionine – low-folate diets, and risk of colon cancer in men, J Natl Cancer Inst, 87, 265-273.
[414] Glynn S.A. Albanes D. Pietinen P. Brown C.C. Rautalahti M. Tangrea J.A. Gunter E.W. Barrett M.J. Virtamo J. Taylor P.R. (1996) Colorectal cancer and folate status: a nested case-control study among male smokers, Cancer Epidemiol Biomarkers Prev, 5, 487-494.
[415] Giovannucci E. Stampfer M.J. Colditz G.A. Hunter D.J. Fuchs C. Rosner B.A. Speizer F.E. Willett W.C. (1998) Multivitamin use, folate, and colon cancer in women in the Nurses' Health Study, Ann Intern Med, 129, 517-524.
[416] Kato I. Dnistrian A.M. Schwartz M. Toniolo P. Koenig K. Shore R.E. Akhmedkhanov A. Zeleniuch-Jacquotte A. Riboli E. (1999) Serum folate, homocysteine and colorectal cancer risk in women: a nested case-control study, Br J Cancer, 79, 1917-1922.
[417] Fuchs C.S. Willett W.C. Colditz G.A. Hunter D.J. Stampfer M.J. Speizer F.E. Giovannucci E.L. (2002) The influence of folate and multivitamin use on the familial risk of colon cancer in women, Cancer Epidemiol Biomarkers Prev, 11, 227-234.
[418] Terry P. Jain M. Miller A.B. Howe G.R. Rohan T.E. (2002) Dietary intake of folic acid and colorectal cancer risk in a cohort of women, Int J Cancer, 97, 864-867.
[419] Zhang S. Hunter D.J. Hankinson S.E. Giovannucci E.L. Rosner B.A. Colditz G.A. Speizer F.E. Willett W.C. (1999) A prospective study of folate intake and the risk of breast cancer, JAMA, 281, 1632-1637.
[420] Rohan T.E. Jain M.G. Howe G.R. Miller A.B. (2000) Dietary folate consumption and breast cancer risk, J Natl Cancer Inst, 92, 266-269.
[421] Cho E. Spiegelman D. Hunter D.J. Chen W.Y. Zhang S.M. Colditz G.A. Willett W.C. (2003) Premenopausal intakes of vitamins A, C, and E, folate, and carotenoids, and risk of breast cancer, Cancer Epidemiol Biomarkers Prev, 12, 713-720.

[422] Sellers T.A. Grabrick D.M. Vierkant R.A. Harnack L. Olson J.E. Vachon C.M. Cerhan J.R. (2004) Does folate intake decrease risk of postmenopausal breast cancer among women with a family history? Cancer Causes Control, 15, 113-120.
[423] Shrubsole M.J. Jin F. Dai Q. Shu X.O. Potter J.D. Hebert J.R. Gao Y.T. Zheng W. (2001) Dietary folate intake and breast cancer risk: results from the Shanghai Breast Cancer Study, Cancer Res, 61, 7136-7141.
[424] Meyer F. White E. (1993) Alcohol and nutrients in relation to colon cancer in middle-aged adults, Am J Epidemiol, 138, 225-236.
[425] Giovannucci E. Stampfer M.J. Colditz G.A. Rimm E.B. Trichopoulos D. Rosner B.A. Speizer F.E. Willett W.C. (1993) Folate, methionine, and alcohol intake and risk of colorectal adenoma, Natl Cancer Inst, 85, 875-884.
[426] Sellers T.A. Kushi L.H. Cerhan J.R. Vierkant R.A. Gapstur S.M. Vachon C.M. Olson J.E. Therneau T.M. Folsom A.R. (2001) Dietary folate intake, alcohol, and risk of breast cancer in a prospective study of postmenopausal women, Epidemiology, 12, 420-428.
[427] Zhang S.M. Willett W.C. Selhub J. Hunter D.J. Giovannucci E.L. Holmes M.D. Colditz G.A. Hankinson S.E. (2003) Plasma folate, vitamin B6, vitamin B12, homocysteine, and risk of breast cancer, J Natl Cancer Inst, 95, 373-380.
[428] Ma J. Stampfer M.J. Giovannucci E. Artigas C. Hunter D.J. Fuchs C. Willett W.C. Selhub J. Hennekens C.H. Rozen R. (1997) Methylenetetrahydrofolate reductase polymorphism, dietary interactions, and risk of colorectal cancer, Cancer Res, 57, 1098-1102.
[429] Ma J. Stampfer M.J. Christensen B. Giovannucci E. Hunter D.J. Chen J. Willett W.C. Selhub J. Hennekens C.H. Gravel R. Rozen R. (1999) A polymorphism of the methionine synthase gene: association with plasma folate, vitamin B12, homocyst(e)ine, and colorectal cancer risk, Cancer Epidemiol Biomarkers Prev, 8, 825-829.
[430] Giovannucci E. Chen J. Smith-Warner S.A. Rimm E.B. Fuchs C.S. Palomeque C. Willett W.C. Hunter D.J. (2003) Methylenetetrahydrofolate reductase, alcohol dehydrogenase, diet, and risk of colorectal adenomas, Cancer Epidemiol Biomarkers Prev, 12, 970-979.
[431] Le Marchand L. Donlon T. Hankin J.H. Kolonel L.N. Wilkens L.R. Seifried A. (2002) B-vitamin intake, metabolic genes, and colorectal cancer risk (United States), Cancer Causes Control, 13, 239-248.
[432] Cravo M.L. Pinto A.G. Chaves P. Cruz J.A. Lage P. Nobre Leitao C. Costa Mira F. (1998) Effect of folate supplementation on DNA methylation of rectal mucosa in patients with colonic adenomas: correlation with nutrient intake, Clin Nutr, 17, 45-49.
[433] Hanchette C.L. Schwartz G.G. (1992) Geographic patterns of prostate cancer mortality. Evidence for a protective effect of ultraviolet radiation, Cancer, 70, 2861-2869.
[434] Lefkowitz E.S. Garland C.F. (19940 Sunlight, vitamin D, and ovarian cancer mortality rates in US women, Int J Epidemiol, 23, 1133-1136.
[435] Gorham E.D. Garland F.C. Garland C.F. (1990) Sunlight and breast cancer incidence in the USSR, Int J Epidemiol, 19, 820-824.
[436] Grant W.B. (2003) Ecologic studies of solar UV-B radiation and cancer mortality rates, Recent Results Cancer Res, 164, 371-377.
[437] Grant W.B. (2002) An estimate of premature cancer mortality in the U.S. due to inadequate doses of solar ultraviolet-B radiation, Cancer, 94, 1867-1875.
[438] Garland C. Shekelle R.B. Barrett-Connor E. Criqui M.H. Rossof A.H. Paul O. (1985) Dietary vitamin D and calcium and risk of colorectal cancer: a 19-year prospective study in men, Lancet, 1, 307-309.
[439] Garland C.F. Comstock G.W. Garland F.C. Helsing K.J. Shaw E.K. Gorham E.D. (1989) Serum 25-hydroxyvitamin D and colon cancer: eight-year prospective study, Lancet, 2, 1176-1178.
[440] Gann P.H. Ma J. Hennekens C.H. Hollis B.W. Haddad J.G. Stampfer M.J. (1996) Circulating vitamin D metabolites in relation to subsequent development of prostate cancer, Cancer Epidemiol Biomarkers Prev, 5, 121-126.
[441] Martinez M.E. Giovannucci E.L. Colditz G.A. Stampfer M.J. Hunter D.J. Speizer F.E. Wing A. Willett W.C. (1996) Calcium, vitamin D, and the occurrence of colorectal cancer among women, J Natl Cancer Inst, 88, 1375-1382.
[442] Tangrea J. Helzlsouer K. Pietinen P. Taylor P. Hollis B. Virtamo J. Albanes D. (1997) Serum levels of vitamin D metabolites and the subsequent risk of colon and rectal cancer in Finnish men, Cancer Causes Control, 8, 615-625.
[443] John E.M. Schwartz G.G. Dreon D.M. Koo J. (1999) Vitamin D and breast cancer risk: the NHANES I Epidemiologic follow-up study, 1971–1975 to 1992. National Health and Nutrition Examination Survey, Cancer Epidemiol Biomarkers Prev, 8, 399-406.
[444] Ahonen M.H. Tenkanen L. Teppo L. Hakama M. Tuohimaa P. (2000) Prostate cancer risk and prediagnostic serum 25-hydroxyvitamin D levels (Finland), Cancer Causes Control, 11, 847-852.

[445] Grau M.V. Baron J.A. Sandler R.S. Haile R.W. Beach M.L. Church T.R. Heber D. (2003) Vitamin D, calcium supplementation, and colorectal adenomas: results of a randomized trial, J Natl Cancer Inst, 95, 1765-1771.
[446] Tuohimaa P. Tenkanen L. Ahonen M. Lumme S. Jellum E. Hallmans G. Stattin P. Harvei S. Hakulinen T. Luostarinen T. Dillner J. Lehtinen M. Hakama M. (2004) Both high and low levels of blood vitamin D are associated with a higher prostate cancer risk: a longitudinal, nested case-control study in the Nordic countries, Int J Cancer, 108, 104-108.
[447] Holick M.F. (2004) Vitamin D: importance in the prevention of cancers, type 1 diabetes, heart disease, and osteoporosis, Am J Clin Nutr, 79, 362-371.
[448] Vieth R. Kimball S. Hu A. Walfish P.G. (2004) Randomized comparison of the effects of the vitamin D3 adequate intake versus 100 mcg (4000 IU) per day on biochemical responses and the wellbeing of patients, Nutr J, 3, 8.
[449] Schwartz G.G. Whitlatch L.W. Chen T.C. Lokeshwar B.L. Holick M.F. (1998) Human prostate cells synthesize 1,25-dihydroxyvitamin D3 from 25-hydroxyvitamin D3, Cancer Epidemiol Biomarkers Prev, 7, 391-395.
[450] Tangpricha V. Flanagan J.N. Whitlatch L.W. Tseng C.C. Chen T.C. Holt P.R. Lipkin M.S. Holick M.F. (2001) 25-hydroxyvitamin D-1alpha-hydroxylase in normal and malignant colon tissue, Lancet, 357, 1673-1674.
[451] Friedrich M. Rafi L. Mitschele T. Tilgen W. Schmidt W. Reichrath J. (2003) Analysis of the vitamin D system in cervical carcinomas, breast cancer and ovarian cancer, Recent Results Cancer Res, 164, 239-246.
[452] Schwartz G.G. Eads D. Rao A. Cramer S.D. Willingham M.C. Chen T.C. Jamieson D.P. Wang L. Burnstein K.L. Holick M.F. Koumenis C. (2004) Pancreatic cancer cells express 25-hydroxyvitamin D-1alpha-hydroxylase and their proliferation is inhibited by the prohormone 25-hydroxyvitamin D3, Carcinogenesis, 25, 1015-1026.
[453] Mawer E.B. Hayes M.E. Heys S.E. Davies M. White A. Stewart M.F. Smith G.N. (1994) Constitutive synthesis of 1,25-dihydroxyvitamin D3 by a human small cell lung cancer cell line, J Clin Endocrinol Metab, 79, 554-560.
[454] Barreto A.M. Schwartz G.G. Woodruff R. Cramer S.D. (2000) 25-Hydroxyvitamin D3, the prohormone of 1,25-dihydroxyvitamin D3, inhibits the proliferation of primary prostatic epithelial cells, Cancer Epidemiol Biomarkers Prev, 9, 265-270.
[455] Holt P.R. Arber N. Halmos B. Forde K. Kissileff H. McGlynn K.A. Moss S.F. Kurihara N. Fan K. Yang K. Lipkin M. (2002) Colonic epithelial cell proliferation decreases with increasing levels of serum 25-hydroxy vitamin D, Cancer Epidemiol Biomarkers Prev, 11, 113-119.
[456] Saffiotti U. (1969) Role of vitamin A in carcinogenesis, Am J Clin Nutr, 22(8), 1088.
[457] Epstein J.H. (1977) Effects of beta-carotene on ultraviolet induced cancer formation in the hairless mouse skin, Photochem Photobiol, 25(2), 211-213.
[458] Bjelke E. (1978) Dietary vitamin A and human lung cancer, Int J Cancer, 15, 561-565.
[459] Mathews-Roth M.M. (1982) Antitumour activity of beta-carotene, canthaxanthin and phytoene, Oncology, 39(1), 33-37.
[460] Hsing A.W. Comstock G.W. Abbey H. Polk B.F. (1990) Serologic precursors of cancer. Retinol, carotenoids, and tocopherol and risk of prostate cancer, J Natl Cancer Inst, 82, 941-946.
[461] Giovannucci E. Ascherio A. Rimm E.B. Stampfer M.J. Colditz G.A. Willett W.C. (1995) Intake of carotenoids and retinol in relation to risk of prostate cancer, J Natl Cancer Inst, 87, 1767-1776.
[462] Giovannucci E. Rimm E.B. Liu Y. Stampfer M.J. Willett W.C. (2002) A prospective study of tomato products, lycopene, and prostate cancer risk, J Natl Cancer Inst, 94, 391-398.
[463] Wu K. Erdman J.W. Jr Schwartz S.J. Platz E.A. Leitzmann M. Clinton S.K. DeGroff V. Willett W.C. Giovannucci E. (2004) Plasma and dietary carotenoids, and the risk of prostate cancer: a nested case-control study, Cancer Epidemiol Biomarkers Prev, 13, 260-269.
[464] Gann P.H. Ma J. Giovannucci E. Willett W. Sacks F.M. Hennekens C.H. Stampfer M.J. (1999) Lower prostate cancer risk in men with elevated plasma lycopene levels: results of a prospective analysis, Cancer Res, 59, 1225-1230.
[465] Kucuk O. Sarkar F.H. Sakr W. Djuric Z. Pollak M.N. Khachik F. Li Y.W. Banerjee M. Grignon D. Bertram J.S. Crissman J.D. Pontes E.J. Wood D.P. Jr (2001) Phase II randomized clinical trial of lycopene supplementation before radical prostatectomy, Cancer Epidemiol Biomarkers Prev, 10, 861-868.
[466] Bowen P. Chen L. Stacewicz-Sapuntzakis M. Duncan C. Sharifi R. Ghosh L. Kim H.S. Christov-Tzelkov K. van Breemen R. (2002) Tomato sauce supplementation and prostate cancer: lycopene accumulation and modulation of biomarkers of carcinogenesis, Exp Biol Med, 227, 886-893.
[467] Hennekens C.H. Buring J.E. Manson J.E. Stampfer M. Rosner B. Cook N.R. Belanger C. LaMotte F. Gaziano J.M. Ridker P.M. Willett W. Peto R. (1996) Lack of effect of long-term supplementation with beta carotene on the incidence of malignant neoplasms and cardiovascular disease, N Engl J Med, 334, 1145-1149.

468 Omenn G.S. Goodman G.E. Thornquist M.D. Balmes J. Cullen M.R. Glass A. Keogh J.P. Meyskens F.L. Valanis B. Williams J.H. Barnhart S. Hammar S. (1996) Effects of a combination of beta carotene and vitamin A on lung cancer and cardiovascular disease, N Engl J Med, 334, 1150-1155.
469 The Alpha-Tocopherol, Beta Carotene Cancer Prevention Study Group (1994) The effect of vitamin E and beta carotene on the incidence of lung cancer and other cancers in male smokers, N Engl J Med, 330, 1029-1035.
470 Hickey S. Roberts H. Cathcart R (2005) Dynamic flow, a new model for ascorbate, Journal of Orthomolecular Medicine, in press.
471 Le Marchand L. Hankin J.H. Kolonel L.N. Beecher G.R. Wilkens L.R. Zhao L.P. (1993) Intake of specific carotenoids and lung cancer risk, Cancer Epidemiol Biomarkers Prev, 2, 183-187.
472 Comstock G.W. Alberg A.J. Huang H.Y. Wu K. Burke A.E. Hoffman S.C. Norkus E.P. Gross M. Cutler R.G. Morris J.S. Spate V.L. Helzlsouer K.J. (1997) The risk of developing lung cancer associated with antioxidants in the blood: ascorbic acid, carotenoids, alpha-tocopherol, selenium, and total peroxyl radical absorbing capacity, Cancer Epidemiol Biomarkers Prev, 6, 907-916.
473 Garcia-Closas R. Agudo A. Gonzalez C.A. Riboli E. (1998) Intake of specific carotenoids and flavonoids and the risk of lung cancer in women in Barcelona, Spain, Nutr Cancer, 32, 154-158.
474 Stefani E.D. Boffetta P. Deneo-Pellegrini H. Mendilaharsu M. Carzoglio J.C. Ronco A. Olivera L. (1999) Dietary antioxidants and lung cancer risk: a case-control study in Uruguay, Nutr Cancer, 34, 100-110.
475 Knekt P. Jarvinen R. Teppo L. Aromaa A. Seppanen R. (1999) Role of various carotenoids in lung cancer prevention, J Natl Cancer Inst, 91, 182-184.
476 Michaud D.S. Feskanich D. Rimm E.B. Colditz G.A. Speizer F.E. Willett W.C. Giovannucci E. (2000) Intake of specific carotenoids and risk of lung cancer in 2 prospective US cohorts, Am J Clin Nutr, 72, 990-997.
477 Yuan J.M. Ross R.K. Chu X.D. Gao Y.T. Yu M.C. (2001) Prediagnostic levels of serum beta-cryptoxanthin and retinol predict smoking-related lung cancer risk in Shanghai, China, Cancer Epidemiol Biomarkers Prev, 10, 767-773.
478 Rohan T.E. Jain M. Howe G.R. Miller A.B. (2002) A cohort study of dietary carotenoids and lung cancer risk in women (Canada), Cancer Causes Control, 13, 231-237.
479 Ito Y. Wakai K. Suzuki K. Tamakoshi A. Seki N. Ando M. Nishino Y. Kondo T. Watanabe Y. Ozasa K. Ohno Y. (2003) Serum carotenoids and mortality from lung cancer: a case-control study nested in the Japan Collaborative Cohort (JACC) study, Cancer Sci, 94, 57-63.
480 Yuan J.M. Stram D.O. Arakawa K. Lee H.P. Yu M.C. (2003) Dietary cryptoxanthin and reduced risk of lung cancer: the Singapore Chinese Health Study, Cancer Epidemiol Biomarkers Prev, 12, 890-898.
481 Mannisto S. Smith-Warner S.A. Spiegelman D. Albanes D. Anderson K. van den Brandt P.A. Cerhan J.R. Colditz G. Feskanich D. Freudenheim J.L. Giovannucci E. Goldbohm R.A. Graham S. Miller A.B. Rohan T.E. Virtamo J. Willett W.C. Hunter D.J. (2004) Dietary carotenoids and risk of lung cancer in a pooled analysis of seven cohort studies, Cancer Epidemiol Biomarkers Prev, 13, 40-48.
482 Smith R.A. Mettlin C.J. Davis K.J. Eyre H. (2000) American Cancer Society guidelines for the early detection of cancer, CA Cancer J Clin, 50(1), 34-49.
483 Miller B.A. Feuer E.J. Hankey B.F. (1991) The increasing incidence of breast cancer since 1982: relevance of early detection, Cancer Causes Control, 2(2), 67-74.
484 Flehinger B.J. Kimmel M. Melamed M.R. (1992) The effect of surgical treatment on survival from early lung cancer. Implications for screening, Chest, 101, 1013-1018.
485 Johansson J.E. Adami H.O. Andersson S.O. Bergstrom R. Holmberg L. Krusemo U.B. (1992) High 10-year survival rate in patients with early, untreated prostatic cancer, JAMA, 267(16), 2191-2196.
486 Johansson J.E. Holmberg L. Johansson S. Bergstrom R. Adami H.O. (1997) Fifteen-year survival in prostate cancer. A prospective, population-based study in Sweden, 277(6), 467-471.
487 Albertsen P.C. Fryback D.G. Storer B.E. Kolon T.F. Fine J. (1995) Long-term survival among men with conservatively treated localized prostate cancer, JAMA, 274(8), 626-631.
488 Chu K.C. Tarone R.E. Chow W.H. Hankey B.F. Ries L.A. (1994) Temporal patterns in colorectal cancer incidence, survival, and mortality from 1950 through 1990, J Natl Cancer Inst, 86(13), 997-1006.
489 Gordon R. (2001) Doctor in the House, (REPRINT), House of Stratus, Ripon, Yorkshire, England.
490 Gordon R. (1983) Great medical disasters, Hutchinson and Co, London, England.
491 Welch H.G. Woloshin S. Schwartz L.M. (2005) Skin biopsy and incidence of melanoma: population based ecological study, BMJ, 331, 481-484.
492 Raffle A.E. Alden B. Quinn M. Babb P.J. Brett M.T. (2003) Outcomes of screening to prevent cancer: analysis of cumulative incidence of cervical abnormality and modelling of cases and deaths prevented, BMJ, 326(7395), 901.
493 Sasieni P.D. (2003) Outcomes of screening to prevent cancer: think of screening as insurance, BMJ, 327(7405), 50

[494] Gigerenzer G. (2003) Calculated risks: how to know when numbers deceive you, Simon & Schuster, New York.
[495] Thompson I.M. Pauler D.K. Goodman P.J. Tangen C.M. Lucia M.S. Parnes H.L. Minasian L.M. Ford L.G. Lippman S.M. Crawford E.D. Crowley J.J. Coltman C.A. (2004) Prevalence of prostate cancer among men with a prostate-specific antigen level < or =4.0 ng per millilitre, NEJM, 50(22), 2239–2246.
[496] Anon (2005) US National Cancer Institute website, reviewed 08/17/2004, http://cis.nci.nih.gov/fact/5_29.htm.
[497] Keetch D.W. Catalona W.J. Smith D.S. (1994) Serial prostatic biopsies in men with persistently elevated serum prostate specific antigen values, The Journal of Urology, 151(6), 1571–1574.
[498] Lafata J.E. Simpkins J. Lamerato L. Poisson L. Divine G. Johnson C.C. (2004) The economic impact of false-positive cancer screens, Cancer Epidemiol Biomarkers Prev, 13(12), 2126-2132.
[499] Moss R.W. (2000) Questioning Chemotherapy, Equinox Press, New York.
[500] Welch H.G. Schwartz L.M. Woloshin S. (2000) Are increasing 5-year survival rates evidence of success against cancer? JAMA, 283(22), 2975-2978.
[501] Greenlee R.T. Hill-Harmon M.B. Murray T. Thun M. (2001) Cancer statistics, CA Cancer J Clin, 51, 15-36.
[502] Ries L.A.G. Eisner M.P. Kosary C.L. Hankey B.F. Miller B.A. Clegg L. Mariotto A. Feuer EJ, Edwards BK (eds). (2004) SEER Cancer Statistics Review, 1975-2001, National Cancer Institute. Bethesda, MD, http://seer.cancer.gov/csr/1975_2001/. Ries L. Eisner M. Kosary C. et al. SEER Cancer Statistics Review, 1973-1997. Bethesda, MD: National Cancer Institute; 2000
[503] Smith R.A. von Eschenbach A.C. Wender R. Levin B. Byers T. Rothenberger D. Brooks D. Creasman W. Cohen C. Runowicz C. Saslow D. Cokkinides V. Eyre H. (2001) American Cancer Society guidelines for the early detection of cancer: update of early detection guidelines for prostate, colorectal, and endometrial cancers. Also: update 2001--testing for early lung cancer detection,CA Cancer J Clin, 51(1), 38-75.
[504] Feinstein A.R. Sosin D.M. Wells C.K. (1985) The Will Rogers phenomenon. Stage migration and new diagnostic techniques as a source of misleading statistics for survival in cancer, N Engl J Med, 312(25), 1604-1608
[505] Woodward W.A. Strom E.A. Tucker S.L. McNeese M.D. Perkins G.H. Schechter N.R. Singletary S.E. Theriault R.L. Hortobagyi G.N. Hunt K.K. Buchholz T.A. (2003) Changes in the 2003 American Joint Committee on Cancer staging for breast cancer dramatically affect stage-specific survival, Clin Oncol, 21(17), 3244-3248.
[506] Simes R.J. (1986) Publication bias: the case for an international registry of clinical trials, J Clin Oncol, 4(10), 1529-1541.
[507] Dickersin K. Chan S. Chalmers T.C. Sacks H.S. Smith H. (1987) Publication bias and clinical trials, Control Clin Trials, 8(4), 343-353.
[508] Easterbrook P.J. Berlin J.A. Gopalan R. Matthews D.R. (1991) Publication bias in clinical research, Lancet, 337(8746), 867-872.
[509] Moynihan R. (2003) Who pays for the pizza? Redefining the relationships between doctors and drug companies, BMJ, 326(7400), 1189-1192.
[510] Healy D.T. (2004) Transparency and trust: figure for ghost written articles was misquoted, BMJ, 329(7478), 1345.
[511] Zuckerman D. (2003) Hype in health reporting: "checkbook science" buys distortion of medical news, Int J Health Serv, 33(2), 383-389.
[512] Taylor G.J. Wainwright P. (2005) Open label extension studies: research or marketing? BMJ, 331, 572-574.
[513] Non-small Cell Lung Cancer Collaborative Group (1995) Chemotherapy in non-small cell lung cancer: a meta-analysis using updated data on individual patients from 52 randomised clinical trials, BMJ, 311, 899-909.
[514] Bush R.S. Jenkin R.D. Allt W.E. Beale F.A. Bean H. Dembo A.J. Pringle J.F. (1978) Definitive evidence for hypoxic cells influencing cure in cancer therapy, Br J Cancer Suppl, 37(3), 302-306.
[515] Beskow C. Agren-Cronqvist A.K. Granath F. Frankendal B. Lewensohn R. (2002) Pathologic complete remission after preoperative intracavitary radiotherapy of cervical cancer stage Ib and IIa is a strong prognostic factor for long-term survival: analysis of the Radiumhemmet data 1989-1991, Int J Gynecol Cancer, 2(2), 158-170.
[516] Zijlstra J.M. Dressel A.J. Mens J.W. Tinteren H. Slotman B.J. Jonkhoff A.R. Huijgens P.C. (2002) Radiation therapy in early stage Hodgkin's disease: long-term results and adverse effects, Hematol J, 3(4), 179-184.
[517] Mauch P.M. Kalish L.A. Marcus K.C. Shulman L.N. Krill E. Tarbell N.J. Silver B. Weinstein H. Come S. Canellos G.P. Coleman C.N. (1995) Long-term survival in Hodgkin's disease, Cancer J Sci Am, 1(1), 33.
[518] Lavey R.S. Eby N.L. Prosnitz L.R. (1990) Impact on second malignancy risk of the combined use of radiation and chemotherapy for lymphomas, Cancer, 66(1), 80-88.
[519] Riccardi A. Mora O. Tinelli C. Valentini D. Brugnatelli S. Spanedda R. De Paoli A. Barbarano L. Di Stasi M. Giordano M. Delfini C. Nicoletti G. Bergonzi C. Rinaldi E. Piccinini L. Ascari E. (2000) Long-term

survival of stage I multiple myeloma given chemotherapy just after diagnosis or at progression of the disease: a multicentre randomized study, British Journal of Cancer, 82, 1254-1260.
520 Abel U. (1992) Chemotherapy of advanced epithelial cancer--a critical review, Biomed Pharmacother, 46(10), 439-452.
521 Ettinger D.S. Finkelstein D.M. Abeloff M.D. Skeel R.T. Stott P.B. Frontiera M.S. Bonomi P.D. (1992) Justification for evaluating new anticancer drugs in selected untreated patients with extensive-stage small-cell lung cancer: an Eastern Cooperative Oncology Group randomized study, J Natl Cancer Inst, 84(14), 1077-1084.
522 Chute J.P. Chen T. Feigal E. Simon R. Johnson B.E. (1999) Twenty years of phase III trials for patients with extensive-stage small-cell lung cancer: perceptible progress, J Clin Oncol, 17(6), 1794-1801.
523 Huisman C. Smit E.F. Giaccone G. Postmus P.E. (2000) Second-line chemotherapy in relapsing or refractory non-small-cell lung cancer: a review, J Clin Oncol, 18(21), 3722-3730.
524 Billingham L.J. Cullen M.H. (2001) The benefits of chemotherapy in patient subgroups with unresectable non-small-cell lung cancer, Annals of Oncology, 12(12), 1671-1675.
525 Simmonds P.C. (2000) Palliative chemotherapy for advanced colorectal cancer: systematic review and meta-analysis, Colorectal Cancer Collaborative Group, BMJ, 321(7260), 531–535.
526 Schulman K.A. Stadtmauer E.A. Reed S.D. Glick H.A. Goldstein L.J. Pines J.M. Jackman J.A. Suzuki S. Styler M.J. Crilley P.A. Klumpp T.R. Mangan K.F. (2003) Economic analysis of conventional-dose chemotherapy compared with high-dose chemotherapy plus autologous hematopoietic stem-cell transplantation for metastatic breast cancer, Bone Marrow Transplant, 31(3), 205-210.
527 Yokomizo A. Ono M. Nanri H. Makino Y. Ohga T. Wada M. Okamoto T. Yodoi J. Kuwano M. Kohno K. (1995) Cellular levels of thioredoxin associated with drug sensitivity to cisplatin, mitomycin C, doxorubicin, and etoposide, Cancer Res, 55(19), 4293-4296.
528 Ferlini C. Scambia G. Marone M. Distefano M. Gaggini C. Ferrandina G. Fattorossi A. Isola G. Benedetti Panici P. Mancuso S. (1999) Tamoxifen induces oxidative stress and apoptosis in oestrogen receptor-negative human cancer cell lines, Br J Cancer, 79(2), 257-263.
529 Pollakis G. Goormaghtigh E. Delmelle M. Lion Y. Ruysschaert J.M. (1984) Adriamycin and derivatives interaction with the mitochondrial membrane: O2 consumption and free radicals formation, Res Commun Chem Pathol Pharmacol, 44(3), 445-459.
530 Mizutani H. Tada-Oikawa S. Hiraku Y. Kojima M. Kawanishi S. (2005) Mechanism of apoptosis induced by doxorubicin through the generation of hydrogen peroxide, Life Sci, 76(13), 1439-1453. Epub 2005 Jan 20.
531 Hudis C.A. Seidman A.D. Crown J.P. Balmaceda C. Freilich R. Gilewski T.A. Hakes T.B. Currie V. Lebwohl D.E. Baselga J. Raptis G. Gollub M. Robles M. Bruno R. Norton L. (1996) Phase II and pharmacologic study of docetaxel as initial chemotherapy for metastatic breast cancer, Journal of Clinical Oncology, 14, 58-65.
532 Gandara D.R. Vokes E. Green M. Bonomi P. Devore R. Comis R. Carbone D. Karp D. Belani C. (2000) Activity of docetaxel in platinum-treated non–small-cell lung cancer: results of a phase II multicenter trial, J Clin Oncol, 18, 131-135.
533 Shepherd F.A. (1999) Chemotherapy for non-small cell lung cancer: Have we reached a new plateau? Semin Oncol, 26, 3-11.
534 Shepherd F.A. Dancey J. Ramlau R. Mattson K. Gralla R. O'Rourke M. Levitan N. Gressot L. Vincent M. Burkes R. Coughlin S. Kim Y. Berille J. (2000) Prospective randomized trial of docetaxel versus best supportive care in patients with non–small-cell lung cancer previously treated with platinum-based chemotherapy, Journal of Clinical Oncology, 18(10), 2095-2103.
535 Hayes D.F. Van Zyl J.A. Hacking A. Goedhals L. Bezwoda W.R. Mailliard J.A. Jones S.E. Vogel C.L. Berris R.F. Shemano I. et al. (1995) Randomized comparison of tamoxifen and two separate doses of toremifene in postmenopausal patients with metastatic breast cancer, J Clin Oncol, 13(10), 2556-2566.
536 Denis LJ, Griffiths K. (2000) Endocrine treatment in prostate cancer, Semin Surg Oncol, 18(1), 52-74.
537 Schoenlein P.V. (1993) Cytotechnology, Molecular cytogenetics of multiple drug resistance, 12(1-3), 63-89.
538 Terzi M. (1974) Chromosomal variation and the origin of drug-resistant mutants in mammalian cell lines, Proc Natl Acad Sci U S A, 71(12), 5027–5031.
539 Koski T. Karhu R. Visakorpi T. Vilpo L. Knuutila S. (2000) Complex chromosomal aberrations in chronic lymphocytic leukemia are associated with cellular drug and irradiation resistance, Vilpo J.Eur J Haematol, 65(1), 32-39.
540 Ziemann C. Burkle A. Kahl G.F. Hirsch-Ernst K.I. (1999) Reactive oxygen species participate in mdr1b mRNA and P-glycoprotein overexpression in primary rat hepatocyte cultures, Carcinogenesis, 20(3), 407-414.
541 Schiller J.T. Lowy D.R. (1996) Papillomavirus-like particles and HPV vaccine development, Semin Cancer Biol, 7(6), 373-382.

[542] Linnebacher M. Gebert J. Rudy W. Woerner S. Yuan Y.P. Bork P. von Knebel Doeberitz M. (2001) Frameshift peptide-derived T-cell epitopes: a source of novel tumor-specific antigens, Int J Cancer, 93(1), 6-11.
[543] Baselga J. Norton L. Albanell J. Kim Y.M. Mendelsohn J. (1998) Recombinant humanized anti-HER2 antibody (Herceptin) enhances the antitumor activity of paclitaxel and doxorubicin against HER2/neu overexpressing human breast cancer xenografts, Cancer Res, 58(13), 2825-2831.
[544] Shak S. (1999) Overview of the trastuzumab (Herceptin) anti-HER2 monoclonal antibody clinical program in HER2-overexpressing metastatic breast cancer. Herceptin Multinational Investigator Study Group, Semin Oncol, 26(4 Suppl 12), 71-77.
[545] Raikow R.B. Acevedo H.F. Krichevsky A. Buffo M.J. Fogarty P. (1987) Flow cytofluorometric analysis of choriogonadotropin-like material on the surface of human and mouse malignant cells, Cancer Detect Prev Suppl, 1, 173-81.
[546] Triozzi P.L. Stevens V.C. (1999) Human chorionic gonadotropin as a target for cancer vaccines, Oncol Rep, 6(1), 7-17.
[547] Acevedo H.F. Kellen J.A. Wong A.C. Gardner H.A. Szalai J.P. (1987) Expression of human choriogonadotropin-like material correlates with metastatic phenotype of R3230 AC rat adenocarcinoma, Cancer Invest, 5(3), 177-185.
[548] Acevedo H.F. Raikow R.B. Powell J.E. Stevens V.C. (1987) Effects of immunization against human choriogonadotropin on the growth of transplanted Lewis lung carcinoma and spontaneous mammary adenocarcinoma in mice, Cancer Detect Prev Suppl, 1, 477-86.
[549] Acevedo H.F. Tong J.Y. Hartsock R.J. (1995) Human chorionic gonadotropin-beta subunit gene expression in cultured human fetal and cancer cells of different types and origins, Cancer, 76(8), 1467-1475. Further comments in Cancer 1995 76(8), 1299-1301 and Cancer, 1996, 78(1), 184-186.
[550] Acevedo H.F. Hartsock R.J. (1996) Metastatic phenotype correlates with high expression of membrane-associated complete beta-human chorionic gonadotropin in vivo, Cancer, 78(11), 2388-2399.
[551] Acevedo H.F. Hartsock R.J. Maroon J.C. (1997) Detection of membrane-associated human chorionic gonadotropin and its subunits on human cultured cancer cells of the nervous system, Cancer Detect Prev, 21(4), 295-303.
[552] Kalantarov G. Acevedo H.F. (1998) Demonstration of dose dependent cytotoxic activity in cancer cells by specific human chorionic gonadotropin monoclonal antibodies, Cancer, 83(4), 783-787.
[553] Moulton H.M. Yoshihara P.H. Mason D.H. Iversen P.L. Triozzi P.L. (2002) Active specific immunotherapy with a beta-human chorionic gonadotropin peptide vaccine in patients with metastatic colorectal cancer: antibody response is associated with improved survival, Clin Cancer Res, 8(7), 2044-2051.
[554] Iversen P.L. Mourich D.V. Moulton H.M. (2003) Monoclonal antibodies to two epitopes of beta-human chorionic gonadotropin for the treatment of cancer, Curr Opin Mol Ther, 5(2), 156-160.
[555] Butler S.A. Staite E.M. Iles R.K. (2003) Reduction of bladder cancer cell growth in response to hCGbeta CTP37 vaccinated mouse serum, Oncol Res, 14(2), 93-100.
[556] He L.Z. Ramakrishna V. Connolly J.E. Wang X.T. Smith P.A. Jones C.L. Valkova-Valchanova M. Arunakumari A. Treml J.F. Goldstein J. Wallace P.K. Keler T. Endres M.J. (2004) A novel human cancer vaccine elicits cellular responses to the tumor-associated antigen, human chorionic gonadotropin beta, Clin Cancer Res, 10(6), 1920-1927.
[557] Beard J. (1906) Trypsin and amylotrypsin in cancer, Medical Record, June 23, 1020.
[558] Saruc M. Standop S. Standop J. Nozawa F. Itami A. Pandey K.K. Batra S.K. Gonzalez N.J. Guesry P. Pour P.M. (2004) Pancreatic enzyme extract improves survival in murine pancreatic cancer, Pancreas, 28(4), 401-12.
[559] Gonzalez N.J. Isaacs L.L. (1999) Evaluation of pancreatic proteolytic enzyme treatment of adenocarcinoma of the pancreas, with nutrition and detoxification support, Nutr Cancer, 33(2), 117-124.
[560] Lane W.I. Comac L. (1992) Sharks Don't Get Cancer : How Shark Cartilage Could Save Your Life, Avery, New York.
[561] Lane W.I. Comac L. (1996) Sharks Still Don't Get Cancer, Avery, New York.
[562] Gonzalez R.P. Soares F.S. Farias R.F. Pessoa C. Leyva A. de Barros Viana G.S. Moraes M.O. (2001) Demonstration of inhibitory effect of oral shark cartilage on basic fibroblast growth factor-induced angiogenesis in the rabbit cornea, Biol Pharm Bull. 24(2), 151-154.
[563] Brem H. Folkman J. (1975) Inhibition of tumor angiogenesis mediated by cartilage, J Exp Med, 141(2), 427-439.
[564] Random Samples (2000) Sharks do get cancer, Science, 288(5464), 259.
[565] Lee A. Langer R. (1983) Shark cartilage contains inhibitors of tumor angiogenesis, Science, 221(4616), 1185-1187.
[566] Langer R. Brem H. Falterman K. Klein M. Folkman J. (1976) Isolations of a cartilage factor that inhibits tumor neovascularization, Science, 193(4247), 70-72.

[567] Moses M.A. Sudhalter J. Langer R. (1990) Identification of an inhibitor of neovascularization from cartilage, Science, 248(4961), 1408-1410.
[568] Liu N. Lapcevich R.K. Underhill C.B. Han Z. Gao F. Swartz G. Plum S.M. Zhang L. Green S.J. (2001) Metastatin: a hyaluronan-binding complex from cartilage that inhibits tumor growth, Cancer Res, 61(3), 1022-1028.
[569] Prudden J.F. (1985) The treatment of human cancer with agents prepared from bovine cartilage, J Biol Response Mod, 4(6), 551-584.
[570] Horsman M.R. Alsner J. Overgaard J. (1998) The effect of shark cartilage extracts on the growth and metastatic spread of the SCCVII carcinoma, Acta Oncologica , 37(5), 441-445.
[571] Barber R. Delahunt B. Grebe S.K. Davis P.F. Thornton A. Slim G.C. (2001) Oral shark cartilage does not abolish carcinogenesis but delays tumor progression in a murine model, Anticancer Res, 21(2A), 1065-1069.
[572] Miller D.R. Anderson G.T. Stark J.J. Granick J.L. Richardson D. (1998) Phase I/II trial of the safety and efficacy of shark cartilage in the treatment of advanced cancer, J Clin Oncol, 16(11), 3649-3655.
[573] Batist G. Patenaude F. Champagne P. Croteau D. Levinton C. Hariton C. Escudier B. Dupont E. (2002) Neovastat (AE-941) in refractory renal cell carcinoma patients: report of a phase II trial with two dose levels, Ann Oncol, 13(8), 1259-1263.
[574] Loprinzi C.L. Levitt R. Barton D.L. Sloan J.A. Atherton P.J. Smith D.J. Dakhil S.R. Moore D.F. Krook J.E. Rowland K.M. Mazurczak M.A. Berg A.R. Kim G.P. (2005) Evaluation of shark cartilage in patients with advanced cancer, Cancer, May 23; [Epub ahead of print].
[575] Sheu J.R. Fu C.C. Tsai M.L. Chung W. (1998) Effect of U-995, a potent shark cartilage-derived angiogenesis inhibitor, on anti-angiogenesis and anti-tumor activities, J.Anticancer Res, 18(6A), 4435-4441.
[576] Gingras D. Renaud A. Mousseau N. Beliveau R. (2000) Shark cartilage extracts as antiangiogenic agents: smart drinks or bitter pills? Cancer Metastasis Rev, 19(1-2), 83-86.
[577] Berbari P. Thibodeau A. Germain L. Saint-Cyr M. Gaudreau P. Elkhouri S. Dupont E. Garrel D.R. Elkouri S. (1999) Antiangiogenic effects of the oral administration of liquid cartilage extract in humans, J Surg Res, 87(1), 108-113.
[578] Cho J. Kim Y. (2002) Sharks: a potential source of antiangiogenic factors and tumor treatments, Mar Biotechnol (NY), 4(6), 521-525.
[579] Hassan Z.M. Feyzi R. Sheikhian A. Bargahi A. Mostafaie A. Mansouri K. Shahrokhi S. Ghazanfari T. Shahabi S. (2005) Low molecular weight fraction of shark cartilage can modulate immune responses and abolish angiogenesis, Int Immunopharmacol, 5(6), 961-970.
[580] Anon (2005) Health News, Shark cartilage cancer treatments are pseudoscience, 11(3), 15.
[581] Cameron E. Pauling L. (1993) Cancer and vitamin C, Camino Books, Philadelphia.
[582] Kwon H.Y. Hong S.P. Hahn D.H. Kim J.H. (2003) Apoptosis induction of Persicae Semen extract in human promyelocytic leukemia (HL-60) cells, Arch Pharm Res, 26(2), 157-161.
[583] Ovejera A.A. Houchens D.P. Barker A.D. Venditti J.M. (1978) Inactivity of DL-amygdalin against human breast and colon tumor xenografts in athymic (nude) mice, Cancer Treat Rep, 62(4), 576-578.
[584] Fukuda T. Ito H. Mukainaka T. Tokuda H. Nishino H. Yoshida T. (2003) Anti-tumor promoting effect of glycosides from Prunus persica seeds, Biol Pharm Bull, 26(2), 271-273.
[585] Chitnis M.P. Adwankar M.K. Amonkar A.J. (1985) Studies on high-dose chemotherapy of amygdalin in murine P388 lymphocytic leukaemia and P815 mast cell leukaemia, J Cancer Res Clin Oncol, 109(3), 208-209.
[586] Koeffler H.P. Lowe L. Golde D.W. (1980) Amygdalin (Laetrile): effect on clonogenic cells from human myeloid leukemia cell lines and normal human marrow, Cancer Treat Rep, 64(1), 105-109.
[587] Wodinsky I. Swiniarski J.K. (1975) Antitumor activity of amygdalin MF (NSC-15780) as a single agent and with beta-glucosidase (NSC-128056) on a spectrum of transplantable rodent tumors, Cancer Chemother Rep, 59(5), 939-950.
[588] Laster W.R. Schabel F.M. (1975) Experimental studies of the antitumor activity of amygdalin MF (NSC-15780) alone and in combination with beta-glucosidase (NSC-128056), Cancer Chemother Rep, 59(5), 951-965.
[589] Hill G.J. Shine T.E. Hill H.Z. Miller C. (1976) Failure of amygdalin to arrest B16 melanoma and BW5147 AKR leukaemia, Cancer Res, 36(6), 2102-2107.
[590] Stock C.C. Martin D.S. Sugiura K. Fugmann R.A. Mountain I.M. Stockert E. Schmid F.A. Tarnowski G.S. (1978) Antitumor tests of amygdalin in spontaneous animal tumor systems, J Surg Oncol, 10(2), 89-123.
[591] Ellison N.M. Byar D.P. Newell G.R. (1978) Special report on Laetrile: the NCI Laetrile Review. Results of the National Cancer Institute's retrospective Laetrile analysis, N Engl J Med, 299(10), 549-552.
[592] Moertel C.G. Fleming T.R. Rubin J. Kvols L.K. Sarna G. Koch R. Currie V.E. Young C.W. Jones S.E. Davignon J.P. (1982) A clinical trial of amygdalin (Laetrile) in the treatment of human cancer, N Engl J Med, 306(4), 201-206.
[593] Schmidt E.S. Newton G.W. Sanders S.M. Lewis J.P. Conn E.E. (1978) Laetrile toxicity studies in dogs, JAMA, 239(10), 943-947.

[594] Sadoff L. Fuchs K. Hollander J. (1978) Rapid death associated with laetrile ingestion, JAMA, 239(15), 1532.
[595] Beamer W.C. Shealy R.M. Prough D.S. (1983) Acute cyanide poisoning from laetrile ingestion, Ann Emerg Med,12(7), 449-451.
[596] Braico K.T. Humbert J.R. Terplan K.L. Lehotay J.M. (1979) Laetrile intoxication. Report of a fatal case, N Engl J Med, 300(5), 238-240.
[597] Hall A.H. Linden C.H. Kulig K.W. Rumack B.H. (1986) Cyanide poisoning from laetrile ingestion: role of nitrite therapy, Pediatrics, 78(2), 269-272.
[598] Moertel C.G. Ames M.M. Kovach J.S. Moyer T.P. Rubin J.R. Tinker J.H. (1981) A pharmacologic and toxicological study of amygdalin, JAMA, 245(6), 591-594.
[599] Holzbecher M.D. Moss M.A. Ellenberger H.A. (1984) The cyanide content of laetrile preparations, apricot, peach and apple seeds, J Toxicol Clin Toxicol, 22(4), 341-347.
[600] Carter J.H. McLafferty M.A. Goldman P. (1980) Role of the gastrointestinal microflora in amygdalin (laetrile)-induced cyanide toxicity, Biochem Pharmacol, 29(3), 301-304.
[601] Dorr R.T. Paxinos J. (1978) The current status of laetrile, Ann Intern Med, 89(3), 389-397.
[602] Ames M.M. Moyer T.P. Kovach J.S. Moertel C.G. Rubin J. (1981) Pharmacology of amygdalin (laetrile) in cancer patients, Cancer Chemother Pharmacol, 6(1), 51-57.
[603] Ames MM, Kovach JS, Flora KP. (1978) Initial pharmacologic studies of amygdalin (laetrile) in man, Res Commun Chem Pathol Pharmacol, 22(1), 175-185.
[604] Willhite C.C. (1982) Congenital malformations induced by laetrile, Science, 215(4539), 1513-1515.
[605] Janssen W.F. (1979) Cancer quackery-the past in the present, Semin Oncol, 6(4), 526-536.
[606] Lerner I.J. (1984) The whys of cancer quackery, Cancer, 53(3 Suppl), 815-819.
[607] Lerner I.J. (1987) Cancer quackery, Psychiatr Med, 5(4), 419-429.
[608] Syrigos K.N. Rowlinson-Busza G. Epenetos A.A. (1998) In vitro cytotoxicity following specific activation of amygdalin by beta-glucosidase conjugated to a bladder cancer-associated monoclonal antibody, Int J Cancer, 78(6), 712-719.
[609] Gonzalez N. (1999) Pancreatic cancer, proteolytic enzyme therapy and detoxification, Clinical Pearls News, November.
[610] Leitzmann M.F. Willett W.C. Rimm E.B. Stampfer M.J. Spiegelman D. Colditz G.A. Giovannucci E. (1999) A prospective study of coffee consumption and the risk of symptomatic gallstone disease in men, JAMA, 281(22), 2106-2112.
[611] Stollerman G.H. Bisno A.L. (eds) (1999) Coffee for gallstones, Hospital Practice, August 15, 128.
[612] Hildenbrand G.L. Hildenbrand L.C. Bradford K. Cavin S.W. (1995) Five-year survival rates of melanoma patients treated by diet therapy after the manner of Gerson: a retrospective review, Altern Ther Health Med, 1, 29-37.
[613] Halliwell B. Gutteridge J.M. (1998) Free Radicals In Biology And Medicine, Oxford University Press, Oxford, England.
[614] Nindl G. (2004) Hydrogen peroxide - from oxidative stressor to redox regulator, featured review, Cellscience Reviews, 1(2).
[615] Szatrowski T.P. Nathan C.F. (1991) Production of large amounts of hydrogen peroxide by human tumor cells, Cancer Res, 51(3), 794-798.
[616] Simizu S. Takada M. Umezawa K. Imoto M. (1998) Requirement of caspase-3(-like) protease-mediated hydrogen peroxide production for apoptosis induced by various anticancer drugs, J Biol Chem, 273(41), 26900-26907.
[617] Hampton M.B. Fadeel B. Orrenius S. (1998) Redox regulation of the caspases during apoptosis, Ann N Y Acad Sci, 854, 328-335.
[618] Editorial (2004) 2nd death in S.C. doctor case, Physician defends use of hydrogen peroxide injections for healing, The Charlotte Observer (online), Mon, November 08.
[619] Anon (1993) Questionable methods of cancer management: hydrogen peroxide and other 'hyperoxygenation' therapies, CA Cancer J Clin, 43(1), 47-56
[620] Barrett S. Cassileth B.R. (1991) Dubious cancer treatment. Tampa, Florida: American Cancer Society, Florida Division, (editors), 60-61.
[621] Schriner S.E. Linford N.J. Martin G.M. Treuting P. Ogburn C.E. Emond M. Coskun P.E. Ladiges W. Wolf N. Van Remmen H. Wallace D.C. Rabinovitch P.S. (2005) Extension of murine lifespan by overexpression of catalase targeted to mitochondria, May 5, 10, 1126/science, 1106653 (Science Express Reports).
[622] Loew O. (1901) Catalase – a new enzyme of general occurrence with special reference to the tobacco plant, US Department of Agriculture report, no 68.
[623] Holman R.A. (1961) The nature and functions of catalase, Mother Earth, Journal of the Soil Association, 6, 607-610.

624 Chance B. Sies H. Boveris A. (1979) Hydroperoxide metabolism in mammalian organs, Physiol Rev, 59(3), 527-605.
625 Holman R.A. (1957) On the fundamental importance of the catalase peroxide mechanism in living cells with particular reference to the problem of cancer, Symposium Internazionale De Medicina Ed Igene Sociale, Trieste.
626 Rose S. Bullock S. (1991) The Chemistry of Life, Penguin, London England.
627 Holman R. (2005) Personal communication.
628 Szatrowski T.R. Nathan C.F. (1991) Production of large amounts of hydrogen peroxide by human tumor cells, Cancer Res, 51(3), 794-798.
629 Oberley LW. (2001) Anticancer therapy by overexpression of superoxide dismutase, Antioxid Redox Signal, 3(3), 461-472.
630 Deshpande N.N. Sorescu D. Seshiah P. Ushio-Fukai M. Akers M. Yin Q. Griendling K.K. (2002) Mechanism of hydrogen peroxide-induced cell cycle arrest in vascular smooth muscle, Antioxid Redox Signal, 4(5), 845-854.
631 Laurent A. Nicco C. Chéreau C. Goulvestre C. Alexandre J. Alves A. Lévy E. Goldwasser F. Panis Y. Soubrane O. Weill B. Batteux F. (2005) Controlling tumor growth by modulating endogenous production of reactive oxygen species, Experimental Therapeutics, Molecular Targets, and Chemical Biology, Cancer Res, 65, 948-956.
632 Parchment R.E. (1993) The implications of a unified theory of programmed cell death, polyamines, oxyradicals and histogenesis in the embryo, Int J Dev Biol, 37(1), 75-83.
633 O'Donnell-Tormey J. DeBoer C.J. Nathan C.F. (1985) Resistance of human tumor cells in vitro to oxidative cytolysis, J Clin Invest, 76(1), 80–86.
634 Brown N.S. Bicknell R. (2001) Hypoxia and oxidative stress in breast cancer: Oxidative stress: its effects on the growth, metastatic potential and response to therapy of breast cancer, Breast Cancer Res, 3(5), 323–327.
635 Benhar M. Engelberg D. Levitzki A. (2002) ROS, stress-activated kinases and stress signaling in cancer, EMBO Reports, 3(5), 420-425.
636 Brar S.S. Corbin Z. Kennedy T.P Hemendinger R. Thornton L. Bommarius B. Arnold R.S. Whorton A.R. Sturrock A.B. Huecksteadt T.P Quinn M.T. Krenitsky K. Ardie K.G. Lambeth J.D. Hoidal J.R (2003) NOX5 NAD(P)H oxidase regulates growth and apoptosis in DU 145 prostate cancer cells, Am J Physiol Cell Physiol, 285, C353-C369.
637 Brar S.S. Kennedy T.P Sturrock A.B. Huecksteadt T.P. Quinn M.T. Whorton A.R. Hoidal J.R. (2002) An NAD(P)H oxidase regulates growth and transcription in melanoma cells, Am J Physiol Cell Physiol, 282(6), C1212-C1224.
638 Nulton-Persson A.C. Szweda L.I. (2001) Modulation of mitochondrial function by hydrogen peroxide, J Biol Chem, 276(26), 23357-23361.
639 Lee S. Yang K. Kwon J. Lee C. Jeong W. Rhee S.G. (2002) Reversible Inactivation of the Tumor Suppressor PTEN by H2O2, J Biol Chem, 277(23), 20336-20342.
640 Lia J. Oberley L.W. Fan M. Colbu N.H. (1998) Inhibition of AP-1 and NF-B by manganese-containing superoxide dismutase in human breast cancer cells, Faseb J, 12, 1713-1723.
641 Manna S.K. Zhang H.J. Yan T. Oberley L.W Aggarwal B.B. (1998) Overexpression of manganese superoxide dismutase suppresses tumor necrosis factor-induced apoptosis and activation of nuclear transcription factor-B and activated protein-1, Biol Chem, 273(21), 13245-13254.
642 Zeisel S.H. (2004) Antioxidants suppress apoptosis, J Nutrition, 134(11), 3179S-3180S.
643 Vaquero E.C. Edderkaoui M. Pandol S.J. Gukovsky I. Gukovskaya A.S. (2004) Reactive Oxygen Species Produced by NAD(P)H Oxidase Inhibit Apoptosis in Pancreatic Cancer Cells, J. Biol. Chem, 279(33), 34643-34654.
644 Pelicano H. Feng L. Zhou Y. Carew J.S. Hileman E.O. Plunkett W. Keating M.J. Huang P. (2003) Inhibition of mitochondrial respiration, J Biol Chem, 278(39), 37832-37839.
645 Oberley T.D. Oberley L.W. (1997) Antioxidant enzyme levels in cancer, Histol Histopathol, 12(2), 525-535.
646 Ambrosone C.B. Freudenheim J.L. Thompson P.A. Bowman E. Vena J.E. Marshall J.R. Graham S. Laughlin R. Nemoto T. Shields P.G. (1999) Manganese Superoxide Dismutase (MnSOD) Genetic polymorphisms, dietary antioxidants, and risk of breast cancer, Cancer Res, 59, 602-606.
647 McCormick W.J. (1952) Ascorbic acid as a chemotherapeutic agent, Arch Pediat, 69, 151-155.
648 Chen Q. Espey M.G. Krishna M.C. Mitchell J.B. Corpe C.P. Buettner G.R. Shacter E. Levine M. (2005) Pharmacologic ascorbic acid concentrations selectively kill cancer cells: Action as a pro-drug to deliver hydrogen peroxide to tissues, Proc Natl Acad Sci U S A,Sep 12; [Epub ahead of print].
649 Youngson R. Schott I. (1995) Medical Blunders, Robinson, London.
650 Fletcher A.E. Breeze E. Shetty P.S. (2003) Antioxidant vitamins and mortality in older persons: findings from the nutrition add-on study to the Medical Research Council Trial of Assessment and Management of Older People in the Community, Am J Clin Nutr, 78, 999-1010.

[651] Lee K.W. Lee H.J. Surh Y.J. Lee C.Y. (2003) Vitamin C and cancer chemoprevention: reappraisal, Am J Clin Nutr, 78, 1074-1078.
[652] Padayatty S.J. Sun H. Wang Y. Riordan H.D. Hewitt S.M. Katz A. Wesley R.A. Levine M. (2004) Vitamin C pharmacokinetics: implications for oral and intravenous use, Ann Intern Med, 140, 533-537.
[653] Levy T.E. (2002) Vitamin C Infectious Diseases and Toxins, Curing the Incurable, Xlibris Press.
[654] Lee K.W. Lee H.J. Kang K.S. Lee C.Y. (2002) Preventive effects of vitamin C on carcinogenesis, The Lancet, 359, 9301.
[655] Upham B.L. Kang K.S. Cho H.Y. Trosko J.E. (1997) Hydrogen peroxide inhibits gap junctional intercellular communication in glutathione sufficient but not glutathione deficient cells, Carcinogenesis, 18, 37-42.
[656] Ramp W.K. Thornton P.A. (1968) The effects of ascorbic acid on the glycolytic and respiratory metabolism of embryonic chick tibias, Calcif Tissue Res , 2, 77-82.
[657] Poydock M.E. (1991) Effect of combined ascorbic acid and B12 on survival of mice implanted with Erlich carcinoma and L1210 leukemia, Am J Clin Nutr, 54, 12615-12655.
[658] Gardorov A.K. Loshkomoeva I.N. (1978) Free radical lipid peroxidation and several ways of regulating it with ascorbic acid, Biofizika, 23, 391-392.
[659] Casciari J.J. Riordan H.D. Miranda-Massari J.R. Gonzalez M.J. (2005) Effects of high dose ascorbate administration on L-10 tumor growth in guinea pigs, P R Health Sci J, 24(2), 145-150.
[660] Szent-Gyorgyi A. (1937) Studies on biological oxidation and some of its catalysts, Banh Verlagbuchandlung, Leipzig.
[661] Leung P.Y. Miyashita K. Young M. Tsao C.S. (1993) Cytotoxic effect of ascorbate and its derivatives on cultured malignant and nonmalignant cell lines, Anticancer Res, 13(2), 475-480.
[662] De Laurenzi V. Melino G. Savini I. Annicchiarico-Petruzzelli M. Finazzi-Agro A. Avigliano L. (1995) Cell death by oxidative stress and ascorbic acid regeneration in human neuroectodermal cell lines, Eur J Cancer, 31A(4), 463-466.
[663] Tsao C. Dunham W. Leung P. (1995) Growth control of human colon tumor xenografts by ascorbic acid, copper, and iron, The Cancer Journal, 8, 157-163.
[664] Girgert R. Vogt Y. Becke D. Bruchelt G. Schweizer P. (1999) Growth inhibition of neuroblastoma cells by lovastatin and L-ascorbic acid is based on different mechanisms, Cancer Lett, 137(2), 167-72.
[665] Paolini M. Pozzeti L. Pedulli G.F. Marchesi E. Cantelli-Forti G. (1999) The nature of prooxidant activity of vitamin C, Life Sci, 64, 273–278.
[666] Gonzalez M.J. (1992) Lipid peroxidation and tumor growth: an inverse relationship, Med Hypotheses, 38, 106-10.
[667] Tsao C.S. Dunham W.B. Leung P.Y. (1988) In vivo antineoplastic activity of ascorbic acid for human mammary tumor, In Vivo, 2(2), 147-150.
[668] Cohen M.H. Kransnow S.H. (1987) Cure of advanced Lewis Lung carcinoma (LL): a new treatment strategy, Proc Am Assoc AA CR, 28, 416.
[669] Kock G.H. and Biaglow J.E. (1978) Toxicity, radiation sensitivity modification and metabolic effects of dehydroascorbate and ascorbate in mammalian cells, J Cell Physiol, 94, 299-306.
[670] Josephy P.D. Palcic B. Skarsgard L.D. (1978) Ascorbate-enhanced cytotoxicity of misonidazole, Nature, 271(5643), 370-372.
[671] Kadiiska M.B. Hanna P.M. Hernandez L. Mason R.P. (1992) In vivo evidence of hydroxyl radical formation after acute copper and ascorbic acid intake: electron spin resonance spin-trapping investigation, Molecular Pharmacol, 42, 723-729.
[672] Margalioth E.J. Schenker J.G. Chevion M. (1983) Copper and zinc levels in normal and malignant tissues, Cancer, 52(5), 868-872.
[673] Rizk S.L. Sky-Peck H.H. (1984) Comparison between concentrations of trace elements in normal and neoplastic human breast tissue, Cancer Research, 44(11), 5390-5394.
[674] Borrello S. De Leo M.E. Galeotti T. (1993) Defective gene expression of MnSOD in cancer cells, Mol Aspects Med, 14(3), 253-258.
[675] Gonzalez M.J. Mora E. Riordan N.H. Riordan H.D. Mojica P. (1998) Rethinking Vitamin C And Cancer: An Update On Nutritional Oncology, Cancer Prevention International, 3, 215-234.
[676] Riordan N.H. Riordan H.D. Jackson J.A. Casciari J.P. (2000) Clinical and experimental experiences with intravenous vitamin C, Journal of Orthomolecular Medicine, 15(4) 201-213.
[677] Hoffer A. Vitamin C and Cancer, Quarry Health Books, Quebec, Canada.
[678] Cathcart R. (2005) Personal communication.
[679] Cathcart R.F. (1981) Vitamin C titrating to bowel tolerance, anascorbemia and acute induced scurvy, Medical Hypotheses, 7, 1359-1376.

680 Riordan H.D. Riordan N.H. Jackson J.A. Casciari J.J. Hunninghake R. Gonzalez M.J. Mora E.M. Miranda-Massari J.R. Rosario N. Rivera A.P (2004) Intravenous vitamin C as a chemotherapy agent: a report on clinical cases, R Health Sci J, 23(2), 115-118.
681 Vera J.C. Rivas C.I. Fischbarg J. Golde D.W. (1993) Mammalian facilitative hexose transporters mediate the transport of dehydroascorbic acid, Nature, 364(6432), 79-82.
682 Rumsey S.C. Kwon O. Xu G.W. Burant C.F. Simpson I. Levine M. (1997) Glucose transporter isoforms GLUT1 and GLUT3 transport dehydroascorbic acid, J Biol Chem, 272(30), 18982-18989.
683 Rumsey S.C. Daruwala R. Al-Hasani H. Zarnowski M.J. Simpson I.A. Levine M. (2000) Dehydroascorbic acid transport by GLUT4 in Xenopus oocytes and isolated rat adipocytes, J Biol Chem, 275(36), 28246-28253.
684 Langemann H. Torhorst J. Kabiersch A. Krenger W. Honegger C.G. (1989) Quantitative determination of water- and lipid-soluble antioxidants in neoplastic and non-neoplastic human breast tissue, Int J Cancer, 43(6), 1169-1173.
685 Malo C. Wilson J.X. (2000) Glucose modulates vitamin C transport in adult human small intestinal brush border membrane vesicles, J Nutr, 130(1), 63-69.
686 Lutsenko E.A. Carcamo J.M. Golde D.W. (2004) A human sodium-dependent vitamin C transporter 2 isoform acts as a dominant-negative inhibitor of ascorbic acid transport, Mol Cell Biol, 24(8), 3150-3156.
687 Baader S.L. Bruchelt G. Trautner M.C. Boschert H. Niethammer D. (1994) Uptake and cytotoxicity of ascorbic acid and dehydroascorbic acid in neuroblastoma (SK-N-SH) and neuroectodermal (SK-N-LO) cells, Anticancer Res, 14(1A), 221-227.
688 Nualart F.J. Rivas C.I. Montecinos V.P. Godoy A.S. Guaiquil V.H. Golde D.W. Vera J.C. (2003) Recycling of vitamin C by a bystander effect, J Biol Chem, 278(12), 10128-10133.
689 Vera J.C. Rivas C.I. Zhang R.H. Farber C.M. Golde D.W. (1994) Human HL-60 myeloid leukemia cells transport dehydroascorbic acid via the glucose transporters and accumulate reduced ascorbic acid, Blood, 84(5), 1628-1634.
690 Agus D.B. Vera J.C. Golde D.W. (1999) Stromal cell oxidation: a mechanism by which tumors obtain vitamin C, Cancer Res, 59(18), 4555-4558.
691 Corti A. Raggi C. Franzini M. Paolicchi A. Pompella A. Casini A.F. (2004) Plasma membrane gamma-glutamyltransferase activity facilitates the uptake of vitamin C in melanoma cells, Free Radic Biol Med, 37(11), 1906-1915.
692 Spielholz C. Golde D.W. Houghton A.N. Nualart F. Vera J.C. (1997) Increased facilitated transport of dehydroascorbic acid without changes in sodium-dependent ascorbate transport in human melanoma cells, Cancer Res, 57(12), 2529-2537.
693 Murata A, Morishige F, Yamaguchi H. (1982) Prolongation of survival times of terminal cancer patients by administration of large doses of ascorbate, International Journal for Vitamin and Nutrition Research, Supplement, 23, 101-113.
694 DeWys W.D. (1982) How to evaluate a new treatment for cancer, Your Patient and Cancer, 2(5), 31-36.
695 Creagan E.T. Moertel C.G. O'Fallon J.R. Schutt A.J. O'Connell M.J. Rubin J. Frytak S. (1979) Failure of high-dose vitamin C (ascorbic acid) therapy to benefit patients with advanced cancer, A controlled trial, N Engl J Med, 301, 687–690.
696 Moss M. (1992) The Cancer Industry, Equinox Press, New York.
697 Pauling L. (1980) Vitamin C therapy and advanced cancer (letter). N Engl J Med, 302, 694.
698 Moertel C.G. Creagan E.T. (1980) Vitamin C therapy and advanced cancer (letter), New England Journal of Medicine, 302, 694-695.
699 Anon (1991) Unproven methods of cancer management, Laetrile, CA Cancer J Clin, 41(3), 187-192.
700 Moertel C.G. (1982) Laetrile, N Engl J Med, 307, 120-121.
701 Moertel C.G. Fleming T.R. Creagan E.T. Rubin J. O'Connell M.J. Ames M.M. (1985) High-dose vitamin C versus placebo in the treatment of patients with advanced cancer who have had no prior chemotherapy. A randomized double-blind comparison, N Engl J Med, 312, 137–141.
702 Padayatty S.J. Levine M. (2000) Reevaluation of ascorbate in cancer treatment: Emerging evidence, open minds and serendipity, Journal of the American College of Nutrition, 19(4), 423-425
703 Hoffer A. Pauling L. (1990) Hardin Jones biostatistical analysis of mortality data for cohorts of cancer patients with a large fraction surviving at the termination of the study, and a comparison of survival times of cancer patients receiving large regular oral doses of vitamin C and other nutrients with similar patients not receiving those doses, Journal of Orthomolecular Medicine, 5, 143-154.
704 Hoffer A. Pauling L. (1993) Hardin Jones biostatistical analysis of mortality data for a second set of cohorts of cancer patients with a large fraction surviving at the termination of the study and a comparison of survival times of cancer patients receiving large regular oral doses of vitamin C and other nutrients with similar patients not receiving these doses, Journal of Orthomolecular Medicine, 8, 1547-1567.
705 Cathcart R. F. (2004) Personal communication.
706 Kumar S. (2005) Personal communication.

[707] De Laurenzi V. Melino G. Savini I. Annicchiarico-Petruzzelli M. Finazzi-Agro A. Avigliano L. (1995) Cell death by oxidative stress and ascorbic acid regeneration in human neuroectodermal cell lines, Eur J Cancer, 31A(4), 463-466.
[708] Meadows G.G. Pierson H.F. Abdallah R.M. (1991) Ascorbate in the treatment of experimental transplanted melanoma, Am J Clin Nutr, 54(6 Suppl), 1284S-1291S.
[709] Riordan H.D. Jackson J.A. Schultz M. (1990) Case study: high-dose intravenous vitamin C in the treatment of a patient with adenocarcinoma of the kidney, J. Ortho Med, 5, 5-7.
[710] Riordan N. Jackson J.A. Riordan H.D. (1996) Intravenous vitamin C in a terminal cancer patient, J. Ortho Med, 11, 80-82.
[711] Riordan N.H. Riordan H.D. Meng X. Li Y. Jackson J.A. (1995) Intravenous ascorbate as a tumor cytotoxic chemotherapeutic agent, Med Hypotheses, 44, 207-213.
[712] Riordan H.D. Riordan N.H. Meng X. Zhong Z. Jackson J.A. Improved microplate fluorometer counting of viable tumor and normal cells, Anticancer Res, 927-932.
[713] Gonzalez M.J. Miranda-Massari J.R. Mora E.M. Jimenez I.Z. Matos M.I. Riordan H.D. Casciari J.J. Riordan N.H. Rodriguez M. Guzman A. (2002) Orthomolecular oncology: a mechanistic view of intravenous ascorbate's chemotherapeutic activity, P R Health Sci J, Mar, 21(1), 39-41.
[714] Raic-Malic S. Svedruzic D. Gazivoda T. Marunovic A. Hergold-Brundic A. Nagl A. Balzarini J. De Clercq E. Mintas M. (200) Synthesis and antitumor activities of novel pyrimidine derivatives of 2, 3-O, O-dibenzyl-6-deoxy-L-ascorbic acid and 4, 5-didehydro-5, 6- dideoxy-L-ascorbic acid. J Med Chem, 43(25), 4806-4811.
[715] Tashiro F. Sugiyama A. Urano Y. Kochi M. (2002) Sodium 5, 6-benzylidene-L-ascorbate induces in vitro neuronal cell differentiation accompanying apoptosis and necrosis, Anticancer Res, 22(3), 1423-1431.
[716] Esposito E. Cervellati F. Menegatti E. Nastruzzi C. Cortesi R. (2002) Spray dried Eudragit microparticles as encapsulation devices for vitamin C, Int J Pharm, 21, 242(1-2), 329-334.
[717] Clement M.V. Ramalingam J. Long L.H. Halliwell B. (2001) The in vitro cytotoxicity of ascorbate depends on the culture medium used to perform the assay and involves hydrogen peroxide, Antioxid Redox Signal, 3(1), 157-163.
[718] Benade L, Howard T, Burk D. (1969) Synergistic killing of Ehrlich ascites carcinoma cells by ascorbate and 3-amino-1, 2, 4, -triazole, Oncology, 23, 33–43.
[719] Bram S Froussard P. Guichard M. Jasmin C. Augery Y. Sinoussi-Barre F. Wray W. (1980) Vitamin C preferential toxicity for malignant melanoma cells, Nature, 284, 629-631.
[720] Helgestad J. Pettersen R. Storm-Mathisen I. Schjerven L. Ulrich K. Smeland E.B. Egeland T. Sorskaard D. Brogger A. Hovig T. (1990) Characterization of a new malignant human T-cell line (PFI-285) sensitive to ascorbic acid, Eur J Haematol, 44, 9-17.
[721] Park C.H. Aniare M. Savin M. A. Hoogstraten B. (1980) Growth suppression of human leukemic cells in vitro by L-ascorbic acid, Cancer Res, 40, 1062-1065.
[722] Yamafuji K. Nakamurar Y. Omura H. Soeda T. Gyotok K. (1971) Anti-tumor potency of ascorbic, dehydroascorbic, or 2, 3-diiketogulonic acid and their action on deoxyribonucleic acid, A Krebsforsch, 76, 1-7.
[723] Yagashita K. Takahashi N. Yamamoto H. Jinnouchi H. Hiyoshi S. Miyakawa T. (1976) Effects of tetraacetyl-bis-dehydroAA, a derivative of ascorbic acid, on Ehrlich cells and HeLa cells (human carcinoma cells), J Nutr Sci Vitaminol, 22, 419427.
[724] Zheng Q.S. Sun X.L. Wang C.H. (2002) Redifferentiation of human gastric cancer cells induced by ascorbic acid and sodium selenite, Biomed Environ Sci, 15(3), 223-232.
[725] Ely J.T.A. (2003) Inadequate levels of essential nutrients in developed nations as a risk factor for disease: a review, Rev Environ Health, 18(2), 111-29.
[726] MyNutrition.co.uk (2004) ONUK Survey – 37,053 respondents.
[727] Vastag B. (2004) Obesity is now on everyone's plate, JAMA, 291, 1186-1188.
[728] Sturm R. (2000) Increases in clinically severe obesity in the United States, 1986–2000, Arch Intern Med, 163, 2146-2148.
[729] Mokdad A.H. Marks J.S. Stroup D.F. Gerberding J.L. (2004) Actual causes of death in the United States, 2000, JAMA, 291, 1238-1245.
[730] Calle E.E. Rodriguez C. Walker-Thurmond K. Thun M.J. (2003) Overweight, obesity, and mortality from cancer in a prospectively studied cohort of U.S. adults, N Engl J Med, 348, 1625-1638.
[731] Wackers F. (2002) It pays to be sweet, Biologist, 49(4), 165-169.
[732] Holldober B. Wilson E.O. (1990) The Ants, Springer, New York.
[733] Yudkin J. (1972) Sweet and Dangerous, Bantam Books, New York.
[734] Spitz D.R. Sim J.E. Ridnour L.A. Galoforo S.S. Lee Y.J. (2000) Glucose deprivation-induced oxidative stress in human tumor cells. A fundamental defect in metabolism? Ann N Y Acad Sci, 899, 349-362.
[735] Hursting S.D. Lavigne J.A. Berrigan D. Perkins S.N. Barrett J.C. (2003) Calorie restriction, aging, and cancer prevention: mechanisms of action and applicability to humans, Annu Rev Med, 54, 131-152.

[736] Dirx M.J. Zeegers M.P. Dagnelie P.C. van den Bogaard T. van den Brandt P.A. (2003) Energy restriction and the risk of spontaneous mammary tumors in mice: a meta-analysis, Int J Cancer, 106, 766-770.
[737] Harvell D.M. Strecker T.E. Xie B. Pennington K.L. McComb R.D. Shull J.D. (2002) Dietary energy restriction inhibits estrogen-induced mammary, but not pituitary, tumorigenesis in the ACI rat, Carcinogenesis, 23, 161-169.
[738] Matsuzaki J. Yamaji R. Kiyomiya K. Kurebe M. Inui H. Nakano Y. (2000) Implanted tumor growth is suppressed and survival is prolonged in sixty percent of food-restricted mice, J Nutr, 130, 111-115.
[739] Michels K.B. Ekbom A. (2004) Caloric restriction and incidence of breast cancer, JAMA, 291, 1226-1230.
[740] Brand-Miller J. Wolever T.M.S. Foster-Powell K. Colagiuri S. (2002) The New Glucose Revolution: The Authoritative Guide to the Glycemic Index--the Dietary Solution for Lifelong Health, Marlowe & Co, New York.
[741] Brand-Miller J. Foster-Powell K. Holt S. Burani J. (2003) The New Glucose Revolution Complete Guide to Glycemic Index Values, Marlowe & Co, New York.
[742] Foster-Powell K. Holt S.H. Brand-Miller J.C. (2002) International table of glycemic index and glycemic load values, Am J Clin Nutr, 76, 5-56.
[743] Augustin L.S. Gallus S. Negri E. La Vecchia C. (2004) Glycemic index, glycemic load and risk of gastric cancer, Ann Oncol, 15, 581-584.
[744] Augustin L.S. Gallus S. Franceschi S. Negri E. Jenkins D.J. Kendall C.W. Dal Maso L. Talamini R. La Vecchia C. (2003) Glycemic index and load and risk of upper aero-digestive tract neoplasms (Italy), Cancer Causes Control, 14, 657-662.
[745] Augustin L.S. Gallus S. Bosetti C. Levi F. Negri E. Franceschi S. Dal Maso L. Jenkins D.J. Kendall C.W. La Vecchia C. (2003) Glycemic index and glycemic load in endometrial cancer, Int J Cancer, 105, 404-407.
[746] Augustin L.S. Polesel J. Bosetti C. Kendall C.W. La Vecchia C. Parpinel M. Conti E. Montella M. Franceschi S. Jenkins D.J. Dal Maso L. (2003) Dietary glycemic index, glycemic load and ovarian cancer risk: a case-control study in Italy, Ann Oncol, 14, 78-84.
[747] Franceschi S. Dal Maso L. Augustin L. Negri E. Parpinel M. Boyle P. Jenkins D.J. La Vecchia C. (2001) Dietary glycemic load and colorectal cancer risk, Ann Oncol, 12, 173-178.
[748] Carroll K.K. (1977) Dietary factors in hormone-dependent cancers, Curr Concepts Nutr, 6, 25–40.
[749] Hems G. (1978) The contributions of diet and childbearing to breast cancer rates, Br J Cancer, 37, 974–982.
[750] Schoen R.E. Tangen C.M. Kuller L.H. Burke G.L. Cushman M. Tracy R.P. Dobs A. Savage P.J. (1999) Increased blood glucose and insulin, body size, and incident colorectal cancer, J Natl Cancer Inst, 91, 1147-1154.
[751] McKeown-Eyssen G. (1994) Epidemiology of colorectal cancer revisited: are serum triglycerides and/or plasma glucose associated with risk? Cancer Epidemiology Biomarkers & Prevention, 3(8), 687-695.
[752] Hu F.B. Manson J.E. Liu S. Hunter D. Colditz G.A. Michels K.B. Speizer F.E. Giovannucci E. (1999) Prospective study of adult onset diabetes mellitus (type 2) and risk of colorectal cancer in women, J Natl Cancer Inst, 91, 542-547.
[753] Colangelo L.A. Gapstur S.M. Gann P.H. Dyer A.R. Liu K. (2002) Colorectal cancer mortality and factors related to the insulin resistance syndrome, Cancer Epidemiol Biomarkers Prev, 11, 385-391.
[754] Coughlin S.S. Calle E.E. Teras L.R. Petrelli J. Thun M.J. (2004) Diabetes mellitus as a predictor of cancer mortality in a large cohort of US adults, Am J Epidemiol, 159, 1160-1167.
[755] La Vecchia C. Negri E. Decarli A. Franceschi S. (1997) Diabetes mellitus and colorectal cancer risk, Cancer Epidemiol Biomarkers Prev, 6, 1007-1010.
[756] Sandhu M.S. Luben R. Khaw K.T. (2001) Self reported non-insulin dependent diabetes, family history, and risk of prevalent colorectal cancer: population based, cross sectional study, J Epidemiol Community Health, 55, 804-805.
[757] Anderson K.E. Anderson E. Mink P.J. Hong C.P. Kushi L.H. Sellers T.A. Lazovich D. Folsom A.R. (2001) Diabetes and endometrial cancer in the Iowa women's health study, Cancer Epidemiol Biomarkers Prev, 10, 611-616.
[758] Calle E.E. Murphy T.K. Rodriguez C. Thun M.J. Heath C.W. Jr (1998) Diabetes mellitus and pancreatic cancer mortality in a prospective cohort of United States adults, Cancer Causes Control, 9, 403-410.
[759] Bostick R.M. Potter J.D. Kushi L.H. Sellers T.A. Steinmetz K.A. McKenzie D.R. Gapstur S.M. Folsom A.R. (1994) Sugar, meat, and fat intake, and non-dietary risk factors for colon cancer incidence in Iowa women (United States), Cancer Causes Control, 5, 38-52.
[760] Frazier A.L. Li L. Cho E. Willett W.C. Colditz G.A. (2004) Adolescent diet and risk of breast cancer, Cancer Causes Control, 15, 73-82.
[761] Higginbotham S. Zhang Z.F. Lee I.M. Cook N.R. Giovannucci E. Buring J.E. Liu S. (2004) Dietary glycemic load and risk of colorectal cancer in the Women's Health Study, J Natl Cancer Inst, 96, 229-233.

[762] Cho E. Spiegelman D. Hunter D.J. Chen W.Y. Colditz G.A. Willett W.C. (2003) Premenopausal dietary carbohydrate, glycemic index, glycemic load, and fiber in relation to risk of breast cancer, Cancer Epidemiol Biomarkers Prev, 12, 1153-1158.
[763] Folsom A.R. Demissie Z. Harnack L. (2003) Glycemic index, glycemic load, and incidence of endometrial cancer: the Iowa women's health study, Nutr Cancer, 46, 119-124.
[764] Higginbotham S. Zhang Z.F. Lee I.M. Cook N.R. Buring J.E. Liu S. (2004) Dietary glycemic load and breast cancer risk in the Women's Health Study, Cancer Epidemiol Biomarkers Prev, 13, 65-70.
[765] Michaud D.S. Liu S. Giovannucci E. Willett W.C. Colditz G.A. Fuchs C.S. (2002) Dietary sugar, glycemic load, and pancreatic cancer risk in a prospective study, J Natl Cancer Inst, 94, 1293-1300.
[766] Holmes M.D. Liu S. Hankinson S.E. Colditz G.A. Hunter D.J. Willett W.C. (2004) Dietary carbohydrates, fiber, and breast cancer risk, Am J Epidemiol, 159, 732-739.
[767] Jonas C.R. McCullough M.L. Teras L.R. Walker-Thurmond K.A. Thun M.J. Calle E.E. (2003) Dietary glycemic index, glycemic load, and risk of incident breast cancer in postmenopausal women, Cancer Epidemiol Biomarkers Prev, 12, 573-577.
[768] Oh K. Willett W.C. Fuchs C.S. Giovannucci E.L. Glycemic index, glycemic load, and carbohydrate intake in relation to risk of distal colorectal adenoma in women, Cancer Epidemiol Biomarkers Prev, 13, 1192-1198.
[769] Terry P.D. Jain M. Miller A.B. Howe G.R. Rohan T.E. (2003) Glycemic load, carbohydrate intake, and risk of colorectal cancer in women: a prospective cohort study, J Natl Cancer Inst, 95, 914-916.
[770] Spitz D.R. Sim J.E. Ridnour L.A. Galoforo S.S. Lee Y.J. (2000) Glucose deprivation-induced oxidative stress in human tumor cells. A fundamental defect in metabolism? Ann N Y Acad Sci, 899, 349-362.
[771] Lee Y.J. Galoforo S.S. Berns C.M. Chen J.C. Davis B.H. Sim J.E. Corry P.M. Spitz D.R. (1998) Glucose deprivation-induced cytotoxicity and alterations in mitogen-activated protein kinase activation are mediated by oxidative stress in multidrug-resistant human breast carcinoma cells, J Biol Chem, 273(9), 5294-5299
[772] Ahmad I.M. Aykin-Burns N. Sim J.E. Walsh S.A. Higashikubo R. Buettner G.R. Venkataraman S. Mackey M.A. Flanagan S.W. Oberley L.W. Spitz D.R. (2005) Mitochondria and H2O2 mediate glucose deprivation-induced stress in human cancer cells, J. Biol. Chem, 280(6) 4254-4263.
[773] Lin X. Zhang F. Bradbury C.F. Kaushal A. Li L. Spitz D.R. Aft R.L. Gius D. (2003) 2-Deoxy-D-Glucose-induced cytotoxicity and radiosensitization in tumor cells is mediated via disruptions in thiol metabolism, Cancer Res, 63, 3413-3417.
[774] Younes M. Lechago L.V. Somoano J.R. Mosharaf M. Lechago J. (1996) Wide expression of the human erythrocyte glucose transporter Glut1 in human cancers, Cancer Res, 56(5), 1164-1167.
[775] Brown R.S. Wahl R.L. (1993) Overexpression of Glut-1 glucose transporter in human breast cancer, An immunohistochemical study, Cancer, 72(10), 2979-2985.
[776] Yamamoto T. Seino Y. Fukumoto H. Koh G. Yano H. Inagaki N. Yamada Y. Inoue K. Manabe T. Imura H. (1990) Over-expression of facilitative glucose transporter genes in human cancer, Biochem Biophys Res Commun, 170(1), 223-230.
[777] Reske S.N. Grillenberger K.G. Glatting G. Port M. Hildebrandt M. Gansauge F. Beger H.G. (1997) Overexpression of glucose transporter 1 and increased FDG uptake in pancreatic carcinoma, J Nucl Med, 38(9), 1344-1348.
[778] Smith T.A. (1999) Facilitative glucose transporter expression in human cancer tissue, Br J Biomed Sci, 56(4), 285-592.
[779] Hoh C.K. Hawkins R.A. Glaspy J.A. Dahlbom M. Tse N.Y. Hoffman E.J. Schiepers C. Choi Y. Rege S. Nitzsche E. (1993) Cancer detection with whole-body PET using 2-[18F]fluoro-2-deoxy-D-glucose, J Comput Assist Tomogr, 17(4), 582-589.
[780] Flamen P. Stroobants S. Van Cutsem E. Dupont P. Bormans G. De Vadder N. Penninckx F. Van Hoe L. Mortelmans L. (1999) Additional value of whole-body positron emission tomography with fluorine-18-2-fluoro-2-deoxy-D-glucose in recurrent colorectal cancer, Journal of Clinical Oncology, 17(3), 894-901.
[781] Vansteenkiste J.F. Stroobants S.G. Dupont P.J. De Leyn P.R. Verbeken E.K. Deneffe G.J. Mortelmans L.A. Demedts M.G. (1999) Prognostic importance of the standardized uptake value on (18)F-fluoro-2-deoxy-glucose-positron emission tomography scan in non-small-cell lung cancer: An analysis of 125 cases, J Clin Oncol, 17(10), 3201-3206.
[782] Lindholm P. Minn H. Leskinen-Kallio S. Bergman J. Ruotsalainen U. Joensuu H. (1993) Influence of the blood glucose concentration on FDG uptake in cancer--a PET study, Journal of Nuclear Medicine, 34(1), 1-6.
[783] Tayek J.A. (1992) A review of cancer cachexia and abnormal glucose metabolism in humans with cancer, J Am College of Nutrition, 11(4), 445-456.
[784] Lundholm K. Edstrom S. Karlberg I. Ekman L. Schersten T. (1982) Glucose turnover, gluconeogenesis from glycerol, and estimation of net glucose cycling in cancer patients, Cancer, 50(6), 1142-1150.
[785] Permert J. Ihse I. Jorfeldt L. von Schenck H. Arnqvist H.J. Larsson (1993) Pancreatic cancer is associated with impaired glucose metabolism, J.Eur J Surg, 159(2), 101-107.

786 Shaw J.H. Wolfe R.R. (1987) Glucose and urea kinetics in patients with early and advanced gastrointestinal cancer: the response to glucose infusion, parenteral feeding, and surgical resection, Surgery, 101(2), 181-191.
787 Smith G.D. Egger M. Shipley M.J. Marmot M.G. (1992) Post-challenge glucose concentration, impaired glucose tolerance, diabetes, and cancer mortality in men, American Journal of Epidemiology, 136(9), 1110-1114.
788 Limburg P.J. Anderson K.E. Johnson T.W. Jacobs D.R. Lazovich D. Hong C. Nicodemus K.K. Folsom A.R. (2005) Diabetes mellitus and subsite-specific colorectal cancer risks in the Iowa women's health study, Cancer Epidemiology Biomarkers and Prevention, 14, 133-137.
789 Jee S.H. Ohrr H. Sull J.W. Yun J.E. Ji M. Samet J.M. (2005) Fasting serum glucose level and cancer risk in Korean men and women, JAMA, 293(2), 194-202.
790 Balkau B. Kahn H.S. Courbon D. Eschwège E. Ducimetière P. (2001) Hyperinsulinemia predicts fatal liver cancer but is inversely associated with fatal cancer at some other sites, The Paris prospective study, Diabetes Care, 24, 843-849.
791 Saydah S.H. Loria C.M. Eberhardt M.S. Brancati F.L. (2003) Abnormal glucose tolerance and the risk of cancer death in the United States, Am J Epidemiol, 157, 1092-1100.
792 Aisenberg A.C. (1961) The Glycolysis and Respiration of Tumors, Academic Press, London.
793 Nakashima R.A. Paggi M.G. Pedersen P.L. (1984) Contributions of glycolysis and oxidative phosphorylation to adenosine 5'-triphosphate production in AS-30D hepatoma cells, Cancer Res, 44(12), 5702-5706.
794 Bustamante E. Pedersen P.L. (1977) High aerobic glycolysis of rat hepatoma cells in culture: role of mitochondrial hexokinase, Proc Nat Acad Sci, 74(9), 3735-3739.
795 Bustamante E. Morris H.P. Pedersen P.L. (1981) Energy metabolism of tumor cells. Requirement for a form of hexokinase with a propensity for mitochondrial binding, J. Biol. Chem, 256(16), 8699-8704.
796 Mathupala S.P. Rempel A. Pedersen P.L. (1995) Glucose catabolism in cancer cells, Isolation, sequence and activity of the promoter for type II hexokinase, A.S.B.M.B, 270(28), 16918-16925.
797 Nakashima R.A. Mangan P.S. Colombini M. Pedersen P.L. (1986) Hexokinase receptor complex in hepatoma mitochondria: evidence from N,N'-dicyclohexylcarbodiimide-labeling studies for the involvement of the pore-forming protein VDAC, Biochemistry, 25(5), 1015-1021.
798 Arora K.K. Pedersen P.L. (1988) Functional significance of mitochondrial bound hexokinase in tumor cell metabolism. Evidence for preferential phosphorylation of glucose by intramitochondrially generated ATP, J. Biol. Chem, 263(33), 17422-17428.
799 Heiden M.G. V. Plas D.R. Rathmell J.C. Fox C.J. Harris M.H. Thompson C.B. (2001) Growth Factors Can Influence Cell Growth and Survival through Effects on Glucose Metabolism, Molecular and Cellular Biology, 21(17), 5899-5912.
800 Floridi A. Paggi M.G. Marcante M.L. Silvestrini B. Caputo A. De Martino C.J. (1981) Lonidamine, a selective inhibitor of aerobic glycolysis of murine tumor cells, Natl Cancer Inst, 66(3), 497-499.
801 Floridi A. Paggi M.G. D'Atri S. De Martino C. Marcante M.L. Silvestrini B. Caputo A. (1981) Effect of lonidamine on the energy metabolism of Ehrlich ascites tumor cells, Cancer Res, 41(11), 4661-4666.
802 Parry D.M. Pedersen P.L. (1983) Intracellular localization and properties of particulate hexokinase in the Novikoff ascites tumor. Evidence for an outer mitochondrial membrane location, J. Biol. Chem, 258(18), 10904-10912.
803 Rose I.A. Warms J.V.B. (1967) Mitochondrial Hexokinase Release, Binding and Location, J. Biol Chem, 242, 1635-1645
804 Arora K.K. Pedersen P.L. (1993) Glucose utilization by tumor cells: the enzyme hexokinase autophosphorylates both its N- and C-terminal halves, Arch Biochem Biophys, 304(2), 515-518.
805 Johansson T. Berrez J.M. Nelson B.D. (1985) Evidence that transcription of the hexokinase gene is increased in a rapidly growing rat hepatoma, Biochem Biophys Res Commun, 133(2), 608-613.
806 Paggi M.G. Fanciulli M. Del Carlo C. Citro G. Bruno T. Floridi A. (1991) Glucose utilization by tumor cells: a post-translational modification of mitochondrial hexokinase may play a regulatory role, Biochem Biophys Res Commun, 178(2), 648-655.
807 Shinohara Y. Ichihara J. Terada H. (1991) Remarkably enhanced expression of the type II hexokinase in rat hepatoma cell line AH130, FEBS Lett, 291(1), 55-57.
808 Nakashima R. A. Paggi, M. G. Scott L. J. Pedersen, P. L. (1988) Purification and characterization of a bindable form of mitochondrial bound hexokinase from the highly glycolytic AS-30D rat hepatoma cell line, Cancer Res, 48, 913-919
809 Thelen A.P. Wilson J.E. (1991) Complete amino acid sequence of the type II isozyme of rat hexokinase, deduced from the cloned cDNA: comparison with a hexokinase from novikoff ascites tumor, Arch Biochem Biophys, 286(2), 645-651.

[810] Printz R.L. Koch S. Potter L.R. O'Doherty R.M. Tiesinga J.J. Moritz S. Granner D.K. (1993) Hexokinase II mRNA and gene structure, regulation by insulin, and evolution [published erratum appears in J Biol Chem 1993 May 5, 268(13), 9936], J. Biol. Chem, 268, 5209-5219.
[811] Krone C.A. Ely J.T.A. (2005) Controlling hyperglycemia as an adjunct to cancer therapy, Integr Cancer Ther, 4(1), 25-31.
[812] Gapstur S.M. Gann P.H. Lowe W. Liu K. Colangelo L. Dyer A. (2000) Abnormal glucose metabolism and pancreatic cancer mortality, JAMA, 283(19), 2552-2558.
[813] Hoffer A. (2005) Personnal communication.
[814] Ely J.T.A. (1988) Protein glycation: ascorbate antagonism, Bulletin Amer Physical Society, 33(3), 296.
[815] Krone C. Ely J.T.A. (2001) Vitamin C and glycohemoglobin revisited, Clinical Chemistry, 47, 148.
[816] Krone C. Ely J.T.A. (2002) Glycohaemoglobin and ascorbic acid, Journal of the New Zealand Medical Association, 115(1160), 156.
[817] Ely J.T.A. (1996) Unrecognized pandemic subclinical diabetes of the affluent nations: causes, cost and prevention, J Ortho Med, 11, 95-99.
[818] Harman D. (2001) Aging: Overview, Annals of the New York Academy of Sciences, 928, 1-21.
[819] Ely J.T.A. (1981) Hyperglycemia and major congenital anomalies, N Engl J Med, 305, 833.
[820] Ely J.T.A. and Krone C. (2002) Aging: predictions of a new perspective on old data, Experimental Biology and Medicine, 227, 939-942.
[821] Ely J.T.A. Santisteban G.A. (1983) Gross fetal malformations associated with maternal hyperglycemia, Anatom Record, 205, 54A.
[822] Cooper M.R. McCall C.E. DeChatelet L.R. (1971) Stimulation of leukocyte hexose monophosphate shunt activity by ascorbic acid, Infect Immun,3 , 851–853.
[823] Mowat A.G. Baum J. (1971) Chemotaxis of polymorphonuclear leukocytes from patients with diabetes mellitus, N Engl J Med, 284, 621-627.
[824] Bagdade J.D. Stewart M. Walters E. (1978) Impaired granulocyte adherence, Diabetes, 27, 677-681.
[825] Bagdade J.D. Root R.K. Bulger R.J. (1974) Impaired leukocyte function in patients with poorly controlled diabetes, Diabetes, 23, 9-15.
[826] Nolan C.M. Beaty H.N. Bagdade J.D. (1978) Further characterization of the impaired bactericidal function of granulocytes in patients with poorly controlled diabetes, Diabetes, 27, 889-894.
[827] Goertzel T. Goertzel B. (1995) Linus Pauling, A life in science and politics, Basic Books, New York.
[828] Robinson A.R. Hunsberger A. Westall F.C. (1994) Suppression of squamous cell carcinoma in hairless mice by dietary nutrient variation, Mechanisms of Ageing and Development, 76, 201-214.
[829] Hunsberger E.M. (1975) How I conquered cancer naturally, Harvest House Publishers, Irvine, California.
[830] Robinson A. (1999) Nutrition and cancer, Oregon Institute of Science and Medicine, www.nutritionandcancer.org.
[831] Ely J.T.A. (1996) Glycemic modulation of tumor tolerance, Journal of Orthomolecular Medicine, 11(1), 23-34.
[832] Cheraskin E. RIngsdorf W.M. Jr. Hutchins K. Setyaadmadja A.T. Wideman G.L. (1968) Effect of diet upon radiation response in cervical carcinoma of the uterus, Acta Cyto, 12, 433-438.
[833] Marks P.A. Bishop J.S. (1957) Glucose metabolism in subjects with neoplastic disease response to insulin and glucose tolerance followup studies, Proc Am Assoc Canc Res, 2, 228-229.
[834] Lupulescu A. (1994) The role of vitamins A, beta-carotene, E and C in cancer cell biology, Int J Vit Nutr Res, 63, 3-14.
[835] Kenny P.A. Bissell M.J. (2003) Tumor reversion: correction of malignant behavior by microenvironmental cues, Int J Cancer, 107(5), 688-695.
[836] Santisteban G.A. Ely J.T.A. Hamel E.E. Read D.H. Kozawa S.M. (1985) Glycemic modulation of tumor tolerance in a mouse model of breast cancer, Biochem Biophys Res Commun, 132(3), 1174-1179.
[837] Wise J. (2003) Hunter-gatherers win profit-sharing deal for obesity drug, Bull World Health Organ, 81(5), 382-382.
[838] MacLean D.B. Luo L.G. (2004) Increased ATP content/production in the hypothalamus may be a signal for energy-sensing of satiety: studies of the anorectic mechanism of a plant steroidal glycoside, Brain Res, 1020(1-2), 1-11.
[839] Mangold T. (2003) Sampling the Kalahari cactus diet, BBC Correpondent, Friday, 30 May, 2003, 09:56 GMT 10:56 UK .
[840] Phytopharm plc (2001) Successful Completion of Proof of Principle Clinical Study of P57 for Obesity, Company press release, 05 December, Phytopharm plc.
[841] Mitchell J.S. Simon-Reuss I. (1947) Combination of some effects of x-radiation and a synthetic vitamin K substitute, Nature, 160, 98-99.
[842] Mitchell J.S. (1948) Clinical trials of tetra-sodium 2-methyl-1: 4-naphthohydroquinone diphosphate, in conjunction with x-ray therapy. Brit J Can, 2, 351-359.

[843] Seegers W.H. Bang N.U. (1967) Blood Clotting Enzymology. New York, Academic Press.
[844] Furie B. Furie B.C. (1990) Molecular basis of vitamin K-dependent gamma-carboxylation, Blood, 75, 1753-1762.
[845] Rannels S.R. Gallaher K.J. Wallin R. Rannels D.E. (1987) Vitamin K-dependent carboxylation of pulmonary surfactant-associated proteins, Proc Natl Acad Sci U S A, 84, 5952-5956.
[846] Suttie J.W. (1993) Synthesis of vitamin K-dependent proteins, Faseb J, 7, 445-452.
[847] Douglas A.S. Robins S.P. Hutchison J.D. (1995) Carboxylation of osteocalcin in post-menopausal osteoporotic women following vitamin K and D supplementation, Bone, 17, 15-20.
[848] Schaafsma A. Muskiet F.A. Storm H. (2000) Vitamin D(3) and vitamin K(1) supplementation of Dutch postmenopausal women with normal and low bone mineral densities: effects on serum 25-hydroxyvitamin D and carboxylated osteocalcin, Eur J Clin Nutr, 54, 626-631.
[849] Shearer M.J. (2000) Role of vitamin K and Gla proteins in the pathophysiology of osteoporosis and vascular calcification, Curr Opin Clin Nutr Metab Care, 3, 433-438.
[850] Wu F.Y. Liao W.C. Chang H.M. (1993) Comparison of antitumor activity of vitamins K1, K2 and K3 on human tumor cells by two (MTT and SRB) cell viability assays, Life Sci, 52, 1797- 1804.
[851] Prasad K.N. Edwards-Prasad J. Sakamoto A. (1981) Vitamin K3 (menadione) inhibits the growth of mammalian tumor cells in culture, Life Sci, 29, 1387-1392.
[852] Chlebowski R.T. Akman S.A. Block J.B. (1985) Vitamin K in the treatment of cancer, Cancer Treat Rev, 12, 49-63.
[853] Noto V. Taper H.S. Jiang Y.H. (1989) Effects of sodium ascorbate (vitamin C) and 2-methyl-1,4-naphthoquinone (vitamin K3) treatment on human tumor cell growth in vitro. I. Synergism of combined vitamin C and K3 action, Cancer, 63, 901-906.
[854] Ngo E.O. Sun T.P. Chang J.Y. (1991) Menadione induced DNA damage in a human tumor cell line, Biochem Pharmacol, 42, 1961-1968.
[855] Wu F.Y. Chang N.T. Chen W.J. Juan C.C. (1993) Vitamin K3-induced cell cycle arrest and apoptotic cell death are accompanied by altered expression of c-fos and c-myc in nasopharyngeal carcinoma cells, Oncogene, 8, 2237-2244.
[856] Su W.C. Sun T.P. Wu F.Y. (1991) The in vitro and in vivo cytotoxicity of menadione (vitamin K3) against rat transplantable hepatoma induced by 3'-methyl-4-dimethyl- aminoazobenzene, Gaoxiong Yi Xue Ke Xue Za Zhi, 7, 454-459.
[857] Gold J. (1986) In vivo synergy of vitamin K3 and methotrexate in tumor-bearing animals, Cancer Treat Rep, 70, 1433-1435.
[858] Wang Z. Wang M. Finn F. Carr B.I. (1995) The growth inhibitory effects of vitamins K and their actions on gene expression, Hepatology, 22, 876-882.
[859] Lamson D.W. Plaza S.M. (2003) The anticancer effects of vitamin K, Alternative Medicine Review, 8(3), 303-318.
[860] Beers M.H. Berkow R. (2004) The Merck Manual of Diagnosis and Therapy, Seventeenth Edition, Centennial Edition, Merck.
[861] Cantoni O. Fiorani M. Cattabeni F. Bellomo G. (1991) DNA breakage caused by hydrogen peroxide produced during the metabolism of 2-methyl-1,4-naphthoquinone (menadione) does not contribute to the cytotoxic action of the quinone, Biochem Pharmacol, 42, S220-S222.
[862] Wang Z. Wang M. Finn F. Carr B.I. (1995) The growth inhibitory effects of vitamins K and their actions on gene expression, Hepatology, 22, 876-882.
[863] Saxena S.P. Israels E.D. Israels L.G. (2001) Novel vitamin K-dependent pathways regulating cell survival, Apoptosis, 6, 57-68.
[864] Costa M.Bellosta P. Basilico C. (1996) Cleavage and release of a soluble form of the receptor tyrosine kinase ARK in vitro and in vivo. J Cell Physiol, 168, 737-744.
[865] Dormady S.P. Zhang X.M. Basch R.S. (2000) Hematopoietic progenitor cells grow on 3T3 fibroblast monolayers that overexpress growth arrestspecific gene-6 (Gas6), Proc Natl Acad Sci USA, 97, 12260-12265.
[866] Wimmel A. Rohner I. Ramaswamy A. Heidtmann H.H. Seitz R. Kraus M. Schuermann M. (1999) Synthesis and secretion of the anticoagulant protein S and coexpression of the Tyro3 receptor in human lung carcinoma cells, Cancer, 86, 43-49.
[867]Bellosta P. Zhang Q. Goff S.P. Basilico C. (1997) Signaling through the ARK tyrosine kinase receptor protects from apoptosis in the absence of growth stimulation, Oncogene, 15, 2387-2397.
[868] Avanzi G.C. Gallicchio M. Bottarel F. et al. (1998) Gas6 inhibits granulocyte adhesion to endothelial cells, Blood, 91, 2334-2340.
[869] Ishimoto Y. Ohashi K. Mizuno K. Nakano T. (2000) Promotion of the uptake of PS liposomes and apoptotic cells by a product of growth arrestspecific gene, Gas6, J Biochem (Tokyo), 127, 411-417.
[870] Chen J. Carey K. Godowski P.J. (1997) Identification of Gas6 as a ligand for Mer, a neural cell adhesion molecule related receptor tyrosine kinase implicated in cellular transformation, Oncogene, 14, 2033-2039.

[871] Deitcher S.R. Erban J.K. Limentani S.A. (1996) Acquired free protein S deficiency associated with multiple myeloma: a case report, Am J Hematol, 51, 319-323.
[872] Thomson R.H. (1971) Naturally Occurring Quinones, Academic Press, New York.
[873] Carr B.I. (1996) Suppression of DCP/PIVKA-2 and alpha-fetoprotein levels in human hepatocellular carcinoma (HCC) by high doses of vitamin K1 (VK1), Hepatology, 348A.
[874] Carr B.I. (1994) A phase I/phase II study of high dose vitamin K(VK) to patients with advanced inoperable hepatocellular carcinoma (HCC): interim analysis, Hepatology, 20, 278A .
[875] Zaniboni A. Biasi L. Graffeo M. et al. (1998) Phase II study of high-dose vitamin K1 in hepatocellular carcinoma: a GISCAD study, ASCO, 17, 1182.
[876] McKee R.W. Binkley S.B. MacCorquodale D.W. et al. (1939) The isolation of vitamins K1 and K2, J Am Chem Soc, 61, 1295.
[877] Davidson R.T. Foley A.L. Engelke J.A. Suttie J.W. (1998) Conversion of dietary phylloquinone to tissue menaquinone-4 in rats is not dependent on gut bacteria, J Nutr,128, 220-223.
[878] Thijssen H.H. Drittij-Reijnders M.J. (1996) Vitamin K status in human tissues: tissue-specific accumulation of phylloquinone and menaquinone-4, Br J Nutr, 75, 121-127.
[879] Akedo Y. Hosoi T. Inoue S. et al. (1992) Vitamin K2 modulates proliferation and function of osteoblastic cells in vitro, Biochem Biophys Res Commun, 187, 814-820.
[880] Yaguchi M. Miyazawa K. Otawa M. et al. (1999) Vitamin K2 therapy for a patient with myelodysplastic syndrome, Leukemia, 13, 144-145.
[881] Miyazawa K. Yaguchi M. Funato K. et al. (2001) Apoptosis/differentiation-inducing effects of vitamin K2 on HL-60 cells: dichotomous nature of vitamin K2 in leukemia cells, Leukemia, 15, 1111-1117.
[882] Sakai I. Hashimoto S. Yoda M. et al. (1994) Novel role of vitamin K2: a potent inducer of differentiation of various human myeloid leukemia cell lines, Biochem Biophys Res Commun, 205, 1305-1310.
[883] Yaguchi M. Miyazawa K. Katagiri T. et al. (1997) Vitamin K2 and its derivatives induce apoptosis in leukemia cells and enhance the effect of all-trans retinoic acid, Leukemia, 11, 779-787.
[884] Yaguchi M. Miyazawa K. Otawa M. et al. (1998) Vitamin K2 selectively induces apoptosis of blastic cells in myelodysplastic syndrome: flow cytometric detection of apoptotic cells using APO2.7 monoclonal antibody, Leukemia, 12, 1392-1397.
[885] Takami A. Nakao S. Ontachi Y. et al. (1999) Successful therapy of myelodysplastic syndrome with menatetrenone, a vitamin K2 analog, Int J Hematol, 69, 24-26.
[886] Fujita H. Tomiyama J. Tanaka T. (1998) Vitamin K2 combined with all-trans retinoic acid induced complete remission of relapsing acute promyelocytic leukaemia, Br J Haematol, 103, 584-585.
[887] Miyazawa K. Nishimaki J. Ohyashiki K. et al. (2000) Vitamin K2 therapy for myelodysplastic syndromes (MDS) and post-MDS acute myeloid leukemia: information through a questionnaire survey of multi-center pilot studies in Japan, Leukemia, 14, 1156-1157.
[888] Jancin B. (2002) Vitamin K cuts hepatocellular CA mortality, Fam Pract News, 32, 16.
[889] Budavari S. O'Neil M.J. Smith A. Heckelman P.E. (1989) The Merck Index, Merck Inc,Rahway, NJ.
[890] Taggart W.V. Matschiner J.T. (1969) Metabolism of menadione-6,7-3H in the rat, Biochemistry, 8, 1141-1146.
[891] Chlebowski R.T. Dietrich M. Akman S. Block J.B. (1985) Vitamin K3 inhibition of malignant murine cell growth and human tumor colony formation, Cancer Treat Rep, 69, 527-532.
[892] Akman S.A. Doroshow J.H. Dietrich M.F. et al. (1987) Synergistic cytotoxicity between menadione and dicumarol vs. murine leukemia L1210, J Pharmacol Exp Ther, 240, 486-491.
[893] Akman S.A. Dietrich M. Chlebowski R. et al. (1985) Modulation of cytotoxicity of menadione sodium bisulfite versus leukemia L1210 by the acid-soluble thiol pool, Cancer Res, 45, 5257-5262.
[894] Waxman S. Bruckner H. (1982) The enhancement of 5-fluorouracil anti-metabolic activity by leucovorin, menadione and alpha-tocopherol, Eur J Cancer Clin Oncol, 18, 685-692.
[895] Su Y.Z. Duarte T.E. Dill P.L. Weisenthal L.M. (1987) Selective enhancement by menadiol of in vitro drug activity in human lymphatic neoplasms, Cancer Treat Rep, 71, 619-625.
[896] Hitomi M. Yokoyama F. Kita Y. Nonomura T. Masaki T. Yoshiji H. Inoue H. Kinekawa F. Kurokohchi K. Uchida N. Watanabe S. Kuriyama S. (2005) Antitumor effects of vitamins K1, K2 and K3 on hepatocellular carcinoma in vitro and in vivo, Int J Oncol, 26(3), 713-720.
[897] Parekh H.K. Mansuri-Torshizi H. Srivastava T.S. Chitnis M.P. (1992) Circumvention of adriamycin resistance: effect of 2-methyl-1,4- naphthoquinone (vitamin K3) on drug cytotoxicity in sensitive and MDR P388 leukemia cells, Cancer Lett, 61, 147-156.
[898] Mitchell J.S. Brinkley D. Haybittle J.L. (1965) Clinical trial of radiosensitizers, including synkavit and oxygen inhaled at atmospheric pressure, Acta Radiol Ther Phys Biol, 3, 329-341.
[899] Taper H.S. Keyeux A. Roberfroid M. (1996) Potentiation of radiotherapy by nontoxic pre-treatment with combined vitamins C and K3 in mice bearing solid transplantable tumor, Anticancer Res, 16, 499-503.
[900] Liao W.C. Wu F.Y. Wu C.W. (2000) Binary/ternary combined effects of vitamin K3 with other

antitumor agents in nasopharyngeal carcinoma CG1 cells, Int J Oncol, 17, 323-328.
901 Margolin K.A. Akman S.A. Leong L.A. et al. (1995) Phase I study of mitomycin C and menadione in advanced solid tumors, Cancer Chemother Pharmacol, 36, 293-298
902 Tetef M. Margolin K. Ahn C. et al. (1995) Mitomycin C and menadione for the treatment of lung cancer: a phase II trial, Invest New Drugs, 13, 157-162.
903 Tetef M. Margolin K. Ahn C. et al. (1995) Mitomycin C and menadione for the treatment of advanced gastrointestinal cancers: a phase II trial, J Cancer Res Clin Oncol, 121, 103-106.
904 Hitomi M, Nonomura T, Yokoyama F, Yoshiji H, Ogawa M, Nakai S, Deguchi A, Masaki T, Inoue H, Kimura Y, Kurokohchi K, Uchida N, Kuriyama S. (2005) In vitro and in vivo antitumor effects of vitamin K5 on hepatocellular carcinoma, Int J Oncol, 26(5), 1337-1344.
905 Sun X. Ross D. (1996) Quinone-induced apoptosis in human colon adenocarcinoma cells via DT-diaphorase mediated bioactivation, Chem Biol Interact, 100(3), 267-276.
906 Emdadul Haque M. Asanuma M. Higashi Y. Miyazaki I. Tanaka K. Ogawa N. (2003) Apoptosis-inducing neurotoxicity of dopamine and its metabolites via reactive quinone generation in neuroblastoma cells, Biochim Biophys Acta, 1619(1), 39-52.
907 Calviello G. Di Nicuolo F. Piccioni E. Marcocci M.E. Serini S. Maggiano N. Jones K.H. Cornwell D.G. Palozza P. (2003) Gamma-Tocopheryl quinone induces apoptosis in cancer cells via caspase-9 activation and cytochrome c release, Carcinogenesis, 24(3), 427-433.
908 Henry T.R. Wallace K.B. (1996) Differential mechanisms of cell killing by redox cycling and arylating quinones, Archives of Toxicology , 70(8), 482-489.
909 Cai J. Jones D.P. (1998) Superoxide in apoptosis. Mitochondrial generation triggered by cytochrome c loss, J Biol Chem, 273(19), 11401-11404.
910 Berlin V. Haseltine W.A. (1981) Reduction of adriamycin to a semiquinone-free radical by NADPH cytochrome P-450 reductase produces DNA cleavage in a reaction mediated by molecular oxygen, J Biol Chem, 256(10), 4747-4756.
911 Dusre L. Rajagopalan S. Eliot H.M. Covey J.M. Sinha B.K. (1990) DNA interstrand cross-link and free radical formation in a human multidrug-resistant cell line from mitomycin C and its analogues, Cancer Res, 50(3), 648-652.
912 Mans D.R. Retel J. van Maanen J.M. Lafleur M.V. van Schaik M.A. Pinedo H.M. Lankelma J. (1990) Role of the semi-quinone free radical of the anti-tumour agent etoposide (VP-16-213) in the inactivation of single- and double-stranded phi X174 DNA, Br J Cancer, 62(1), 54-60.
913 Kagan V.E. Yalowich J.C. Borisenko G.G. Tyurina Y.Y. Tyurin V.A. Thampatty P. Fabisiak J.P. (1999) Mechanism-based chemopreventive strategies against etoposide-induced acute myeloid leukemia: free radical/antioxidant approach, Mol Pharmacol, 56(3), 494-506.
914 Gant T.W. Rao D.N. Mason R.P. Cohen G.M. (1988) Redox cycling and sulphydryl arylation; their relative importance in the mechanism of quinone cytotoxicity to isolated hepatocytes, Chem Biol Interact, 65, 157-173.
915 Ross D. Thor H. Orrenius S. Moldeus P. (1985) Interaction of menadione (2-methyl-1,4-naphthoquinone) with glutathione, Chem Biol Interact, 55, 177-184.
916 Brown P.C. Dulik D.M. Jones T.W. (1991) The toxicity of menadione (2-methyl-1,4-naphthoquinone) and two thioether conjugates studied with isolated renal epithelial cells, Arch Biochem Biophys, 285, 187-196.
917 Nutter L.M. Ngo E.O. Fisher G.R. Gutierrez P.L. (1992) DNA strand scission and free radical production in menadione-treated cells. Correlation with cytotoxicity and role of NADPH quinine acceptor oxidoreductase, J Biol Chem, 267, 2474-2479.
918 Ross D. Thor H. Threadgill M.D. et al. (1986) The role of oxidative processes in the cytotoxicity of substituted 1,4- naphthoquinones in isolated hepatocytes, Arch Biochem Biophys, 248, 460-466.
919 Sun J.S. Tsuang Y.H. Huang W.C. et al. (1997) Menadione induced cytotoxicity to rat osteoblasts, Cell Mol Life Sci, 53, 967-976.
920 Byczkowski J.Z. Gessner T. (1988) Inhibition of the redox cycling of vitamin K3 (menadione) in mouse liver microsomes, Int J Biochem, 20(10), 1073-1079.
921 Nishikawa Y. Carr B.I. Wang M. et al. (1995) Growth inhibition of hepatoma cells induced by vitamin K and its analogs, J Biol Chem, 270, 28304-28310.
922 Juan C. Markovits J. Sun T. Wu F. (1996) Antitumor drug vitamin K3 inhibits both p34CDC2 kinase and protein tyrosine phosphatase by binding to the sulfhydryl groups of the enzymes, Proc AACR, 37, 2427.
923 Chung J.H. Seo D.C. Chung S.H. et al. (1997) Metabolism and cytotoxicity of menadione and its metabolite in rat platelets, Toxicol Appl Pharmacol, 142, 378-385.
924 Thor H. Smith M.T. Hartzell P. et al. (1982) The metabolism of menadione (2-methyl-1,4-naphthoquinone) by isolated hepatocytes. A study of the implications of oxidative stress in intact cells, J Biol Chem , 257, 12419-12425.

[925] Nutter L.M. Cheng A.L. Hung H.L. et al. (1991) Menadione: spectrum of anticancer activity and effects on nucleotide metabolism in human neoplastic cell lines, Biochem Pharmacol, 41, 1283-1292.
[926] Wilson I. Wardman P. Lin T.S. Sartorelli A.C. (1987) Reactivity of thiols towards derivatives of 2- and 6-methyl-1,4- naphthoquinone bioreductive alkylating agents, Chem Biol Interact, 61, 229-240.
[927] Morrison H Jernstrom B. Nordenskjold M. et al. (1984) Induction of DNA damage by menadione (2-methyl-1,4-naphthoquinone) in primary cultures of rat hepatocytes, Biochem Pharmacol, 33, 1763-1769.
[928] Ham S.W. Song J.H. Kim H.I. Bull S.J. (2000) Mechanism of cell cycle arrest by menadione, Korean Chem. Soc, 21(12) 1173-1174.
[929] Nishikawa Y. Carr B.I. Wang M. Kar S. Finn F. Dowd P. Zheng Z.B. Kerns J. Naganathan S. (1995) Growth Inhibition of Hepatoma Cells Induced by Vitamin K and Its Analogs, J Biol Chem, 270(47), 28304-28310.
[930] Nishikawa Y. Wang Z. Kerns J. et al. (1999) Inhibition of hepatoma cell growth in vitro by arylating and non-arylating K vitamin analogs. Significance of protein tyrosine phosphatase inhibition, J Biol Chem, 274, 34803-34810.
[931] Sata N. Klonowski-Stumpe H. Han B. et al. (1997) Menadione induces both necrosis and apoptosis in rat pancreatic acinar AR4-2J cells, Free Radic Biol Med, 23, 844-850.
[932] De Loecker W. Janssens J. Bonte J. Taper H.S. (1993) Effects of sodium ascorbate (vitamin C) and 2-methyl-1, 4-naphthoquinone (vitamin K3) treatment on human tumor cell growth in vitro II, Synergism with combined chemotherapy action, Anticancer Res, Jan-Feb, 13(1), 103-106.
[933] Gilloteaux J. Jamison J.M. Venugopal M. Giammar D. Summers J.L. (1995) Scanning electron microscopy and transmission electron microscopy aspects of synergistic antitumor activity of vitamin C - vitamin K3 combinations against human prostatic carcinoma cells, Scanning Microsc, Mar, 9(1), 159-173.
[934] Venugopal M. Jamison J.M. Gilloteaux J. Koch J.A. Summers M. Giammar D. Sowick C. Summers J.L. (1996) Synergistic antitumor activity of vitamins C and K3 on human urologic tumor cell lines, Life Sci, 59(17), 1389-1400.
[935] Gilloteaux J. Jamison J.M. Ervin E. et al. (1998) Scanning electron microscopy and transmission electron microscopy aspects of the synergistic antitumor activity of vitamin C/vitamin K3 combinations against human T24 bladder carcinoma: another kind of cell death? Scanning, 20, 208-209.
[936] Jamison M. Gilloteaux J. Venugopal M. Koch J.A. Sowick C. Shah R. Summers J.L. (1996) Flow cytometric and ultrastructural aspects of the synergistic antitumor activity of vitamin C-vitamin K3 combinations against human prostatic carcinoma cells, Tissue Cell, 28(6), 687-701.
[937] Juan C.C. Wu F.Y. (1993) Vitamin K3 inhibits growth of human hepatoma HepG2 cells by decreasing activities of both p34CDC2 kinase and phosphatase, Biochem Biophys Res Commun, 190, 907-913.
[938] Wu F.Y. Sun T.P. (1999) Vitamin K3 induces cell cycle arrest and cell death by inhibiting CDC25 phosphatase, Eur J Cancer, 35, 1388-1393.
[939] Ham S.W. Park H.J. Lim D.H. (1997) Studies on menadione as an inhibitor of the cdc25 phosphatase. Bioorg Chem, 25, 33-36.
[940] Chyan C. Wu F. (1999) Vitamin K3 induces cell death by inhibiting dual specificity phosphatases, Proc AACR, 40, 80.
[941] Bouchard C. Staller P. Eilers M. (1998) Control of cell proliferation by Myc, Trends Cell Biol, 8, 202-206.
[942] Cole M.D. McMahon S.B. The Myc oncoprotein: a critical evaluation of transactivation and target gene regulation, Oncogene, 18, 2916-2924.
[943] Bouzahzah B. Nishikawa Y. Simon D. Carr B.I. (1995) Growth control and gene expression in a new hepatocellular carcinoma cell line, Hep40: inhibitory actions of vitamin K, J Cell Physiol, 165, 459-467.
[944] Hoffman B. Liebermann D.A. (1998) The protooncogene c-myc and apoptosis, Oncogene, 17, 3351-3357.
[945] Evan G.I. Wyllie A.H. Gilbert C.S. Littlewood T.D. Land H. Brooks M. Waters C.M. Penn L.Z. Hancock D.C. (1992) Induction of apoptosis in fibroblasts by c-myc protein, Cell, 69, 119-128.
[946] Caricchio R. Kovalenko D. Kaufmann W.K. (1999) Cohen PL. Apoptosis provoked by the oxidative stress inducer menadione (Vitamin K(3)) is mediated by the Fas/Fas ligand system, Clin Immunol, 93, 65-74
[947] Gilloteaux J. Jamison J.M. Arnold D. Ervin E. Eckroat L. Docherty J.J. Neal D. Summers J.L. (1998) Cancer cell necrosis by autoschizis: synergism of antitumor activity of vitamin C: vitamin K3 on human bladder carcinoma T24 cells, Scanning, 20(8), 564-575.
[948] Sakagami H. Satoh K. Hakeda Y. Kumegawa M. (2000) Apoptosis-inducing activity of vitamin C and vitamin K, Cell Mol Biol (Noisy-le-grand), 46(1), 129-143.
[949] Jamison J.M. Gilloteaux J. Taper H.S. Calderon P.B. Summers J.L. (2002) Autoschizis: a novel cell death, Biochem Pharmacol, 15, 63(10), 1773-1783.
[950] Gilloteaux J. Jamison J.M. Arnold D. Summers J.L. (2001) Autoschizis: another cell death for cancer cells induced by oxidative stress, Ital J Anat Embryol, 106(2 Suppl 1), 79-92.
[951] Okayasu H. Ishihara M. Satoh K. Sakagami H. (2001) Cytotoxic activity of vitamins K1, K2 and K3 against human oral tumor cell lines, Anticancer Res, 21(4A), 2387-2392.

[952] Verrax J. Cadrobbi J. Delvaux M. Jamison J.M. Gilloteaux J. Summers J.L. Taper H.S. Calderon P.B. (2003) The association of vitamins C and K3 kills cancer cells mainly by autoschizis, a novel form of cell death. Basis for their potential use as coadjuvants in anticancer therapy, Invited review, European Journal of Medicinal Chemistry, 38, 451-457.
[953] Calderon P.B. Cadrobbi J. Marques C. Hong-Ngoc N. Jamison J.M. Gilloteaux J. Summers J.L. Taper H.S. (2002) Potential therapeutic application of the association of vitamins C and K(3) in cancer treatment, Curr Med Chem, 9(24), 2271-2285.
[954] Taper H.S. Jamison J.M. Gilloteaux J. Gwin C.A. Gordon T. Summers J.L. (2001) In vivo reactivation of DNases in implanted human prostate tumors after administration of a vitamin C/K(3) combination, J Histochem Cytochem, Jan, 49(1), 109-120.
[955] Lasalvia-Prisco E. Cucchi S. Vazquez J. Lasalvia-Galante E. Golomar W. Gordon W. (2003) Serum markers variation consistent with autoschizis induced by ascorbic acid-menadione in patients with prostate cancer, Med Oncol, 20(1), 45-52.
[956] Gilloteaux J. Jamison J.M. Arnold D. Taper H.S. Summers J.L. (2001) Ultrastructural aspects of autoschizis: a new cancer cell death induced by the synergistic action of ascorbate/menadione on human bladder carcinoma cells, Ultrastruct Pathol, 25(3), 183-192.
[957] Taper HS, de Gerlache J, Lans M, Roberfroid M. (1987) Non-toxic potentiation of cancer chemotherapy by combined C and K3 vitamin pre-treatment, Int J Cancer, Oct 15, 40(4), 575-579.
[958] Taper H.S. Roberfroid M. (1992) Non-toxic sensitization of cancer chemotherapy by combined vitamin C and K3 pretreatment in a mouse tumor resistant to oncovin, Anticancer Res, 12(5), 1651-1654.
[959] Zhang W. Negoro T. Satoh K. Jiang Y. Hashimoto K. Kikuchi H. Nishikawa H. Miyata T. Yamamoto Y. Nakano K. Yasumoto E. Nakayachi T. Mineno K. Satoh T. Sakagami H. (2001) Synergistic cytotoxic action of vitamin C and vitamin K3, Anticancer Res, 21(5), 3439-3444.
[960] Jiang Y. Satoh K. Aratsu C. Kobayashi N. Unten S. Kakuta H. Kikuchi H. Nishikawa H. Ochiai K. Sakagami H. (2001) Combination effect of lignin F and natural products, Anticancer Res, Mar-Apr, 21(2A), 965-970.
[961] Venugopal M. Jamison J.M. Gilloteaux J. et al. (1996) Synergistic antitumour activity of vitamins C and K3 against human prostate carcinoma cell lines, Cell Biol Int, 20, 787-797.
[962] Ervin E. Jamison J.M. Gilloteaux J, et al. (1998) Characterization of the early events in vitamin C and K3-induced death of human bladder tumor cells, Scanning, 20, 210-211.
[963] Jamison J.M. Gilloteaux J. Taper H.S. Summers J.L. (2001) Evaluation of the in vitro and in vivo antitumor activities of vitamin C and K-3 combinations against human prostate cancer, J Nutr, 131, 158S-160S.
[964] Carreau J.P. (1979) Biosynthesis of lipoic acid via unsaturated fatty acids, Methods Enzymol, 62, 152-158.
[965] Berkson B. (1998) The Alpha Lipoic Acid Breakthrough, Three Rivers press (Random House), New York.
[966] Parish R.C. Doering P.L. (1986) Treatment of Amanita mushroom poisoning: a review, Vet Hum Toxicol, 28(4), 318-222.
[967] Kramer K. Packer L. (2001) R-alpha-lipoic acid, In Nutraceuticals in Health and Disease Prevention, Eds Kramer K. Hoppe P. Packer L. Marcel Dekker, New York, 129-164.
[968] Lodge J.K. Youn H.D. Handelman G.J. et al. (1997) Natural sources of lipoic acid: determination of lipoyllysine released from protease-digested tissues by high performance liquid chromatography incorporating electrochemical detection, J Appl Nutr, 49(1 & 2), 3-11.
[969] Hermann R. Niebch G. Borbe H.O. et al. (1996) Enantioselective pharmacokinetics and bioavailability of different racemic a-lipoic acid formulations in healthy volunteers, Eur J Pharm Sci, 4, 167-174.
[970] Teichert J. Kern J. Tritschler H.J. Ulrich H. Preiss R. (1998) Investigations on the pharmacokinetics of alpha-lipoic acid in healthy volunteers, Int J Clin Pharmacol Ther, 36(12), 625-628.
[971] Packer L. Kraemer K. Rimbach G. (2001) Molecular aspects of lipoic acid in the prevention of diabetes complications, Nutrition, 17(10), 888-895.
[972] Marangon K. Devaraj S. Tirosh O. Packer L. Jialal I. (1999) Comparison of the effect of alpha-lipoic acid and alpha-tocopherol supplementation on measures of oxidative stress, Free Rad Biol Med, 27(9-10), 1114-1121.
[973] Baur A. Harrer T. Peukert M. et al. (1991) Alpha-lipoic acid is an effective inhibitor of human immuno-deficiency virus (HIV-1) replication, Klin Wochenschr, 69, 722-724.
[974] Packer L. (1998) Alpha-Lipoic acid: a metabolic antioxidant which regulates NF-kappa B signal transduction and protects against oxidative injury, Drug Metab Rev, 30(2), 245-275.
[975] Jacob S, Rett K, Henriksen EJ, Haring HU. (1999) Thioctic acid - effects on insulin sensitivity and glucose-metabolism, Biofactors, 10(2-3), 169-174.
[976] Thorne Research (1998) Monograph alpha lipoic acid, Alternative Medicine Review, 3 (4), 308-310
[977] Gal E.M. (1965) Reversal of selective toxicity of α-lipoic acid by thiamine in thiamine-deficient rats, Nature, 205, 535.

[978] Patrick L. (2002) Mercury toxicity and antioxidants: Part I: role of glutathione and alpha-lipoic acid in the treatment of mercury toxicity, Alternative Medicine Review, 7(6), 456-471
[979] Suh J.H. Shigeno E.T. Morrow J.D. Cox B. Rocha A.E. Frei B. Hagen T.M. (20010 Oxidative stress in the aging rat heart is reversed by dietary supplementation with (R)-(alpha)-lipoic acid, Faseb J, 15(3), 700-706.
[980] Ou P. Nourooz-Zadeh J. Tritschler H.J. Wolff S. (1996) Activation of aldose reductase in rat lens and metal-ion chelation by aldose reductase inhibitors and lipoic acid, Free Radic Res, 25, 337-346.
[981] Sigel H. Prijs B. McCormick D.B. Shih J.C.H. (1978) Stability of binary and ternary complexes of a-lipoate and lipoate derivatives with Mn^{2+} , Cu^{2+}, and Zn^{2+} in solution, Arch Biochem Biophys, 187, 208-214.
[982] Grunert R.R. (1960) The effect of DL α-lipoic acid on heavy-metal intoxication in mice and dogs, Arch Biochem Biophys, 86, 190-194.
[983] Bustamante J. Lodge J.K. Marcocci L. Tritschler H.J. Packer L. Rihn B.H. (1998) Alpha-lipoic acid in liver metabolism and disease, Free Radic Biol Med, 24(6), 1023-1039.
[984] Miura K. Clarkson T.W. (1993) Reduced methylmercury accumulation in a methylmercuryresistant rat pheochromocytoma PC12 cell line, Toxicol Appl Pharmacol, 118, 39-45.
[985] Muller L. Menzel H. (1990) Studies on the efficacy of lipoate and dihydrolipoate in the alteration of cadmium toxicity in isolated hepatocytes, Biochem Biophys Acta, 1052, 386-391.
[986] Keith R.L. Setiarahardjo I. Fernando Q. et al. (1997) Utilization of renal slices to evaluate the efficacy of chelating agents for removing mercury from the kidney, Toxicology, 116, 67-75.
[987] Biewenga G.P. Haenen G.R. Bast A. (1997) The pharmacology of the antioxidant lipoic acid, Gen Pharmacol, 29(3), 315-331.
[988] Bast A. Haenen G.R. (2001) Lipoic acid: a multifunctional nutraceutical, In: Kramer K, Hoppe P, Packer L, eds. Nutraceuticals in Health and Disease Prevention, Marcel Dekker, Inc, New York.
[989] Packer L. Witt E. Tritschler H.J. (1995) Alpha-Lipoic acid as a biological antioxidant, Free Radic Biol Med, 19, 227-250.
[990] Handelman G.J. Han D. Tritschler H. Packer L. (1994) α-Lipoic acid reduction by mammalian cells to the dithiol form and release into the culture medium, Biochem Pharmacol, 47, 1725-1730.
[991] Packer L. Kraemer K. Rimbach G. (2001) Molecular aspects of lipoic acid in the prevention of diabetes complications, Nutrition, 17(10), 888-895.
[992] Kagan V. Serbinova E. Packer L. (1990) Antioxidant effects of ubiquinones in microsomes and mitochondria are mediated by tocopherol recycling, Biochem Biophys Res Comm, 169, 851-857.
[993] Busse E. Zimmer G. Schopohl B. et al. (1992) Influence of alpha-lipoic acid on intracellular glutathione in vitro and in vivo, Arzneimittel-Forschung, 42, 829-831.
[994] Scholich H. Murphy M.E. Sies H. (1989) Antioxidant activity of dihydrolipoate against microsomal lipid peroxidation and its dependence on α-tocopherol. Biochem Biophys Acta, 1001, 256-261.
[995] Scheer B. Zimmer G. (1993) Dihydrolipoic acid prevents hypoxic/reoxygenation and peroxidative damage in rat mitochondria, Arch Biochem Biophys, 302, 385-390.
[996] Assadnazari H. Zimmer G. Freisleben H.J. et al. (1993) Cardioprotective efficiency of dihydrolipoic acid in working rat hearts during hypoxia and reoxygenation, P nuclear magnetic resonance investigations, Arzneimittel-Forschung, 43, 425-432.
[997] Prehn J.H. Karkoutly C. Nuglisch J. et al. (1992) Dihydrolipoate reduces neuronal injury after cerebral ischemia, J Cereb Blood Flow Metab, 12, 78-87.
[998] Panigrahi M. Sadguna Y. Shivakumar B.R. et al. (1996) Alpha-Lipoic acid protects against reperfusion injury following cerebral ischemia in rats, Brain Res, 717, 184-188.
[999] Cao X. Phillis J.W. (1995) The free radical scavenger, alpha-lipoic acid, protects against cerebral ischemia-reperfusion injury in gerbils, Free Rad Res, 23, 365-370.
[1000] Haramaki N. Assadnazari H. Zimmer G. et al. (1995) The influence of vitamin E and dihydrolipoic acid on cardiac energy and glutathione status under hypoxia-reoxygenation, Biochem Mol Biol Int, 37, 591-597.
[1001] Biewenga G.P. Veening-Griffioen D.H. Nicastia A.J. Haenen G.R. Bast A. (1998) Effects of dihydrolipoic acid on peptide methionine sulfoxide reductase, Implications for antioxidant drugs, Arzneimittelforschung, 48(2), 144-148.
[1002] Hinz M, Krappmann D, Eichten A, Heder A, Scheidereit C, Strauss M. (1999) NF-kappaB function in growth control: regulation of cyclin D1 expression and G0/G1-to-S-phase transition, Mol Cell Biol, 19(4), 2690-2698.
[1003] Zhang W.J. Frei B. (2001) Alpha-lipoic acid inhibits TNF-alpha-induced NF-kappaB activation and adhesion molecule expression in human aortic endothelial cells, Faseb J, 15(13), 2423-2432.
[1004] Mizuno M. Packer L. (1994) Effects of alpha-lipoic acid and dihydrolipoic acid on expression of proto-oncogene c-fos, Biochem Biophys Res Commun, 200(2), 1136-1142.
[1005] Ziegler D. Reljanovic M. Mehnert H. Gries F.A. (1999) Alpha-lipoic acid in the treatment of diabetic polyneuropathy in Germany: current evidence from clinical trials, Exp Clin Endocrinol Diabetes, 107(7), 421-430.

[1006] Maitra I. Serbinova E. Tritschler H.J. Packer L. (1996) Stereospecific effects of R-lipoic acid on buthionine sulfoximine-induced cataract formation in newborn rats, Biochem Biophys Res Commun, 221(2), 422-429.
[1007] Rett K. Wicklmayr E. Maerker P. Russ D. Nehrdich D. Hermann R. (1995) Effect of acute infusion of thioctic acid on oxidative and non-oxidative metabolism in obese subjects with NIDDM, Diabetologia, 38, A41.
[1008] Jacob S. Henriksen E.J. Schiemann A.L. Simon I. Clancy D.E. Tritschler H.J. Jung W.I. Augustin H.J. Dietze G.J. (1995) Enhancement of glucose disposal in patients with type 2 diabetes by alpha lipoic acid, Arzneimittel-Forschung, 45, 872-874.
[1009] Konrad T. Vicini P. Kusterer K. et al. (1999) Alpha-Lipoic acid treatment decreases serum lactate and pyruvate concentrations and improves glucose effectiveness in lean and obese patients with type 2 diabetes, Diabetes Care, 22(2), 280-287.
[1010] Jacob S. Henriksen E.J. Tritschler H.J. et al. (1996) Improvement of insulin-stimulated glucose-disposal in type 2 diabetes after repeated parenteral administration of thioctic acid, Exp Clin Endocrinol Diabetes, 104, 284-288.
[1011] Streeper R.S. Henriksen E.J. Jacob S. Hokama J.Y. Fogt D.L. Tritschler H.J. (1997) Differential effects of lipoic acid stereoisomers on glucose metabolism in insulin-resistant skeletal muscle, Am J Physiol, 273(1 Pt 1), E185-191.
[1012] Estrada D.E. Ewart H.S. Tsakiridis T. Volchuk A. Ramlal T. Tritschler H. Klip A. (1996) Stimulation of glucose uptake by the natural coenzyme alpha-lipoic acid/thioctic acid: participation of elements of the insulin signaling pathway, Diabetes, 45(12), 1798-1804.
[1013] Borcea V. Nourooz-Zadeh J, Wolff S.P. et al. (1999) Alpha-Lipoic acid decreases oxidative stress even in diabetic patients with poor glycemic control and albuminuria, Free Radic Biol Med, 26(11-12), 1495-1500.
[1014] Androne L. Gavan N.A. Veresiu I.A. Orasan R. (2000) In vivo effect of lipoic acid on lipid peroxidation in patients with diabetic neuropathy, In Vivo, 14(2), 327-330.
[1015] Ames B.N. (1998) Micronutrients prevent cancer and delay aging, Toxicol Lett, 102-103, 5-18.
[1016] Culy C.R. Clemett D. Wiseman L.R. (2000) Oxaliplatin: A review of its pharmacological properties and clinical efficacy in metastatic colorectal cancer and its potential in other malignancies, Drugs, 60, 895-924.
[1017] Ziegler D. Hanefeld M. Ruhnau K.J. Hasche H. Lobisch M. Schutte K. Kerum G. Malessa R. (1999) Treatment of symptomatic diabetic polyneuropathy with the antioxidant alpha-lipoic acid: a 7-month multicenter randomized controlled trial (ALADIN III Study). ALADIN III Study Group. Alpha-Lipoic Acid in Diabetic Neuropathy, Diabetes Care, 22(8), 1296-1301.
[1018] Gedlicka C. Scheithauer W. Schull B. Kornek G.V. (2002) Effective treatment of oxaliplatin-induced cumulative polyneuropathy with alpha-lipoic acid, J Clin Oncol, 20(15), 3359-3361.
[1019] Gedlicka C. Kornek G.V. Schmid K. Scheithauer W. (2003) Amelioration of docetaxel/cisplatin induced polyneuropathy by alpha-lipoic acid, Ann Oncol, 14(2), 339-340.
[1020] Somani S.M. Husain K. Whitworth C. Trammell G.L. Malafa M. Rybak L.P. (2000) Dose-dependent protection by lipoic acid against cisplatin-induced nephrotoxicity in rats: antioxidant defense system, Pharmacol Toxicol, 86(5), 234-241.
[1021] Dovinova I. Novotny L. Rauko P. Kvasnicka P. (1999) Combined effect of lipoic acid and doxorubicin in murine leukaemia, Neoplasma, 46(4), 237-241.
[1022] Gatei M. Shkedy D. Khanna K.K. Uziel T. Shiloh Y. Pandita T.K. Lavin M.F. Rotman G. (2001) Ataxia-telangiectasia: chronic activation of damage-responsive functions is reduced by alpha-lipoic acid, Oncogene, 20(3), 289-294.
[1023] van de Mark K. Chen J.S. Steliou K. Perrine S.P. Faller D.V. (2003) Alpha-lipoic acid induces p27Kip-dependent cell cycle arrest in non-transformed cell lines and apoptosis in tumor cell lines, J Cell Physiol, 194(3), 325-340.
[1024] Scott B.C. Aruoma O.I. Evans P.J. O'Neill C. Van der Vliet A. Cross C.E. Tritschler H. Halliwell B. (1994) Lipoic and dihydrolipoic acids as antioxidants. A critical evaluation, Free Radic Res, 20(2), 119-133.
[1025] Wenzel U. Nickel A. Daniel H. (2005) Alpha-lipoic acid induces apoptosis in human colon cancer cells by increasing mitochondrial respiration with a concomitant O2-.-generation, Apoptosis, 10(2), 359-368.
[1026] Pack R.A. Hardy K. Madigan M.C. Hunt N.H. (2002) Differential effects of the antioxidant alpha-lipoic acid on the proliferation of mitogen-stimulated peripheral blood lymphocytes and leukaemic T cells, Mol Immunol, 38(10), 733-745.
[1027] Sen C.K. Sashwati R. Packer L. (1999) Fas mediated apoptosis of human Jurkat T-cells: intracellular events and potentiation by redox-active alpha-lipoic acid, Cell Death Differ, 6(5), 481-491.
[1028] Casciari J.J. Riordan N.H. Schmidt T.L. Meng X.L. Jackson J.A. Riordan H.D. (2001) Cytotoxicity of ascorbate, lipoic acid, and other antioxidants in hollow fibre in vitro tumours, British Journal of Cancer, 84, 11, 1544-1550.
[1029] Savvov V.I. Karpov L.M. (1982) Characteristics of 35S-lipoic acid absorption by the blood cells in breast cancer, Vopr Onkol, 28(7), 11-13.

[1030] Savvov V.I. Karpov L.M. (1978) Pyruvate oxidation and S35-lipoic acid fixation by breast tumor homogenates, Vopr Onkol, 24(8), 97-99.
[1031] Karpov L.M. Dvuzhil'naia E.D. Savvov V.I. Anisimov V.D. (1975) Permeability of the mitochondrial membranes of the organs of white rats innoculated with Walker carcinoma to lipoic acid and thiamine labelled with S35, Vopr Onkol, 21(8), 69-73.
[1032] Karpov L.M. Rozanov A. Savvov V.I. (1973) Effect of Walker carcinoma on the distribution of S35-lipoic acid in the body of white rats, Vopr Onkol, 19(12), 57-61.
[1033] Karpov L.M. Dvuzhil'naia E.D. Savvov V.I. Phan Van Thuy (1977) S35 lipoic acid distribution and its effect on pyruvate dehydrogenase activity in rats with Walker carcinoma, Vopr Onkol, 23(10), 87-90.
[1034] Prasad K.N. Kumar B. Yan X.D. Hanson A.J. Cole W.C. (2003) Alpha-tocopheryl succinate, the most effective form of vitamin E for adjuvant cancer treatment: a review, J Am Coll Nutr, 22(2), 108-117.
[1035] Prasad K.N. Ramanujam S. Gaudreau D. (1979) Vitamin E induces morphological differentiation and increases the effect of ionizing radiation on neuroblastoma cells in culture, Proc Soc Exp Biol Med, 161, 570 –573.
[1036] Prasad K.N. Edwards-Prasad J. Ramanujam S. Sakamoto A. (1980) Vitamin E increases the growth inhibitory and differentiating effects of tumor therapeutic agents on neuroblastoma and glioma cells in culture, Proc Soc Exp Biol Med, 164, 158 –163.
[1037] Prasad K.N. Edwards-Prasad J. (1982) Effects of tocopherol (vitamin E) acid succinate on morphological alterations and growth inhibition in melanoma cells in culture, Cancer Res, 42, 550 –555.
[1038] Turley J.M. Fu T. Ruscetti F.W. Mikovits J.A. Bertolette D.C. Birchenall-Roberts M.C. (1997) Vitamin E succinate induces Fas-mediated apoptosis in estrogen receptor-negative human breast cancer cells, Cancer Res, 57(5), 881-890.
[1039] Zhao B. Yu W. Qian M. Simmons-Menchaca M. Brown P. Birrer M.J. Sanders B.G. Kline K. (1997) Involvement of activator protein-1 (AP-1) in induction of apoptosis by vitamin E succinate in human breast cancer cells, Mol Carcinog, 19(3), 180-190.
[1040] Yu W. Heim K. Qian M. Simmons-Menchaca M. Sanders B.G. Kline K. (1997) Evidence for role of transforming growth factor-beta in RRR-alpha-tocopheryl succinate-induced apoptosis of human MDA-MB-435 breast cancer cells, Nutr Cancer, 27(3), 267-278.
[1041] Pussinen P.J. Lindner H. Glatter O. Reicher H. Kostner G.M. Wintersperger A. Malle E. Sattler W. (2000) Lipoprotein-associated alpha-tocopheryl-succinate inhibits cell growth and induces apoptosis in human MCF-7 and HBL-100 breast cancer cells, Biochim Biophys Acta, 1485(2-3), 129-144.
[1042] Malafa M.P. Neitzel L.T. (2000) Vitamin E succinate promotes breast cancer tumor dormancy, J Surg Res, 93(1), 163-170.
[1043] Barnett K.T. Fokum F.D. Malafa M.P. (2002) Vitamin E succinate inhibits colon cancer liver metastases, J Surg Res, 106(2), 292-298.
[1044] Israel K. Yu W. Sanders B.G. Kline K. (2000) Vitamin E succinate induces apoptosis in human prostate cancer cells: role for Fas in vitamin E succinate-triggered apoptosis, Nutr Cancer, 36(1), 90-100.
[1045] Kogure K. Hama S. Manabe S. Tokumura A. Fukuzawa K. (2002) High cytotoxicity of alpha-tocopheryl hemisuccinate to cancer cells is due to failure of their antioxidative defense systems, Cancer Lett, 186(2), 151-156
[1046] Kang Y. Lee E. Choi M. Ku j. Kim S.H. Park Y. Lim S. (2004) Role of reactive oxygen species in the induction of apoptosis by alpha-tocopheryl succinate, Int J Cancer, 112(3), 385-392.
[1047] Wu Y. Zu K. Ni J. Yeh S. Kasi D. James N.S. Chemler S. Ip C. (2004) Cellular and molecular effects of alpha-tocopheryloxybutyrate: lessons for the design of vitamin E analog for cancer prevention, Anticancer Res, 24(6), 3795-3802.
[1048] Neuzil J. Weber T. Schroder A. Lu M. Ostermann G. Gellert N. Mayne G.C. Olejnicka B. Negre-Salvayre A. Sticha M. Coffey R.J. Weber C. (2001) Induction of cancer cell apoptosis by alpha-tocopheryl succinate: molecular pathways and structural requirements, Faseb J, 15(2), 403-415.
[1049] Weber T. Lu M. Andera L. Lahm H. Gellert N. Fariss M.W. Korinek V. Sattler W. Ucker D.S. Terman A. Schroder A. Erl W. Brunk U.T. Coffey R.J. Weber C. Neuzil J. (2002) Vitamin E succinate is a potent novel antineoplastic agent with high selectivity and cooperativity with tumor necrosis factor-related apoptosis-inducing ligand (Apo2 ligand) in vivo, Clin Cancer Res, 8(3), 863-869.
[1050] Neuzil J. Weber T. Gellert N. Weber C. (2001) Selective cancer cell killing by alpha-tocopheryl succinate, Brit J Cancer, 84, 87-89.
[1051] Israel K. Sanders B.G. Kline K. (1995) RRR-a-Tocopheryl Succinate inhibits the proliferation of human prostatic tumor cells with defective cell cycle/differentiation pathway, Nutr Cancer, 24, 161-169.
[1052] Yu W. Sanders B.G. Kline K. (1997) RRR-a-tocopheryl succinate inhibits EL4 thymic lymphoma cell growth by inducing apoptosis and DNA synthesis arrest, Nutr Cancer, 27, 92-101

1053 Simmons-Menchaca M. Qian M. Yu W. Sanders B.G. Kline K. (1995) RRR-a-Tocophoryl succinate inhibits DNA synthesis and enhances the production and secretion of biologically active transforming growth factor-b by avian retrovirus-transformed lymphoid cells, Nutr Cancer, 24, 171-185.
1054 Wu K. Zhao Y. Liu B.H. Li Y. Liu F. Guo J. Yu W. (2002) RRR-a-tocopheryl succinate inhibits human gastric cancer SGC-7901 cell growth by inducing apoptosis and DNA synthesis arrest, World J Gastroenterol, 8, 26-30.
1055 Kim S.J. Bang O.S. Lee Y.S. Kang S.S. (1998) Production of inducible nitric oxide is required for monocytic differentiation of U937 cells induced by vitamin E-succinate, J Cell Sci, 111, 435-441.
1056 You H. Yu W. Sanders B.G. Kline K. (2001) RRR-a-tocopheryl succinate induces MDA-MB-435 and MCF-7 human breast cancer cells to undergo differentiation, Cell Growth Differ, 12, 471-480.
1057 Yu W. Sanders B.G. Kline K. (1996) Modulation of murine EL-4 thymic lymphoma cell proliferation and cytokine production by Vitamin E succinate, Nutr Cancer, 25, 137-149.
1058 Shklar G. Schwartz J. Trickler D.P. Niukian K. (1987) Regression by vitamin E of experimental oral cancer, J Natl Cancer Inst, 78(5), 987-992.
1059 Malafa MP, Fokum FD, Mowlavi A, Abusief M, King M. (2002) Vitamin E inhibits melanoma growth in mice, Surgery, 131(1), 85-91.
1060 Neuzil J. (2003) Vitamin E succinate and cancer treatment: a vitamin E prototype for selective antitumour activity, Br J Cancer, 89(10), 1822-1826.
1061 Jiang Q. Wong J. Ames B.N. (2004) Gamma-tocopherol induces apoptosis in androgen-responsive LNCaP prostate cancer cells via Caspase-dependent and independent mechanisms, Ann N Y Acad Sci, 1031, 399-400.
1062 Shah S. Sylvester P.W. (2004) Tocotrienol-induced caspase-8 activation is unrelated to death receptor apoptotic signaling in neoplastic mammary epithelial cells, Exp Biol Med (Maywood), 229(8), 745-755.
1063 Agarwal M.K. Agarwal M.L. Athar M. Gupta S. (2004) Tocotrienol-rich fraction of palm oil activates p53, modulates Bax/Bcl2 ratio and induces apoptosis independent of cell cycle association, Cell Cycle, 3(2), 205-211.
1064 Sylvester P.W. Shah S.J. (2005) Mechanisms mediating the antiproliferative and apoptotic effects of vitamin E in mammary cancer cells, Front Biosci, 10, 699-709.
1065 Sakai M. Okabe M. Yamasaki M. Tachibana H. Yamada K. (2004) Induction of apoptosis by tocotrienol in rat hepatoma dRLh-84 cells, Anticancer Res, 24(3a), 1683-1688.
1066 Prasad K.N. Hernandez C. Edwards-Prasad J. Nelson J. Borus T. Robinson W.A. (1994) Modification of the effect of tamoxifen, cis-platin, DTIC, and interferon-alpha 2b on human melanoma cells in culture by a mixture of vitamins, Nutr Cancer, 22(3), 233-245.
1067 Prasad K.N. Kumar R. (1996) Effect of individual and multiple antioxidant vitamins on growth and morphology of human nontumorigenic and tumorigenic parotid acinar cells in culture, Nutr Cancer, 26,11–19.
1068 Folkers K. Osterborg A. Nylander M. Morita M. Mellstedt H. (1997) Activities of vitamin Q10 in animal models and a serious deficiency in patients with cancer, Biochem Biophys Res Commun, 234(2), 296-299.
1069 Palan P.R. Mikhail M.S. Shaban D.W. Romney S.L. (2003) Plasma concentrations of coenzyme Q10 and tocopherols in cervical intraepithelial neoplasia and cervical cancer, Eur J Cancer Prev, 12(4), 321-326.
1070 Folkers K. (1996) Relevance of the biosynthesis of coenzyme Q10 and of the four bases of DNA as a rationale for the molecular causes of cancer and a therapy, Biochem Biophys Res Commun, 224(2), 358-361.
1071 Langsjoen H. Langsjoen P. Langsjoen P. Willis R. Folkers K. (1994) Usefulness of coenzyme Q10 in clinical cardiology: a long-term study, Mol Aspects Med, 15 Suppl, s165-175.
1072 Langsjoen P. Langsjoen P. Willis R. Folkers K. (1994) Treatment of essential hypertension with coenzyme Q10, Mol Aspects Med, 15 Suppl, S265-272.
1073 Iarussi D. Auricchio U. Agretto A. Murano A. Giuliano M. Casale F. Indolfi P. Iacono A. (1994) Protective effect of coenzyme Q10 on anthracyclines cardiotoxicity: control study in children with acute lymphoblastic leukemia and non-Hodgkin lymphoma, Mol Aspects Med, 15 Suppl, s207-212.
1074 Olson R.E. (2001) Karl August Folkers (1906-1997), J Nutr, 131(9), 2227-2230.
1075 Lockwood K. Moesgaard S. Hanioka T. Folkers K. (1994) Apparent partial remission of breast cancer in 'high risk' patients supplemented with nutritional antioxidants, essential fatty acids and coenzyme Q10, Mol Aspects Med, 15 Suppl, s231-240.
1076 Lockwood K. Moesgaard S. Yamamoto T. Folkers K. (1995) Progress on therapy of breast cancer with vitamin Q10 and the regression of metastases, Biochem Biophys Res Commun, 212(1), 172-177.
1077 Folkers K. Brown R. Judy W.V. Morita M. (1993) Survival of cancer patients on therapy with coenzyme Q10, Biochem Biophys Res Commun, 192(1), 241-245.
1078 Hodges S. Hertz N. Lockwood K. Lister R. (1999) CoQ10: could it have a role in cancer management? Biofactors, 9(2-4), 365-370.
1079 Damkier A. Jensen A.B. Rose C. (1994) Use of Q10 in cancer patients, 156(6), 813-818.

[1080] Lopez-Lluch G. Barroso M.P. Martin S.F. Fernandez-Ayala D.J. Gomez-Diaz C. Villalba J.M. Navas P. (1999) Role of plasma membrane coenzyme Q on the regulation of apoptosis, Biofactors, 9(2-4), 171-177.
[1081] Crane F.L. (2001) Biochemical functions of coenzyme Q10, Am Coll Nutr, 20(6), 591-598.
[1082] Brandt U. (1999) Proton translocation in the respiratory chain involving ubiquinone--a hypothetical semiquinone switch mechanism for complex I, Biofactors, 9(2-4), 95-101.
[1083] Yu C.A. Zhang L. Deng K.P. Tian H. Xia D. Kim H. Deisenhofer J. Yu L. (1999) Structure and reaction mechanisms of multifunctional mitochondrial cytochrome bc1 complex, Biofactors, 9(2-4), 103-109.
[1084] Crane F.L. (2000) New functions for coenzyme Q, Protoplasma, 213, 127–133.
[1085] McLennan H.R. Degli Esposti M. (2000) The contribution of mitochondrial respiratory complexes to the production of reactive oxygen species, J Bioenerg Biomembr, 32(2), 153-162.
[1086] Hansford R.G. Hogue B.A. Mildaziene V. (1997) Dependence of H2O2 formation by rat heart mitochondria on substrate availability and donor age, J Bioenerg Biomembr, 29(1), 89-95.
[1087] Nohl H. Gille L. Schönheit K. Lin Y. (1996) Conditions allowing redox cycling ubisemiquinone in mitochondria to establish a direct couple with molecular oxygen, Free Radical Bio Med, 20, 207–213.
[1088] Wang G.S. Olsson J.M. Eriksson L.C. Stål P. (2000) Diet restriction increases ubiquinone contents and inhibits progression of hepatocellular carcinoma in the rat, Scand J. Gastr, 35(1), 83 – 89.
[1089] Luzi C. Brisdelli F. Cinque B. Cifone G. Bozzi A. (2004) Differential sensitivity to resveratrol-induced apoptosis of human chronic myeloid (K562) and acute lymphoblastic (HSB-2) leukemia cells, Biochem Pharmacol, 68(10), 2019-2030.
[1090] Mahyar-Roemer M. Katsen A. Mestres P. Roemer K. (2001) Resveratrol induces colon tumor cell apoptosis independently of p53 and precede by epithelial differentiation, mitochondrial proliferation and membrane potential collapse, Int J Cancer, 94(5), 615-622.
[1091] Kang J.H. Park Y.H. Choi S.W. Yang E.K. Lee W.J. (2003) Resveratrol derivatives potently induce apoptosis in human promyelocytic leukemia cells, Exp Mol Med, 35(6), 467-474.
[1092] Larrosa M. Tomas-Barberan F.A. Espin J.C. (2004) The grape and wine polyphenol piceatannol is a potent inducer of apoptosis in human SK-Mel-28 melanoma cells, Eur J Nutr, 43(5), 275-284.
[1093] Laux M.T. Aregullin M. Berry J.P. Flanders J.A. Rodriguez E. (2004) Identification of a p53-dependent pathway in the induction of apoptosis of human breast cancer cells by the natural product, resveratrol, J Altern Complement Med, 10(2), 235-239.
[1094] Kim Y.A. Choi B.T. Lee Y.T. Park D.I. Rhee S.H. Park K.Y. Choi Y.H. (2004) Resveratrol inhibits cell proliferation and induces apoptosis of human breast carcinoma MCF-7 cells, Oncol Rep, 11(2), 441-446.
[1095] Scarlatti F. Sala G. Somenzi G. Signorelli P. Sacchi N. Ghidoni R. (2003) Resveratrol induces growth inhibition and apoptosis in metastatic breast cancer cells via de novo ceramide signalling, Faseb J, 17(15), 2339-2341.
[1096] Lu R. Serrero G. (1999) Resveratrol, a natural product derived from grape, exhibits antiestrogenic activity and inhibits the growth of human breast cancer cells, J Cell Physiol, 179(3), 297-304.
[1097] Kim Y.A. Rhee S.H. Park K.Y. Choi Y.H. (2003) Antiproliferative effect of resveratrol in human prostate carcinoma cells, J Med Food, 6(4), 273-280.
[1098] Aggarwal B.B. Bhardwaj A. Aggarwal R.S. Seeram N.P. Shishodia S. Takada Y. (2004) Role of resveratrol in prevention and therapy of cancer: preclinical and clinical studies, Anticancer Res, 24(5A), 2783-2840.
[1099] Joe A.K. Liu H. Suzui M. Vural M.E. Xiao D. Weinstein I.B. (2002) Resveratrol induces growth inhibition, S-phase arrest, apoptosis, and changes in biomarker expression in several human cancer cell lines, Clin Cancer Res, 8(3), 893-903.
[1100] Bernhard D. Tinhofer I. Tonko M. HuÈbl H. Ausserlechner M.J. 2, Greil R. Kofler R. Csordas A. (2000) Resveratrol causes arrest in the S-phase prior to Fas-independent apoptosis in CEM-C7H2 acute leukaemia, Cell Death and Differentiation, 7, 834-842.
[1101] Atten M.J. Godoy-Romero E. Attar B.M. Milson T. Zopel M. Holian O. (2005) Resveratrol regulates cellular PKC alpha and delta to inhibit growth and induce apoptosis in gastric cancer cells, Invest New Drugs, 23(2), 111-119.
[1102] Chen Y. Tseng S.H. Lai H.S. Chen W.J. (2004) Resveratrol-induced cellular apoptosis and cell cycle arrest in neuroblastoma cells and antitumor effects on neuroblastoma in mice, Surgery, 136(1), 57-66.
[1103] Tseng S.H. Lin S.M. Chen J.C. Su Y.H. Huang H.Y. Chen C.K. Lin P.Y. Chen Y. (2004) Resveratrol suppresses the angiogenesis and tumor growth of gliomas in rats, Clin Cancer Res, 10(6), 2190-2202.
[1104] Liu H.S. Pan C.E. Yang W. Liu X.M. (2003) Antitumor and immunomodulatory activity of resveratrol on experimentally implanted tumor of H22 in Balb/c mice, World J Gastroenterol, 9(7), 1474-1476
[1105] Scifo C. Cardile V. Russo A. Consoli R. Vancheri C. Capasso F. Vanella A. Renis M. (2004) Resveratrol and propolis as necrosis or apoptosis inducers in human prostate carcinoma cells, Oncol Res, 14(9), 415-426.
[1106] Fuggetta M.P. D'Atri S. Lanzilli G. Tricarico M. Cannavo E. Zambruno G. Falchetti R. Ravagnan G. (2004) In vitro antitumour activity of resveratrol in human melanoma cells sensitive or resistant to temozolomide, Melanoma Res, 14(3), 189-196.

1107 Surh Y.J. Hurh Y.J. Kang J.Y. Lee E. Kong G. Lee S.J. (1999) Resveratrol, an antioxidant present in red wine, induces apoptosis in human promyelocytic leukemia (HL-60) cells, Cancer Lett, 140(1-2), 1-10.
1108 Tsan M.F. White J.E. Maheshwari J.G. Chikkappa G. (2002) Anti-leukemia effect of resveratrol, Leuk Lymphoma, 43(5), 983-987.
1109 Roman V. Billard C. Kern C. Ferry-Dumazet H. Izard J.C. Mohammad R. Mossalayi D.M. Kolb J.P. (2002) Analysis of resveratrol-induced apoptosis in human B-cell chronic leukaemia, Br J Haematol, 117(4), 842-851.
1110 Lee E.J. Min H.Y. Joo Park H. Chung H.J. Kim S. Nam Han Y. Lee S.K. (2004) G2/M cell cycle arrest and induction of apoptosis by a stilbenoid, 3,4,5-trimethoxy-4'-bromo-cis-stilbene, in human lung cancer cells, Life Sci, 75(23), 2829-2839.
1111 Gautam S.C. Xu Y.X. Dumaguin M. Janakiraman N. Chapman R.A. (2000) Resveratrol selectively inhibits leukemia cells: a prospective agent for ex vivo bone marrow purging, Bone Marrow Transplant, 25(6), 639-645.
1112 Asensi M. Medina I. Ortega A. Carretero J. Bano M.C. Obrador E. Estrela J.M. (2002) Inhibition of cancer growth by resveratrol is related to its low bioavailability, Free Radic Biol Med, 33(3), 387-398.
1113 Gao X. Xu Y.X. Divine G. Janakiraman N. Chapman R.A. Gautam S.C. (2002) Disparate in vitro and in vivo antileukemic effects of resveratrol, a natural polyphenolic compound found in grapes, J Nutr, 132(7), 2076-2081.
1114 Rimando A.M. Cuendet M. Desmarchelier C. Mehta R.G. Pezzuto J.M. Duke S.O. (2002) Cancer chemopreventive and antioxidant activities of pterostilbene, a naturally occurring analogue of resveratrol, J Agric Food Chem, 50(12), 3453-3457.
1115 Werneke U. Earl J. Seydel C. Horn O. Crichton P. Fannon D. (2004) Potential health risks of complementary alternative medicines in cancer patients, Br J Cancer, 90(2), 408-413.
1116 Pinn G. (2001) Herbal medicine in oncology, Aust Fam Physician, 30(6), 575-580.
1117 Arditti F.D. Rabinkov A. Miron T. Reisner Y. Berrebi A. Wilchek M. Mirelman D. (2005) Apoptotic killing of B-chronic lymphocytic leukemia tumor cells by allicin generated in situ using a rituximab-alliinase conjugate, Mol. Cancer Ther, 4, 325-332.
1118 Miron T. Mironchik M. Mirelman D. Wilchek M. Rabinkov A. (2003) Inhibition of tumor growth by a novel approach: In situ allicin generation using targeted alliinase delivery, Mol. Cancer Ther, 2, 1295-1301.
1119 Kucuk O. Sarkar F.H. Djuric Z. Sakr W. Pollak M.N. Khachik F. Banerjee M. Bertram J.S. Wood D.P. Jr (2002) Effects of lycopene supplementation in patients with localized prostate cancer, Exp Biol Med, 227, 881–885.
1120 Dorai T. Aggarwal B.B. (2004) Role of chemopreventive agents in cancer therapy, Cancer Lett, 215(2), 129-140.
1121 Sun S.Y. Hail N. Jr Lotan R. (2004) Apoptosis as a novel target for cancer chemoprevention, J Natl Cancer Inst, 96(9), 662-672.
1122 Kimura Y. (2005) New anticancer agents: in vitro and in vivo evaluation of the antitumor and antimetastatic actions of various compounds isolated from medicinal plants, In Vivo, 19(1), 37-60.
1123 Olsson M.E. Gustavsson K.E. Andersson S. Nilsson A. Duan R.D. (2004) Inhibition of cancer cell proliferation in vitro by fruit and berry extracts and correlations with antioxidant levels, J Agric Food Chem, 52(24), 7264-7271.
1124 Chang W.H. Liu J.J. Chen C.H. Huang T.S. Lu F.J. (2002) Growth inhibition and induction of apoptosis in MCF-7 breast cancer cells by fermented soy milk, Nutr Cancer, 43(2), 214-226.
1125 Malik A. Afaq F. Sarfaraz S. Adhami V.M. Syed D.N. Mukhtar H. (2005) Pomegranate fruit juice for chemoprevention and chemotherapy of prostate cancer, Proc Natl Acad Sci U S A, Sep 28; [Epub ahead of print].
1126 Liontas A. Yeger H. (2004) Curcumin and resveratrol induce apoptosis and nuclear translocation and activation of p53 in human neuroblastoma, Anticancer Res, 24(2B), 987-998.
1127 Syng-Ai C. Kumari A.L. Khar A. (2004) Effect of curcumin on normal and tumor cells: role of glutathione and bcl-2, Mol Cancer Ther, 3(9), 1101-1108.
1128 Jaruga E. Salvioli S. Dobrucki J. Chrul S. Bandorowicz-Pikula J. Sikora E. Franceschi C. Cossarizza A. Bartosz G. (1998) Apoptosis-like, reversible changes in plasma membrane asymmetry and permeability, and transient modifications in mitochondrial membrane potential induced by curcumin in rat thymocytes, FEBS Lett, 433(3), 287-293.
1129 Bhaumik S. Anjum R. Rangaraj N. Pardhasaradhi B.V. Khar A. (1999) Curcumin mediated apoptosis in AK-5 tumor cells involves the production of reactive oxygen intermediates, FEBS Lett, 456(2):311-314.
1130 Marko D. Puppel N. Tjaden Z. Jakobs S. Pahlke G. (2004) The substitution pattern of anthocyanidins affects different cellular signaling cascades regulating cell proliferation, Mol Nutr Food Res, 48(4), 318-325.

[1131] Afaq F. Saleem M. Krueger C.G. Reed J.D. Mukhtar H. (2005) Anthocyanin- and hydrolyzable tannin-rich pomegranate fruit extract modulates MAPK and NF-kappaB pathways and inhibits skin tumorigenesis in CD-1 mice, Int J Cancer, 113(3), 423-433.
[1132] Fimognari C. Berti F. Nusse M. Cantelli-Forti G. Hrelia P. (2004) Induction of apoptosis in two human leukemia cell lines as well as differentiation in human promyelocytic cells by cyanidin-3-O-beta-glucopyranoside, Biochem Pharmacol, 67(11), 2047-2056.
[1133] Fang J. Lu J. Holmgren A. (2005) Thioredoxin reductase is irreversibly modified by curcumin: a novel molecular mechanism for its anticancer activity, J Biol Chem, Epub ahead of print.
[1134] Karunagaran D. Rashmi R. Kumar T.R. (2005) Induction of apoptosis by curcumin and its implications for cancer therapy, Curr Cancer Drug Targets, 5(2), 117-129.
[1135] Aime S. Barge A. Cabella C. Crich S.G. Gianolio E. (2004) Targeting cells with MR imaging probes based on paramagnetic Gd(III) chelates, Curr Pharm Biotechnol, 5(6), 509-518.
[1136] Artemov D. (2003) Molecular magnetic resonance imaging with targeted contrast agents, J Cell Biochem, 90(3), 518-524.
[1137] Young S.W. Qing F. Harriman A. Sessler J.L. Dow W.C. Mody T.D. Hemmi G.W. Hao Y. Miller R.A. (1996) Gadolinium(III) texaphyrin: a tumor selective radiation sensitizer that is detectable by MRI, Proc Natl Acad Sci U S A, 93(13), 6610-6615.
[1138] Rosenthal D.I. Nurenberg P. Becerra C.R. Frenkel E.P. Carbone D.P. Lum B.L. Miller R. Engel J. Young S. Miles D. Renschler M.F. (1999) A phase I single-dose trial of gadolinium texaphyrin (Gd-Tex), a tumor selective radiation sensitizer detectable by magnetic resonance imaging, Clin Cancer Res, 5(4), 739-745.
[1139] Carde P. Timmerman R. Mehta M.P. Koprowski C.D. Ford J. Tishler R.B. Miles D. Miller R.A. Renschler M.F. (2001) Multicenter phase Ib/II trial of the radiation enhancer motexafin gadolinium in patients with brain metastases, J Clin Oncol, 19(7), 2074-2083.
[1140] Young S.W. Sidhu M.K. Qing F. Muller H.H. Neuder M. Zanassi G. Mody T.D. Hemmi G. Dow W. Mutch J.D. et al. (1994) Preclinical evaluation of gadolinium (III) texaphyrin complex. A new paramagnetic contrast agent for magnetic resonance imaging, Invest Radiol, 29(3), 330-338.
[1141] Xu S. Zakian K. Thaler H. Matei C. Alfieri A. Chen Y. Koutcher J.A. (2001) Effects of Motexafin gadolinium on tumor metabolism and radiation sensitivity, Int J Radiat Oncol Biol Phys, 49(5), 1381-1390.
[1142] Carde P. Timmerman R. Mehta M.P. Koprowski C.D. Ford J. Tishler R.B. Miles D. Miller R.A. Renschler M.F. (2001) Multicenter phase Ib/II trial of the radiation enhancer motexafin gadolinium in patients with brain metastases, J Clin Oncol, 19(7), 2074-1083.
[1143] Mehta M.P. Suh J.H. (2004) Novel radiosensitizers for tumors of the central nervous system, Curr Opin Investig Drugs, 5(12), 1284-1291.
[1144] Magda D. Lepp C. Gerasimchuk N. Lee I. Sessler J.L. Lin A. Biaglow J.E. Miller R.A. (2001) Redox cycling by motexafin gadolinium enhances cellular response to ionizing radiation by forming reactive oxygen species, Int J Radiat Oncol Biol Phys, 51(4), 1025-1036.
[1145] Evens A.M. Lecane P. Magda D. Prachand S. Singhal S. Nelson J. Miller R.A. Gartenhaus R.B. Gordon L.I. (2005) Motexafin gadolinium generates reactive oxygen species and induces apoptosis in sensitive and highly resistant multiple myeloma cells, Blood, 105(3), 1265-1273. Epub 2004 Sep 23.
[1146] Renschler M.F. (2004) The emerging role of reactive oxygen species in cancer therapy, Eur J Cancer, 40(13), 1934-1940.
[1147] Biaglow J.E. (2005) The thioredoxin reductase/thioredoxin system: novel redox targets for cancer therapy, Cancer Biol Ther, 4(1), 6-13. Epub 2004 Jan 8.
[1148] Evens A.M. (2004) Motexafin gadolinium: a redox-active tumor selective agent for the treatment of cancer, Curr Opin Oncol, 16(6), 576-580.

78841860R00166

Made in the USA
San Bernardino, CA
09 June 2018